Lichen-Derived Products

Scrivener Publishing
100 Cummings Center, Suite 541J
Beverly, MA 01915-6106

Emerging Trends in Medicinal and Pharmaceutical Chemistry

Series Editor: Shahid-ul-Islam

The Emerging Trends in Medicinal Chemistry and Pharmacology Series is intended to provide recent trends, the state -of-the-art, and advancements particularly in the rapidly growing fields of drug design and synthesis, medicinal natural products, phytochemistry, pharmacology and applications. With a focus on generating means to combat different human diseases, the series addresses novel strategies and advanced methodology to circumvent the invasion from microbial infections and to ameliorate the effects caused by dreadful diseases. Each volume from the series will provide high-level research books covering theoretical and experimental approaches of medicinal natural products, antimicrobial drugs, chemotherapeutic agents, anticancer agents, phytochemistry and pharmacology. The volumes will be written by international scientists for a broad readership researchers and students in biology, chemistry, biochemistry, medicinal science, chemical and biomedical engineering.

Publishers at Scrivener
Martin Scrivener (martin@scrivenerpublishing.com)
Phillip Carmical (pcarmical@scrivenerpublishing.com)

Lichen-Derived Products

Extraction and Applications

Edited by
Mohd Yusuf

Scrivener
Publishing

WILEY

This edition first published 2020 by John Wiley & Sons, Inc., 111 River Street, Hoboken, NJ 07030, USA and Scrivener Publishing LLC, 100 Cummings Center, Suite 541J, Beverly, MA 01915, USA

For more information about Scrivener publications please visit www.scrivenerpublishing.com.

Wiley Global Headquarters
111 River Street, Hoboken, NJ 07030, USA

For details of our global editorial offices, customer services, and more information about Wiley products visit us at www.wiley.com.

Library of Congress Cataloging-in-Publication Data

ISBN 978-1-119-59171-9

Cover image: Pixabay.Com
Cover design by Russell Richardson

Set in size of 11pt and Minion Pro by Manila Typesetting Company, Makati, Philippines

Contents

Preface

Naturally derived products and their derivatives have had a strong influence ever since ancient times and continue to dominate nearly every field of application today, ranging from food, medicine, agriculture, textiles, healthcare, etc., due to their enthusiastic performance. To meet the demands of the current era, scientists need to communicate the relevance of their work in the field with specific arguments for eco-friendly alternatives. With increased awareness of product safety and closer attention being paid to the possible adverse impacts of synthetic ready-to-use products on the ecosystem, during the last few decades research has continued to be conducted on more and more new alternatives based on naturally derived functional materials. So, since a sustainable world requires the utilization of renewable materials or resources for use in a wide range of applications, there needs to be a multidisciplinary effort from a technological perspective to perform further research toward new opportunities to develop new sustainable products.

The purpose of this handbook is to provide reference material that includes current developments along with a future outlook on the topic. The handbook is divided into two sections: "Morphological Overview and Extraction Prospects" and "Trends and Applications". Part I contains four chapters that provide an overview and systematically discuss the physical morphology, suitability, and extraction aspects of lichens and their secondary metabolites. Part II includes eight chapters that give in-depth insights on recent and valuable applications of lichen and their obtained products in several applied sectors, including ethnopharmacology, therapeutics, paper and dye, nutraceuticals, cosmetics, herbal industries, etc.

Overall, *Lichen-Derived Products: Extraction and Applications* highlights the importance of the subject. The book offers valuable resource material that is equally beneficial for undergraduate and postgraduate students, educators, and researchers involved in natural product research, phytochemistry, and pharmacological research.

My sincere thanks go to the eminent authors for their priceless contributions that provide excellent informative coverage. Valuable comments and suggestions from our readers on the presented material are welcome, which will help to improve future volumes.

I express my appreciation to Dr. M. I. Khan, Principal, YMD College, Nuh, Haryana, for his worthy suggestions and moral support. A deep sense of gratitude goes to my research guide Dr. F. Mohammad (former Professor, Department of Chemistry, Faculty of Natural Sciences, Jamia Millia Islamia, New Delhi, India) and my colleague and friend Dr. M. Shahid (Marie Quire Fellow, University of Glasgow, Glasgow, UK) for their discussions on the topic, careful advice, critique, and valuable suggestions. Also, I am thankful to my parents (Mohd Yasin and Mrs. Sakina Yasin) and spouse (Mrs. Salma Yusuf) for their moral support and careful efforts and patience.

Particularly, I would like to express my sincere thanks to Scrivener Publishing, especially Martin Scrivener, for inviting me to compile this volume.

Mohd Yusuf
(Editor)
Head
Department of Chemistry
YMD College, M. D. University, Nuh
Haryana 122107 India
May 2020

PART 1

MORPHOLOGICAL OVERVIEW AND EXTRACTION PROSPECTS

1

An Overview on Lichen's Morphology, Chemistry of Derived Products, and Current Scenario

Mohd Yusuf

Department of Chemistry, YMD College, M. D. University, Nuh, Haryana, India

Abstract

Lichens, the mutual symbionts of a fungus with microalga and/or a cyanobacterium, are important constituents of many ecosystems. A large number of natural products and secondary metabolites have been obtained from lichens, especially, dyes, drugs, and other constituents. In the present scenario, these natural products/secondary metabolites have shown impressive range of biological activities such as antimicrobial, anticancer, antifungal, antioxidant, antiviral, anti-inflammatory, analgesic, antipyretic activities, and many more. Traditionally, lichens as whole and their extracts were used to treat various diseases and ailments around the world. This chapter deals with the discussion on lichen's morphology and chemistry of derived products.

Keywords: Lichen, morphology, biological activity, antibacterial, antifungal, dyes

1.1 Introduction

Lichens are an inevitable segment of all ecosystems due to their symbiotic nature not only fixation of essential nutrients from the air to the soil but also production of fresh oxygen into air. Lichens are slowest growing, unique in nature and shape, stable, and self-supporting symbiotic organisms propagate on living organisms such as shrubs, tortoise's carapaces, trunks and woods of trees, etc., and mounted surfaces like soil, bricks,

Email: yusuf1020@gmail.com

Mohd Yusuf (ed.) Lichen-Derived Products: Extraction and Applications, (3–26) © 2020 Scrivener Publishing LLC

and many more, in an association of a fungi with algal isolate [1, 2]. It is estimated that lichens are survived with approximately 25,000 species in which 98% are Ascomycotes fungal partners and remaining 2% are cyanobacteria partners [3]. Old documentations reveal the uses of lichens and their extracts/obtained products have been utilized, as one of the best traditional medicines for the treatment of several human (wounds, burns, fissures, impotency, etc.), plants (rots and worts), and animal diseases (disfertility, wounds, skin issues) [4, 5]. Lichen found to exhibit manifold biological activities in various screenings, for example, antimicrobial, anticancer, antifungal, antioxidant, antiviral, anti-inflammatory, analgesic, antipyretic, growth inhibitory, and enzyme inhibitory activities, etc. [1–8]. Extracts of many lichen species of the genus *Usnea* were found usable as analgesic remedy in several countries of Asia, Europe, Africa, and Argentina [9].

The plant-like portion of the lichens is morphologically multifarious and generally often called as thallus. Primarily, the main three categories of lichens were recognized already such as crustose, foliose, and fruticose [1]. Furthermore, lichenologists, chiefly Baron, Dobson, Smith, and their co-workers described comprehensive diversities of thallus types (Table 1.1). Lichens inherently synthesize various types of secondary metabolites. There are more than 1,000 lichen-derived substances/biomolecules that have been identified through analytical and experimental methods [2]. The discussion on lichen's thallus morphology and the chemistry of their derived products with their considerable applications is highlighted herein this chapter.

1.2 The Lichen's Body: Thallus Morphology

Lichens are found growing on high peaks from the sea to the mountains such as wood, trunks, barks of trees, crustacean to the soil, rocks, walls, and many more places either in the sunlight or in the dark grooves. Abundantly, they propagate in clean environment like, forest ecosystem, and in a wide range of locations and climates from the Polar Regions to the tropics. Thus, they act as environmental indicator. Lichen thallus is the vegetative part and morphologically diversified in nature with branched, proliferated, tabulate, leafy, fibrous, tiny plates, layers of powdery granules, or without any specified thallus [10]. The main three categories of lichens were recognized already such as crustose, foliose, and fruticose [9]. Furthermore, lichenologists, chiefly Baron, Dobson, Smith, and their co-workers described comprehensive diversities of thallus types (Table 1.1) [5–10].

Table 1.1 Morphological diversities in lichen thalli.

Lichen types	Thallus morphology		
Foliose	 *Evernia prunastri*	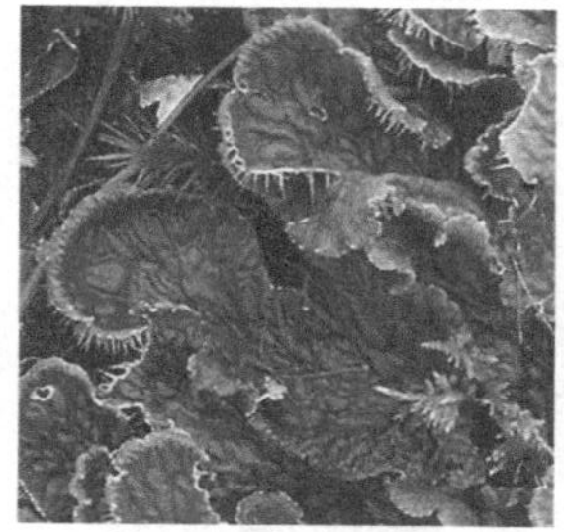 *Peltigera membranacea*	 *Parmotrema perlatum*
Fruticose	 *Usnea subfloridana*	 *Roccella phycopsis*	 *Ramalina polymorpha*

(*Continued*)

Table 1.1 Morphological diversities in lichen thalli. (*Continued*)

Lichen types	**Thallus morphology**		
Squamulose			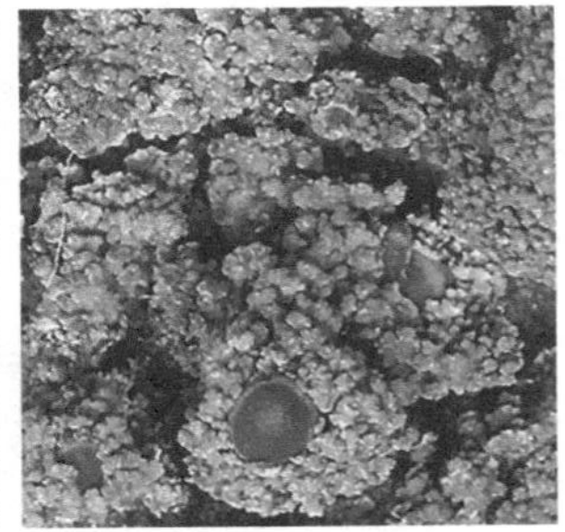
	Squamarina cartilaginea	*Cladonia subcervicornis*	*Vahliella leucophaea*
Crustose	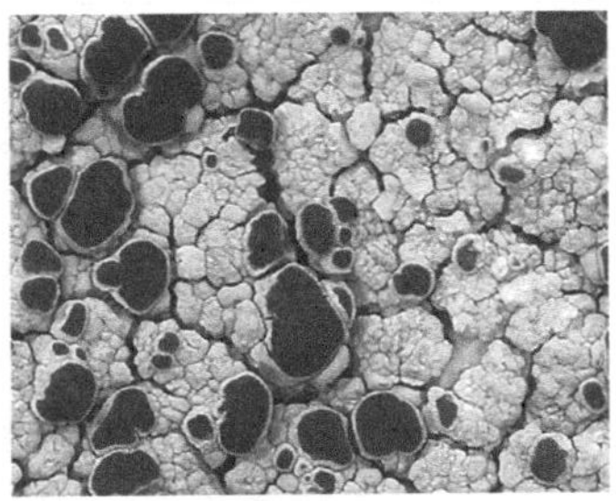	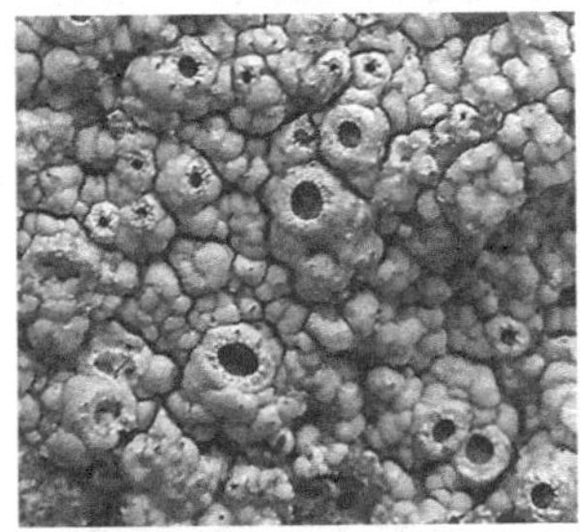	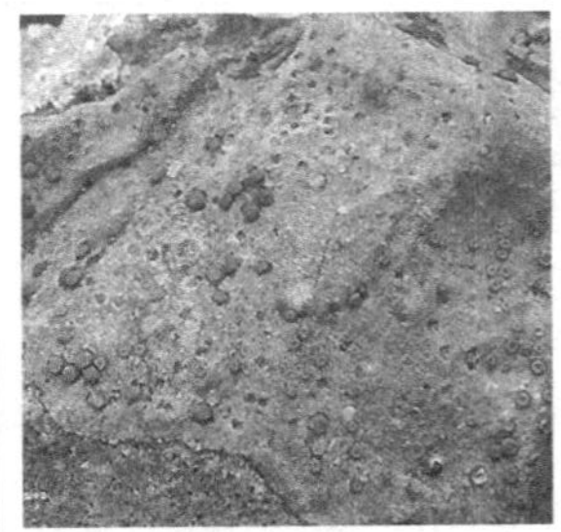
	Ophioparma ventosa	*Diploschistes scruposus*	*Caloplaca ochracea*

(*Continued*)

Table 1.1 Morphological diversities in lichen thalli. (*Continued*)

Lichen types	**Thallus morphology**		
Placodioid lichens	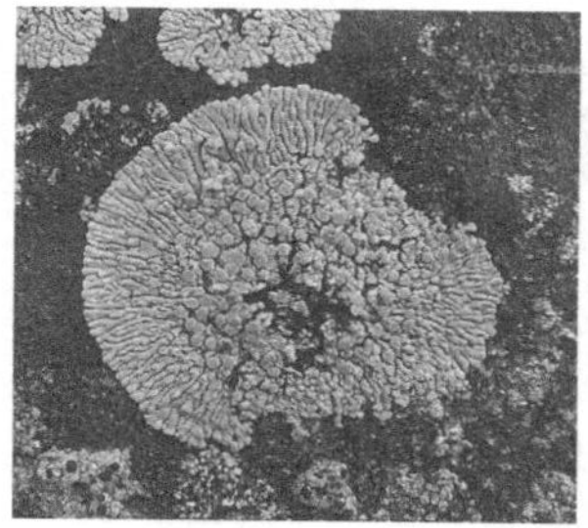	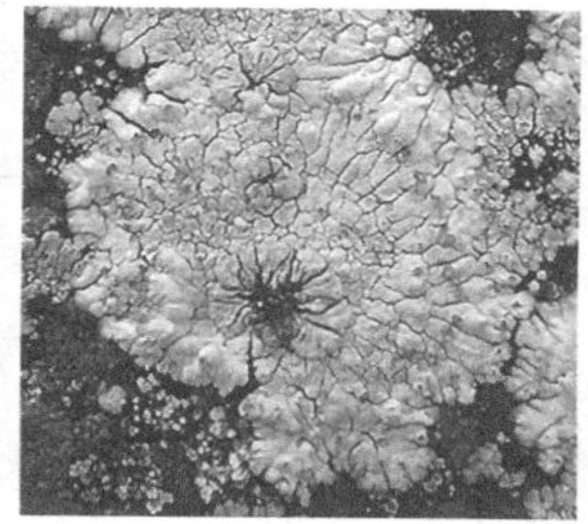	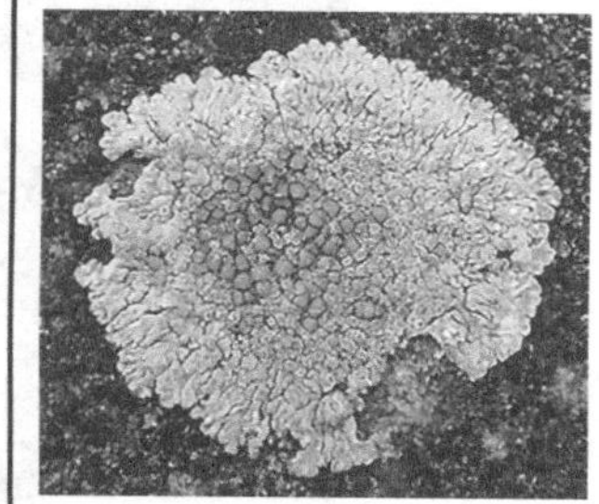
	Caloplaca thallincola	*Placopsis lambii*	*Lecanora muralis*
Leprose			
	Lepraria incana	*Caloplaca chrysodeta*	*Chrysothrix candelaris*

(*Continued*)

Table 1.1 Morphological diversities in lichen thalli. (*Continued*)

Lichen types	**Thallus morphology**		
Filamentous			
	Cystocoleus ebeneus	*Ephebe lanata*	*Polychidium muscicola*
Without any definite shape or lichens without any visible thallus	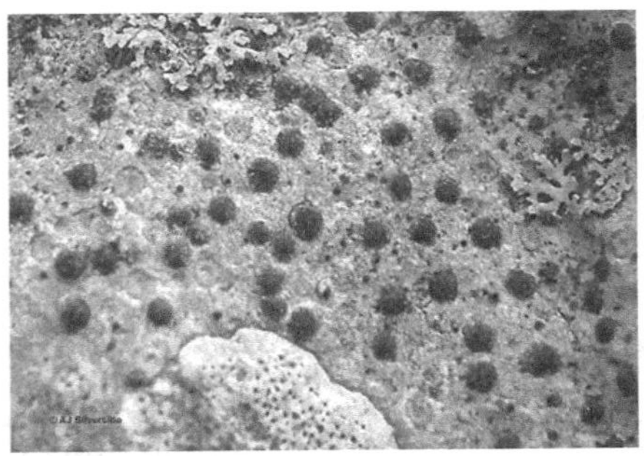	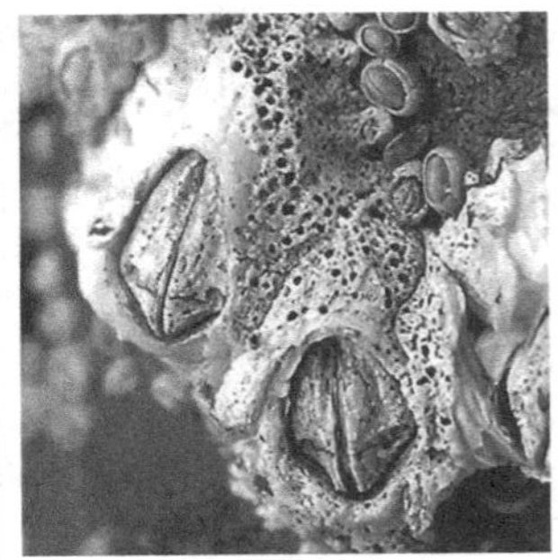	
	Acrocordia conoidea	*Collemopsidium foveolatum*	

1.2.1 Foliose

Foliose lichens mostly grow up in stratum with variable upper and lower surfaces. The thallus are somewhat flat but are convoluted or in leafy form and also can stand in erect mode away from the substrate like a swarm with multiple branches. Examples are *Evernia prunastri*, *Pseudevernia furfuracea*, *Evernia prunastri*, *Peltigera membranacea*, and *Parmotrema perlatum*.

1.2.2 Fruiticose

Fruiticose lichens are tickering in air like pendulum and have either a flat or a branched thallus covered by a solo outer cortex. For example, flattened branches—*Ramalina sp.*, and with branched aerial lobes such as *Evernia sp.* Other examples are *Usnea subfloridana* and *Roccella phycopsis*.

1.2.3 Squamulose Lichens

The squamulose lichens are specific arrangements of squamules (overlapping scales) which are heavy in weight, oblate, and succinct in size. Anatomically, squamulose lichens indicate variation to the smaller foliose lichens in which the presence of lower cortex is either not and or is highly differentiated. In case of *Catapyrenium* and *Placidium* species, medulla part may be sometimes colored. More examples are *Squamarina cartilaginea*, *Cladonia subcervicornis*, and *Vahliella leucophaea*.

1.2.4 Crustose

Crustose lichens mostly grow and spread over the crust such as soil, bark, and trunks of trees, and rocks with much variation like yellow, red, and grey-green in color-appearance, but found abundantly in grey-green color. Such lichens are firmly attached to the surface and are sighted in a bag-like texture. Examples are *Ophioparma ventosa*, *Diploschistes scruposus*, and *Caloplaca ochracea*.

1.2.5 Placodioid/Crustose-Placodioid/Squamulose-Placodioid

This is a true variant of crustose and squamulose that alternatively became placodioid. Some placodioid species are *Caloplaca thallincola*, *Placopsis lambii*, and *Lecanora muralis*. This type of thallus does not attach too much with the base and have a lower cortex. Also, the lobe margin extends into narrow and spread after.

1.2.6 Leprose

The surface of this type of thallus made up of lumps or granules like structure in which algal and fungal hyphae are present. Examples are *Lepraria incana*, *Caloplaca chrysodeta*, and *Chrysothrix candelaris*.

1.2.7 Filamentous Lichens

Filamentous lichens are also known as fibrous lichens owe to its shape. The morphology of the lichen generally seems dark green in appearance. At inner side algal cells like fiber are situated and the fungal hypae around the fiber form an outer sheath, for example, *Trentepohlia* or trichome-forming cyanobacteria. *Ephebe lanata* is a type of cyanolichens, which is blackish green in color and seen as smooth shrub or bear. Other common examples given are *Cystocoleus ebeneus* and *Polychidium muscicola*.

1.2.8 Without any Definite Shape or Lichens Without any Visible Thallus

Rarely, in this type of Lichens, no thallus are definitely visible. It can be understood in two ways, either it disappears very quickly or it gets immersed in the substrate in which both algae cells and fungal hyphae are present together. Examples of this group are *Acrocordia conoidea* and *Collemopsidium foveolatum*.

1.3 Chemistry of Lichen-Derived Products

Asahina [11], a Japanese scientist, explored diversified chemistry of lichens, with respect, concerning the initial work of two great botanists and entomologists of 18th centuary, William Nylander from Finland and Friedrich Wilhelm Zopf from Germany [12]. In the mid of 19th century, biogenic origin and phylogenetic significance of lichens have accumulated steadily worldwide, and lichen-derived biomolecules have attracted much attention. Consequently, Shibata, another Japanese researcher, performed exhausted work on the biosynthesis process for lichen-derived metabolites. Toward step ahead, they also introduced the process of microcrystallization with respect to the chemistry of lichen-derived compounds [13]. According to Molnar *et al.* (2010), about 1,050 lichen-derived substances have been identified out of which more than 700 were characterized by their chemical structure, mostly bears ester group [2, 14]. Biosynthetically, lichen metabolites can be categorized into four types (Figure 1.1) [15].

(i) Polyketide pathway shows the formation of the maximum lichen-derived compounds such as fatty acids and related compounds, phenolic compounds, etc.,
(ii) Shikimic acid pathway generally responsible for the generation of pulvinic acid derivatives and terphenylquinones,
(iii) Mevalonic acid pathway produces chiefly steroids and terpenes, and
(iv) Photosynnthesis generates sugars.

Lichen metabolites may be categorized into the following classes according to chemical structures (Figure 1.2):

(i) Aliphatics (acids, zeorin like compounds, polyols),
(ii) Carbohydrates (mono, di, and polysaccharides), and
(iii) Aromatics (almost constituents).

Aliphatic metabolites found are acids, polyols, and zeorin like compounds, whereas carbohydrates are the types of mono and complex saccharides. Based on the studies [4–7], the major class aromatics is widely distributed involving, phenolic compounds (orcinol and derivatives), lactones (protolichesterinic acid, nephrosterinic acid), quinines (parietin),

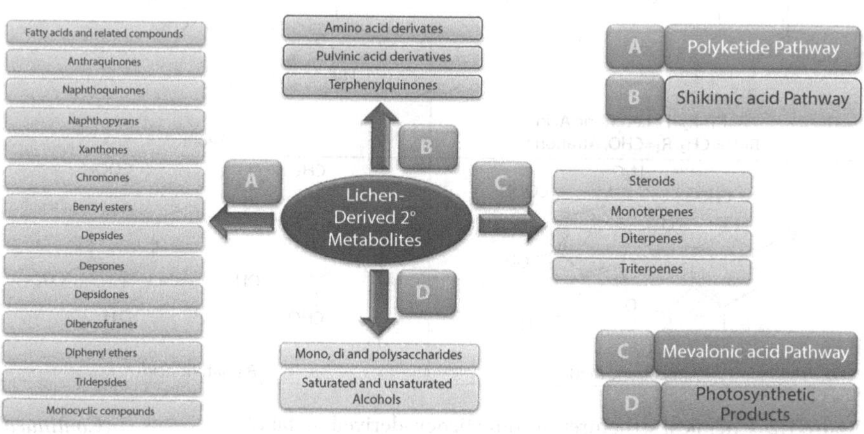

Figure 1.1 Different biosynthetic pathways of Lichen's secondary metabolites/substances.

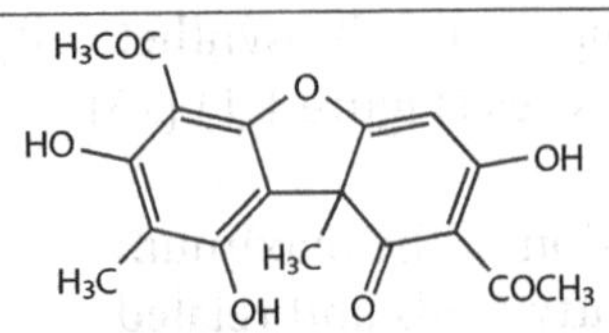

Usnic acid

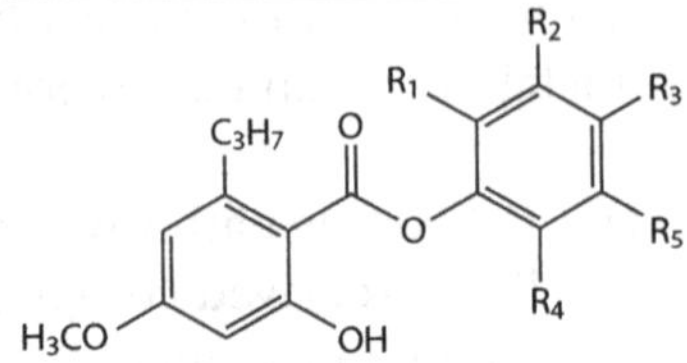

R_1=OH, R_2=COOH, R_3=C_3H_7, R_4=OCH_3, R_5=H Sekikaic Acid

R_1=H, R_2=OH, R_3=COOH, R_4=H, R_5=C_3H_7 Divaricatic Acid

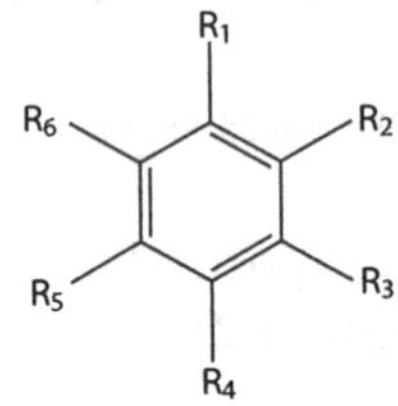

R_1=CH_3, R_2=R_4=R_6=H, R_3=R_5=OH Orcinol

R_1=R_4=CH_3, R_2=R_6=H, R_3=R_5=OH β-Orcinol

R_1=$COOCH_3$, R_2=R_5=CH_3, R_3=H, R_4=R_6=OH Methyl orsellinate

R_1=COOPh, R_2=R_5=CH_3, R_3=H, R_4=R_6=OH Phenyl orsellinate

R_1=COOH, R_2=R_5=CH_3, R_3=H, R_4=R_6=OH Orsellinic acid

R_1=R_4=CH_3, R_2=CH_2OH, R_3= OH, R_5=OCH_3, R_6=H Rhizonyl alcohol

R_1=R_4=CH_3, R_2=CHO, R_3=OH, R_5=OCH_3, R_6=H Rhizonyl aldehyde

R_1=$COOC_2H_5$, R_2=R_5=CH_3, R_3=H, R_4=R_6=OH Ethyl orsellinate

R_1=R_5=OH, R_2=$COOC_2H_5$, R_3=CH_3, R_4=H, R_6=CHO Ethyl heamatomate

R_1=R_4=CH_3, R_2=$COOCH_3$, R_3=R_5=OH, R_6=H Methyl-β-orcinolcarboxylate

R_1=R_5=OH, R_2=$COOCH_3$, R_3=R_6=CH_3, R_4=H, R_5=OH Atraric acid

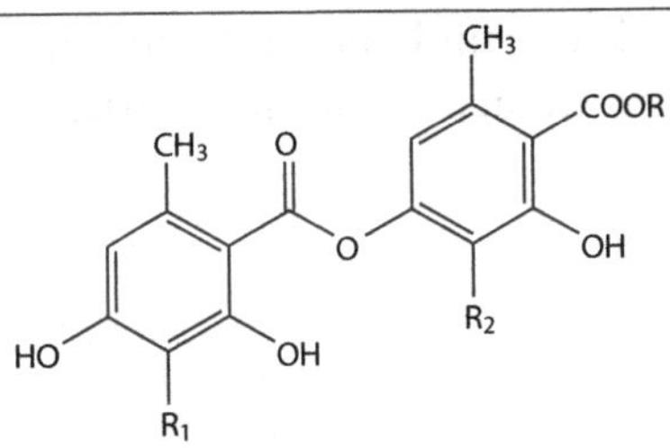

R=R_1=R_2=H; Lecanoric Acid

R=R_2=CH_3, R_1=CHO; Atranorin

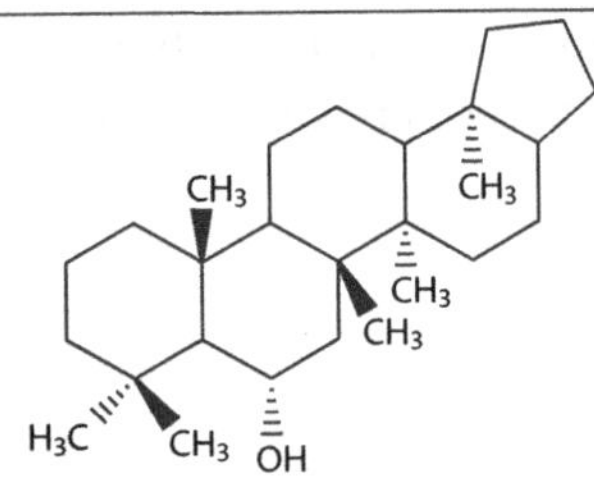

Zeorin

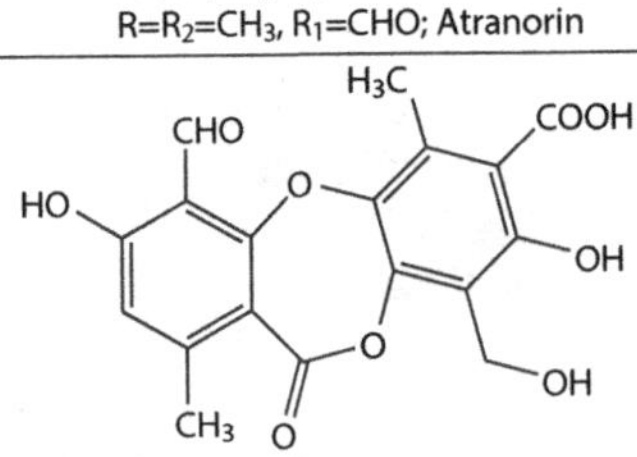

Protocetraric Acid

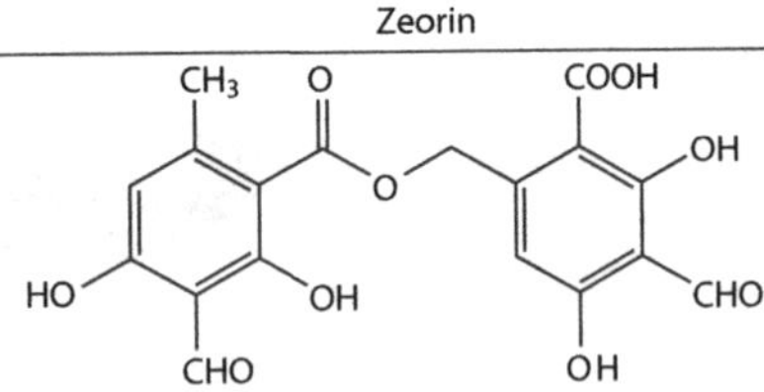

Barbatolic acid

Figure 1.2 Chemical structures of some lichen-derived metabolites. *(Continued)*

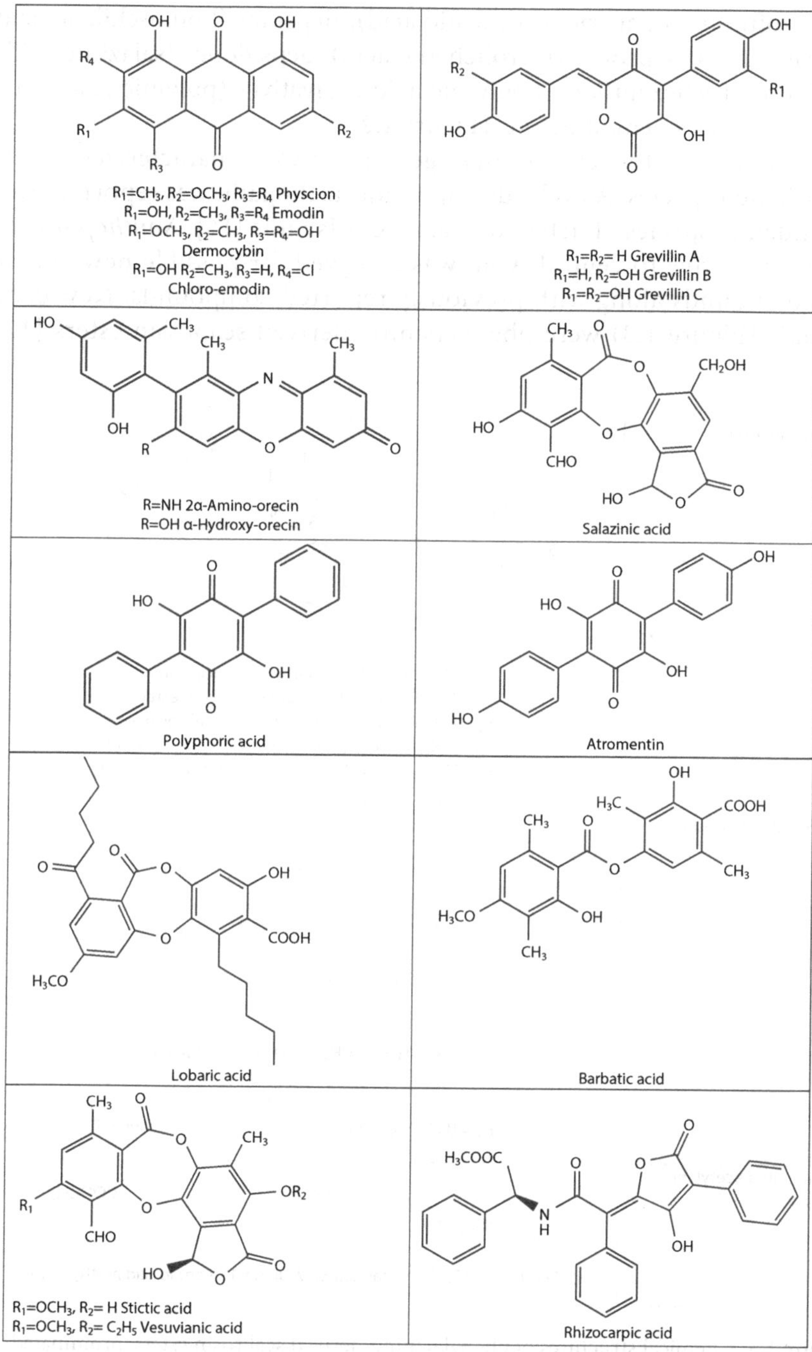

Figure 1.2 (*Continued*) Chemical structures of some lichen-derived metabolites.

dibenzofurans (pannaric acid, usnic acid), depsides (homosekikaic acid, barbatic acid), depsones (picrolichenic acid), depsidones (salazinic acid), xanthones (lichexanthone), pulvinic acid derivatives (pulvinic acid), quinones (physcion, emodin), etc. (Figure 1.2).

Moreover, phytochemicals in large number were characterized in separate lichen species. A cyclic depsipeptide, arthogalin was reported from an endemic species of lichen of Galapagos Islands [16]. *Xanthoparmelia scabrosa*, native to central Asia, was analyzed for possible new chemical compounds along with previously reported compounds. New compounds (Figure 1.3) were phenylalanine-derived scabrosin esters [17],

Scabrocin esters	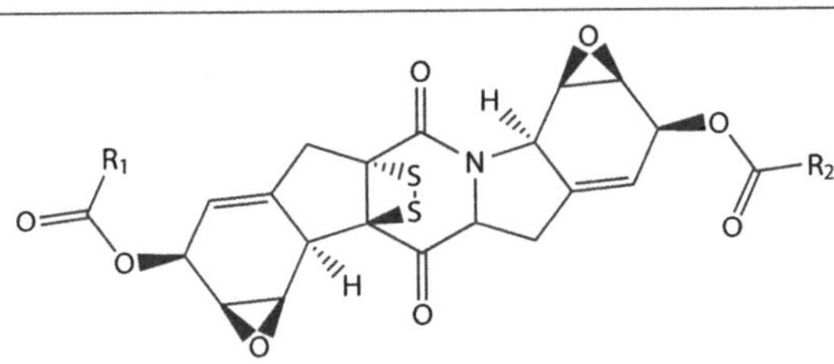
	$R_1 = R_2 = CH_3$ Dimethyl scabrocin ester $R_1 = R_2 = CH_3H_7$ Dipropyl scabrocin ester $R_1 = CH_3$, $R_2 = CH_3H_7$ Methylpropyl scabrocin ester $R_1 = CH_3$, $R_2 = C_5H_{11}$ Methylpentyl scabrocin ester $R_1 = CH_3H_7$, $R_2 = C_5H_{11}$ Propylpentyl scabrocin ester $R_1 = R_2 = C_5H_{11}$ Dipentyl scabrocin ester
Brominated depsidones	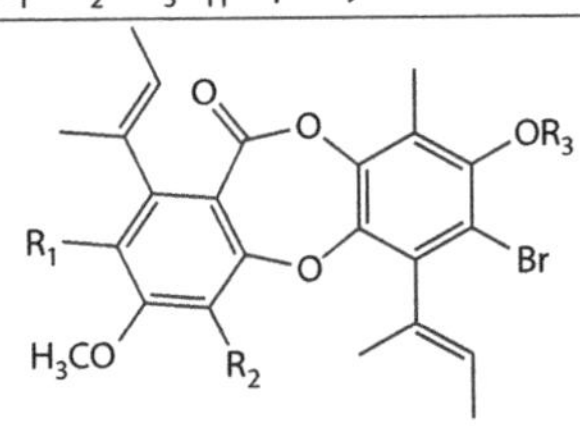
	$R_1 = H$, $R_2 = CH_3$, $R_3 = H$ Acarogobein A $R_1 = Br$, $R_2 = CHO$, $R_3 =$ 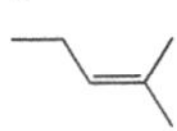 Acarogobein B
Brominated acetylenic fatty acids	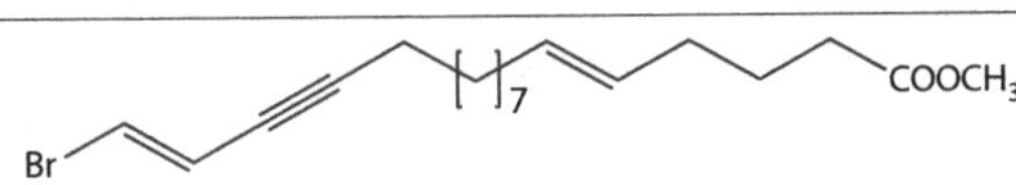 18-Bromo-(5E,17E)-octadeca-5,17-diene-15-yonoic acid methyl ester

Figure 1.3 Chemical structures of phenylalanine-derived scabrosin esters, brominated depsidones, brominated acetylenic fatty acids, monotetrahydrofuranic acetogenins and mono and di-prenylated xanthosides. *(Continued)*

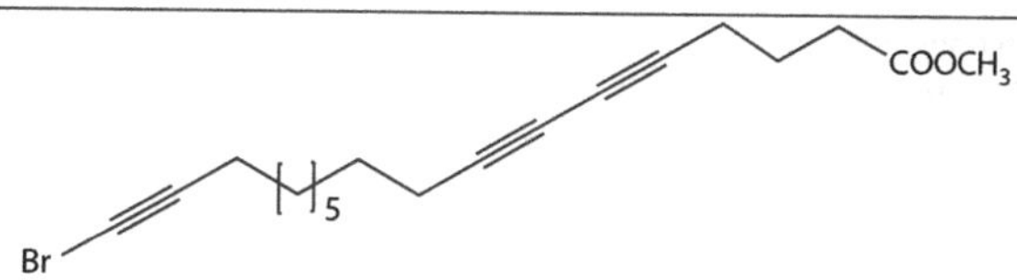

8-Bromo-octadeca-5,7,17-triyonoic acid methyl ester

16,18-Dibromo-(5E,17E)-octadeca-5,17-diene-5,7-diyonoic acid methyl ester

18,18-Dibromo-17-octadecene-5,7-diyonoic acid methyl ester

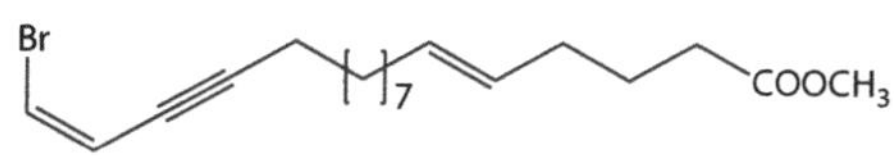

18-Bromo-(5E,17Z)-octadeca-5,7-diene-15-yonoic acid methyl ester

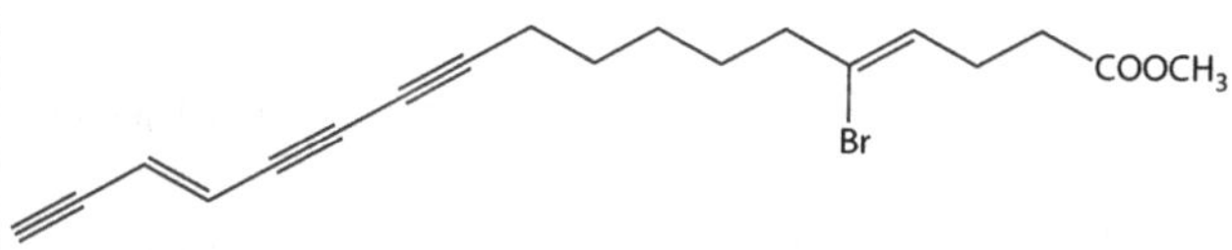

18-Bromo-(5E,15Z)-octadeca-5,5-diene-11,13,17-triyonoic acid methyl ester

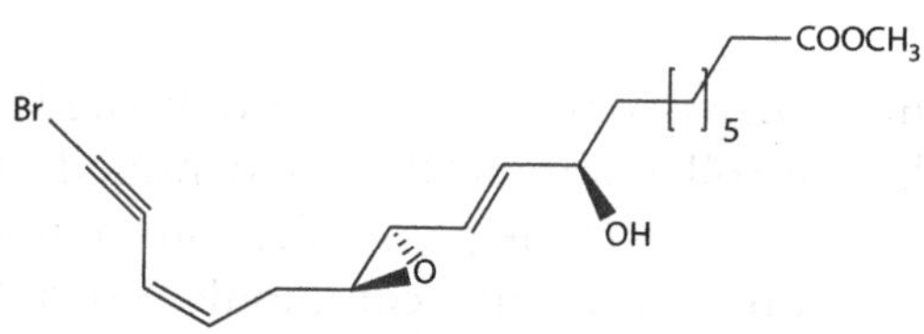

18-Bromo-9-hydroxy-12,13-*trans*-epoxy-(10E,15Z)-octadeca-10,15-diene-17-yonoic acid methyl ester

Figure 1.3 (*Continued*) Chemical structures of phenylalanine-derived scabrosin esters, brominated depsidones, brominated acetylenic fatty acids, monotetrahydrofuranic acetogenins and mono and di-prenylated xanthosides.

Monotetrahydrofuranic acetogenins	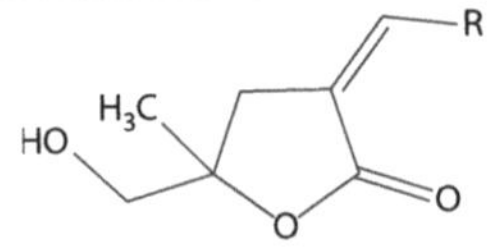
	R = 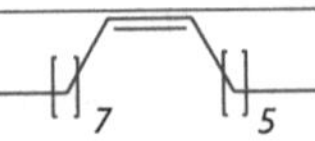 Tornabeatin A R =  Tornabeatin B R = 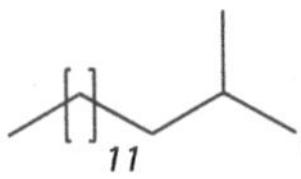 Tornabeatin C R = 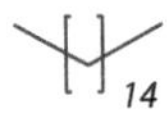 Tornabeatin D
Mono and di-prenylated xanthosides	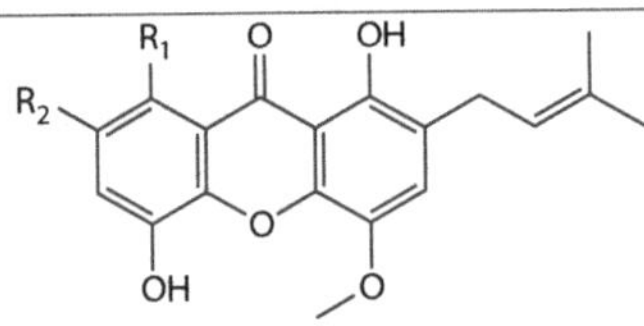
	R_1 = H, R_2 = β-D-Glu Umbilicaxanthoside A R_1 = 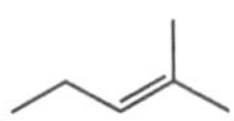, R_2 = β-D-Glu(1-4)-β-D-Glu Umbilicaxanthoside B

Figure 1.3 (*Continued*) Chemical structures of phenylalanine-derived scabrosin esters, brominated depsidones, brominated acetylenic fatty acids, monotetrahydrofuranic acetogenins and mono and di-prenylated xanthosides.

brominated depsidones [18], and brominated acetylenic fatty acids [9]. Rezanka and collaborators [19–22] identified γ-lactones with long-chain fatty acids, macrolactone glycoside, and monotetrahydrofuranic acetogenins (Figure 1.3) in some commonly grown lichen species of Central Asia. Xanthosides (mono- and di-prenylated xanthone glucosides) (Figure 1.3) that are isolated from the lichen, *Umbilicaria proboscidea*, by Torres *et al.* (2004), are found in Schrader of the Ural Series of mountains [23].

Besides many halogenated xanthone classes have been identified such as usneaxanthones A, Arthothelins are obtained several genus of *Lecanora* such as *L. broccha*, *L. straminea*, *L. flavopallescens*, *L. ingae*, *L. bolanderi*, *L. sulphurata L. andrewii*, *L. flavidopallens*, *L. pruinos and L. pinguis*, and the genus *Lecidella*, for example, *L. quema*, *L. asema*, *L. vorax*, *L. meiococca*, and *L. subalpicida*. The compounds of similar class also isolated from *Buellia sp.*, *Melanaria melanospora*, *Arthothelium pacificum*, and *Dimelaena australiensis*. Other class of Asemones was identified in *Micarea isabellina*, *M. austroternaria*, *Lecanora broccha*, *Buellia sp.*, *Lecidella asema*, *L. subalpicida*, and *Pertusaria pycnothelia* (Figure 1.4) [24–27]. Furthermore, Ratnayake and coworkers reported some heterocyclic polyketides from *Kibdelosporangium* sp. with class Isokibdelones (Figure 1.5) [28].

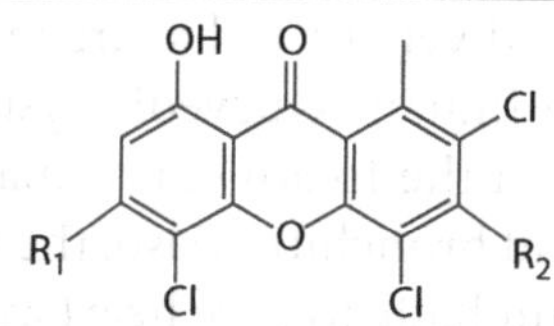

R_1 = OH, R_2 = OH Asemone

R_1 = OCH_3, R_2 = OH 3-*O*-methylasemone

R_1 = OH, R_2 = OCH_3 6-*O*-methylasemone

Figure 1.4 Chemical structure of Asemones.

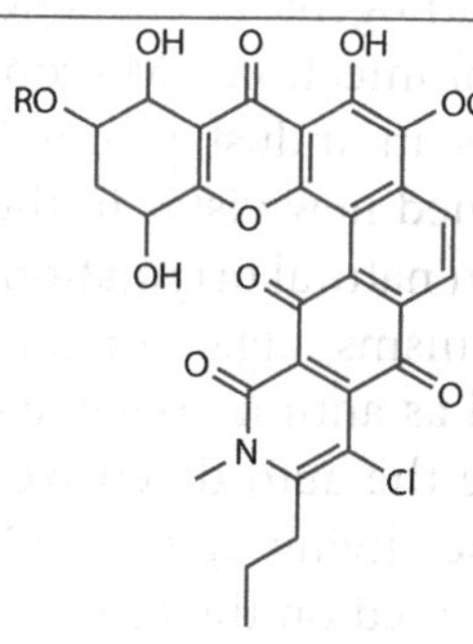	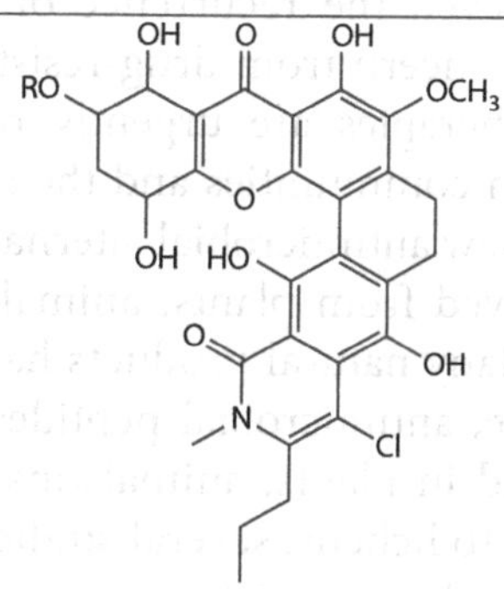
R = H Isokibdelone A R = *Rhambose* Isokibdelone A rhamnoside	Isokibdelone C

Figure 1.5 Chemical structure of Isokibdelones.

1.4 Current Scenario of Lichens and Derived Products Thereof

From ancient times, several species of lichens have been used for diverse purposes. Chiefly, the aqueous extract of purple pigments obtained from *Roccella species* was applied to colorable articles and clothings to make them colorful. This extract is used as pH indicator. Also, brown pigments from *Evernia*, *Ochrolechia*, and *Parmelia* genus were called *crottal* by Romans [29]. Lichens have found able to possess environmental biodetection property. This ability attributed to their sensible behavior toward many air pollutants, for example, oxides of sulfur and nitrogen and specific heavy metal ions [30].

Lichens are one of the important flora of the Asia continent having a vast topographical and climatic diversity and a large varieties have been identified in India [31]. In the Indian Ayurvedic system of medicine, lichens have been found effective in the treatment of many diseases like asthma, leprosy, wound-healing, and bronchitis. Also, the medication/recipe based on lichens was observed emphatic to cure heart as well as spleen size problems. In addition, lichens are consecutively used in the Unani system of medicines to treat a number of curable and incurable ailments, especially, heart problems and stomach-related disorders—inflammation of liver and spleen, diarrhea, and vomiting. Some species are also used as blood purifier and as a cosmetic ingredient [32].

Lichens have been used and recognized for centuries as folk medicines by native peoples of Indians, Chinese, Americans, Europeans, and Haitian in the curing of various ailments.

However, the recurrence of advanced pathogens is challenging and global concern from drug-resistant microbial infections. Microbial resistance therapies are urgently require efforts in industry and scientific research communities and the need has opened new vistas in the discovery of new antimicrobial alternatives based on natural origination that can be derived from plants, animals, microorganisms, algae, or mushrooms [33]. Many natural products have been used as antimicrobial agents. For example, antimicrobial peptides (AMPs) are the auto-defensive proteins reported in plants, animals/insects, and other living systems [34]. With respect to lichens, several studies were performed on the bioactive profile and found to be satisfactory results. The lichenologists express much attention on the lichens and their derived secondary metabolites on account of the inherited valuable bioactive properties. In this regard, about 50% certain species of the known lichens have shown excellent antimicrobial potential [35]. Some species are capable of possessing similar functionality as of vaccines. For instance, *Parmelia sulcata* and *Peltigera apthosa* are found very effective to treat rabies and thrush disease serially [36]. Recent researches demonstrated that many lichen species are found to have antimicrobial, antiprotozoal, anticancer, antiviral, antioxidant, antipyretic, anti-inflammatory, and antidiabetic properties [3–10, 33–36].

Minimum Inhibitory Concentrations (MIC) values for MeCOMe and MeOH extracts about 34 species of Lichens were investigated in detail against four isolates of pathogenic bacteria by Shrestha *et al.* (2014), reproduced in Table 1.2. In this study, most of the tested extracts most of the selected lichen extracts showed good antibacterial potential on the tested pathogenic isolates, *viz.*, *S. aureus* (0), *S. aureus* (COL), and *P. aeruginosa*, but showed no significant activity against *E. coli* strains. The antibacterial potential in considerable manner toward *E. coli* strain was observed for the extracts of *Letharia* species (*L. vulpine* and *L. columbiana*) as well as *Vulpicida canadensis.* In overall, MeCOMe extracts showed superior activity than MeOH extracts [37]. Furthermore, lichen-derived isolated compounds have been tested against several pathogenic organisms. In contrast, two carboxylic acids, namely, isousnic acid and protolichesterinic acid, derived from various lichen species saliently possessed bio-functionality for *Chlorella vulgaris* and *C. sorokiniana*. In addition, few compounds (usnic acid and evernyl) showed no effectiveness against *C. sorokiniana* and *Scenedesmus subspicatus*, but hold activeness for *C. vulgaris* isolates [34–36]. The antifungal effectiveness has been explored for various extracts of some of the lichen species such as *Ramalina roesleri*, *Usnea longissima*, *Parmelia reticulata*, and *Stereocaulon himalayense*. As a result, the extracts observed to have antifungal activity to a significant extent [38]. In addition, lichen-derived products were evaluated for the management of pests and their herbicidal as well as fungicidal activities have been recognized [39].

In Mediterranean civilizations, the most common colors violet and purple were obtained from mollusks and lichens with minute disadvantage of poor light-fastness abilities [40, 41].

Some specific mordants (especially salts of Cu, Fe, Cr, Al, Sn, Pb, etc.) are essentially pH dependant which give desirous colors such as purple, variables of red and orange, with certain dyes and pigments [42, 43]. *Rocella tinctoria* and *R. fuciformis* were most popular lichens which were documented to occur in the Cape Verde Islands (West Coast of Africa), South and Central America, East Indies, Madagascar, and Europe. In France, the lichen *Pertusaria dealbescens* Erichs was reported to have dyeing potential. Consequently, few species of the genus *Rocella* and *Ochrolechia tartarea* Linn. were also indicated the utilization of derived dyes and pigments in Scandinavian and Celtic countries [42, 44–46].

Table 1.2 *In vitro* MIC values (mg/ml) for various lichen species with respect to tested bacterial strains. (Shrestha 2014 Reproduced from Ref. No. [37] with permission. Copyright© Taylor & Francis, 2014).

Bacterial strains	***E. coli***		***P. aeruginosa***		***S. aureus***		**MRSA**	
Selected lichen species	**A**	**M**	**A**	**M**	**A**	**M**	**A**	**M**
Alectoria imshaugii	–	–	31.25	15.6	62.5	62.5	250	125
A. sarmentosa	–	–	31.25	15.6	62.5	31.25	125	62.5
Bryoria fuscescens	–	–	125	–	125	–	500	–
Cladonia furcata	–	–	250	–	500	–	500	–
Evernia prunastri	–	–	31.25	125	62.5	250	125	500
E. strum catawbiense	–	–	250	–	125	–	500	–
Flavocetraria nivalis	–	–	31.25	15.6	62.5	31.25	500	62.5
Hypogymnia physodes	–	–	62.5	250	62.5	62.5	62.5	250
Letharia columbiana	250	–	125	–	125	500	31.25	125
L. vulpina	125	500	125	500	3.9	15.6	31.25	125
Lobaria pulmonaria	–	–	–	–	–	–	–	–
Masonhalea richardsonii	–	–	250	500	125	125	125	250

(*Continued*)

Table 1.2 *In vitro* MIC values (mg/ml) for various lichen species with respect to tested bacterial strains. (Shrestha 2014 Reproduced from Ref. No. [37] with permission. Copyright© Taylor & Francis, 2014). (*Continued*)

Bacterial strains	***E. coli***		***P. aeruginosa***		***S. aureus***		**MRSA**	
Selected lichen species	**A**	**M**	**A**	**M**	**A**	**M**	**A**	**M**
Parmelia sulcata	-	–	125	–	250	–	125	–
Parmotrema reticulatum	–	–	250	–	250	–	250	–
Peltigera aphthosa	–	–	250	–	–	–	–	–
Platismatia glauca	–	–	250	–	500	–	500	–
Ramalina sinensis	–	–	15.6	250	15.6	–	62.5	–
Rhizoplaca chrysoleuca	–	–	7.8	125	7.8	250	15.6	500
R. haydenii	–	–	3.9	31.25	15.6	31.25	15.6	62.5
R. idahoensis	–	–	7.8	31.25	15.6	125	15.6	125
R. marginalis	–	–	7.8	125	7.8	250	7.8	500
R. melanophthalma	–	–	15.6	62.5	15.6	125	31.25	250
R. peltata	–	–	15.6	62.5	31.25	250	31.25	250
Sphaerophorus globosus	–	–	7.8	31.25	62.5	500	62.5	500
Thamnolia vermicularis	–	–	31.25	125	125	500	500	500

(*Continued*)

Table 1.2 *In vitro* MIC values (mg/ml) for various lichen species with respect to tested bacterial strains. (Shrestha 2014 Reproduced from Ref. No. [37] with permission. Copyright© Taylor & Francis, 2014). (*Continued*)

Bacterial strains	***E. coli***		***P. aeruginosa***		***S. aureus***		**MRSA**	
Selected lichen species	**A**	**M**	**A**	**M**	**A**	**M**	**A**	**M**
Tuckermannopsis ciliaris	–	–	125	–	62.5	250	250	500
Umbilicaria americana	–	–	500	–	500	–	–	–
U. mammulata	–	–	500	–	–	–	–	–
Usnea hirta	–	–	3.9	15.6	7.8	31.25	7.8	62.5
U. strigosa	–	–	3.9	31.25	7.8	32.5	15.6	250
Vulpicida canadensis	250	–	15.6	15.6	15.6	62.5	31.25	125
Xanthoparmelia chlorochroa	–	–	3.9	62.5	7.8	62.5	31.25	–
X. coloradoensis	–	–	7.8	62.5	7.8	250	15.6	500
X. wyomingica	–	–	15.6	15.6	15.6	62.5	62.5	500

A, acetone extract; M, methanolic extract; – , not considerable.

Neurodegenerative diseases are the diseases associated with the central nervous system that alter chiefly shape and functionality which strike on various dynamics of our body, for instance, talking, breathing, movement, heat activities, etc. Some lichen species have the potential to treat the mentioned diseases [47]. In our traditional medicine system as well as in the new system of medicine, nature-derived products have proved to be useful for medical and incurable diseases. Thus, nature-derived products have shown curing abilities to treat diabetic and neurological disorders [33–36, 47].

Enzyme inhibition activity was reported not only in derived secondary metabolites but also have been observed in solvent-based extracts from various species of lichenized fungi responsible for the effectiveness to have medicinal properties. For example, enzyme inhibition activity has been found effective toward trypsin, lipase, amylase, urease, tyrosinase, β-glucoronidase, cyclooxygenase, prolyl endopeptidase, lipoxygenase, aromatase, monoamine oxidase, thioredoxin reductase, xanthine oxidase, and many more [36–38]. Lichen-derived compounds such as usnic acid and its derivatives like physodic acid, lobaric acid, atranorin, evernic acid, zeorin, protolichestrinic acid, salazinic acid, and methyl β-orcinolcarboxylate were found to have inhibitors of some disease responsive enzymes [35, 36, 48]. However, lichens contain bioactive constituents in majority and they are under considerable attention within scientific community. Yet, more researches are needed on a larger scale to execute new dimension on lichen epidemics.

1.5 Conclusion and Future Outlook

Traditionally, lichens as whole and their extracts were used to treat various diseases and to dye the clothings around the globe since thousands of years. Lichens, having variable morphology and growing nature in clean ecosystem, possess indicating arrangements to the various environmental factors. They show sensitivity to the change in a small way. This seems to be because their chemical diversity is found different due to climatic phase such as variable temperature, altitude, and climate, and so, are called as universal environmental indicators. Many active derived ingredients from lichen species are found "lead" chemical structures and possess diverse biological properties such as antimicrobial, antiprotozoal, anticancer, antiviral, antioxidant, anti-inflammatory, and antidiabetic activities. Additionally, lichens cannot be potentially sidelined for pest related management, and they can be a better alternative to the synthetic counterparts. However, lichens pay much attention between scientific communities, yet they must be explored further so that they may be utilized for the social as well as economic benefits to our society to increase the natural product-based economic enhancements.

References

1. Mitrovic, T., Stamenkovic, S., Cvelkovic, V., Nikolic, M., Tosic, S., Stojicic, D., Lichen as a source of biologically active compound. *Biol. Nyssana*, 2, 1–6, 2011.
2. Molnár, K. and Farkas, E., Current results on biological activities of lichen secondary metabolites: A review. *Zeitschrift für Nat.*, C, 65, 3–4, 157–173, 2010.
3. Seaward, M.R.D., Enviromental role of lichen, in: *Lichen Biology*, II ed., pp. 274–298, Cambridge University Press, Cambridge, 2008.
4. Huneck, S., The significance of lichens and their metabolites. *Naturwissenschaften*, 86, 559–570, 1999.
5. Huneck, S., New Results on the Chemistry of Lichen Substances, in: *Progress in the chemistry of organic natural products*, W. Herz, H. Falk, G.W. Kirby, R.E. Moore (Eds.), pp. 81–95, Springer, Wien, New York, 2001.
6. Baron, G., *Understanding Lichens: A Beginner's Guide Lichens*, Richmond Publishing Co, Slough, UK, 1999.
7. Ekman, S. and Tønsberg, T., Most species of Lepraria and Leproloma form a monophyletic group closely related to Stereocaulon. *Mycol. Res.*, 106, 1262–1276, 2002.
8. Smith, C.W., Aptroot, A., Coppins, B.J., Fletcher, A., Gilbert, O.L., James, P.W., Wolseley, P.A. (Eds.), *The lichens of Great Britain and Ireland*, British Lichen Society, London, 2009.
9. Rezanka, T. and Dembitsky, V.M., Fatty acids of lichen species from tian shan mountains. *Phytochem.*, 51, 963–968, 1999.
10. Lichen thallus types http://www.lichens.lastdragon.org/faq/lichenthallustypes.html (Retrieved on 10.12.2019).
11. Asdńna, Y., Mikrochemischer nachweis der flechtenstoffe. *Int. J. Japan. Bot.*, 12, 516–525, 1936.
12. Zopf, W., Zur Kenntniss der Flechtensloffe. *Ann. Chem.*, 284, 107–132, 1895.
13. Shibata, S., Yasuhiko Asahina (1880–1975) and his studies on lichenology and chemistry of lichen metabolites. *Bryol.*, 103, 4, 710–719, 2000.
14. Dandapat, M. and Paul, S., Secondary metabolites from lichen *Usnea longissima* and its pharmacological relevance. *Pharmacog. Res.*, 11, 2, 103–109, 2019.
15. Boustie, J. and Grube, M., Lichens—A promising source of bioactive secondary metabolites. *Plant Genetic Resour.*, 3, 2, 273–287, 2005.
16. Huneck, S., Himmelreich, U., Nicholson, G., Arthogalin, a cyclic depsipeptide from the lichen Arthothelium galapagoense. *Zeitschrift für Nat. B.*, 50, 7, 1101–1103, 1995.
17. Ernst-Russell, M.A., Chai, C.L.L., Hurne, A.M., Waring, P., Hockless, D.C.R., Elix, J.A., *Aust. J. Chem.*, 52, 279–283, 1999.

18. Řezanka, T. and Guschina, I.A., Brominated depsidones from *Acarospora gobiensis*, a lichen of Central Asia. *J. Nat. Prod.*, 62, 12, 1675–1677, 1999, 1999.
19. Řezanka, T., Temina, M., Hanuš, L., Dembitsky, V.M., The tornabeatins, four tetrahydro-2-furanone derivatives from the lichenized ascomycete Tornabea scutellifera (With.) JR Laundon. *Phytochem.*, 65, 18, 2605–2612, 2004.
20. Řezanka, T. and Guschina, I.A., Glycosidic compounds of murolic, protoconstipatic and allo-murolic acids from lichens of Central Asia. *Phytochem.*, 54, 6, 635–645, 2000.
21. Řezanka, T. and Guschina, I.A., Glycoside esters from lichens of central Asia. *Phytochem.*, 58, 3, 509–516, 2001.
22. Řezanka, T., Jáchymová, J., Dembitsky, V.M., Prenylated xanthone glucosides from Ural's lichen Umbilicaria proboscidea. *Phytochem.*, 62, 4, 607–612, 2003.
23. Torres, A., Hochberg, M., Pergament, I., Smoum, R., Niddam, V., Dembitsky, V.M., Temina, M., Dor, I., Lev, O., Srebnik, M., Enk, C.D., A new UV-B absorbing mycosporine with photo protective activity from the lichenized ascomycete Collema cristatum. *Eur. J. Biochem.*, 271, 4, 780–784, 2004.
24. Santesson, J., Chemical Studies on Lichens. 12. A New Lichen Xanthone from *Lecanora reuteri*. *Acta Chem. Scand.*, 22, 1698–1712, 1968.
25. Masters, K.S. and Bräse, S., Xanthones from fungi, lichens, and bacteria: The natural products and their synthesis. *Chem. Rev.*, 112, 7, 3717–3776, 2012.
26. Calcott, M.J., Ackerley, D.F., Knight, A., Keyzers, R.A., Owen, J.G., Secondary metabolism in the lichen symbiosis. *Chem. Soc. Rev.*, 47, 5, 730–1760, 2018.
27. Tuong, T.L., Aree, T., Do, L.T., Nguyen, P.K., Wonganan, P., Chavasiri, W., Dimeric tetrahydroxanthones from the lichen. *Usnea aciculifera. Fitoterapia*, 137, 104194, 2019.
28. Ratnayake, R., Lacey, E., Tennant, S., Gill, J.H., Capon, R.J., Isokibdelones: Novel heterocyclic polyketides from a Kibdelosporangium sp. *Org. Lett.*, 8, 23, 5267–5270, 2006.
29. Muggia, L., Schmitt, I., Grube, M., Lichen as treasure chest of natural products. *SIM News*, 59, 85–97, 2009.
30. Yousuf, S. and Choudhary, M.I., Lichens: Chemistry and biological activities, in: *Studies in natural products chemistry*, vol. 43, F.R.S. Atta-ur-Rahman (Ed.), pp. 223–259, Elsevier, Oxford, UK, 2014.
31. Upreti, D.K., Bajpai, R., Nayaka, S., Lichenology: Current research in India, in: *Plant biology and biotechnology*, B. Bahadur, M. Venkat Rajam, L. Sahijram, K. Krishnamurthy (Eds.), pp. 263–280 2015, Springer, New Delhi, India, 2015.
32. Kirtikar, K.R. and Basu, B.D., Archaeological Survey of India, Central Archaeological Library, New Delhi, in: *Indian Medicinal Plants Vol. 4*, Lalit Mohan Basu Publication, Allahabad, India, 1936.
33. Pisoschi, A.M., Pop, A., Georgescu, C., Turcuş, V., Olah, N.K., Mathe, E., An overview of natural antimicrobials role in food. *Eur. J. Med. Chem.*, 143, 922–935, 2018.

34. Tang, S.S., Prodhan, Z.H., Biswas, S.K., Le, C.F., Sekaran, S.D., Antimicrobial peptides from different plant sources: Isolation, characterisation, and purification. *Phytochem.*, 154, 94–105, 2018.
35. Kosanić, M. and Ranković, B., Lichen secondary metabolites as potential antibiotic agents, in: *Lichen Secondary Metabolites*, B. Ranković (Ed.), pp. 99–127, Springer, Cham, 2019.
36. Cheng, G., Dai, M., Ahmed, S., Hao, H., Wang, X., Yuan, Z., Antimicrobial drugs in fighting against antimicrobial resistance. *Front. Microbiol.*, 7, 470, 2016.
37. Shrestha, G., Raphael, J., Leavitt, S.D., St. Clair, L.L., *In vitro* evaluation of the antibacterial activity of extracts from 34 species of North American lichens. *Pharmaceut. Biol.*, 52, 10, 1262–1266, 2014.
38. Goel, M., Sharma, P.K., Dureja, P., Rani, A., Uniyal, P.L., Antifungal activity of extracts of the lichens *Parmelia reticulata*, *Ramalina roesleri*, *Usnea longissima* and *Stereocaulon himalayense*. *Arch. Phytopathol. Plant Prot.*, 44, 13, 1300–1311, 2011.
39. Dayan, F.E. and Romagni, J.G., Lichens as a potential source of pesticides. *Pesticide Outlook*, 12, 6, 229–232, 2001.
40. Cardon, D., *Natural Dyes: Sources, Tradition, Technology and Science*, Archetype Publications, London, 2007.
41. Casselman, K.D., *Lichen Dyes and Dyeing: The New Source Book*, Dover Publications, Mineola, New York, 2001.
42. Ferreira, E.S., Hulme, A.N., McNab, H., Quye, A., The natural constituents of historical textile dyes. *Chem. Soc. Rev.*, 33, 6, 329–336, 2004.
43. Casselman, K.D., Magic, mystery and mayhem: Lichen dyes old and new, in: *Publication of the Natural Dyes Product Research and Development Project*, EVTEK Institute of Art and Design, Vantaa, 2002.
44. Rather, L.J., Jameel, S., Ganie, S.A., Bhat, K.A., Lichen Derived Natural Colorants: History, Extraction, and Applications, in: *Handbook of Renewable Materials for Coloration and Finishing*, Yusuf, M. (Ed.), pp. 103–114, 2018.
45. Shukla, P., Upreti, D.K., Nayaka, S., Tiwari, P., Natural dyes from Himalayan lichens. *Ind. J. Trad. Knowl.*, 13, 1, 195–201, 2014.
46. Gupta, V.K., Fundamentals of Natural Dyes and Its Application on Textile Substrates, in: *Chemistry and Technology of Natural and Synthetic Dyes and Pigments*, A.K. Samanta, N.S. Awwad, H.M. Algarni (Eds.), pp. 2019 Intech Open Ltd., London, UK, 2019.
47. Kosanić, M. and Ranković, B., Antineurodegenerative and Antidiabetic Activity of Lichens, in: *Lichen Secondary Metabolites*, B. Ranković (Ed.), pp. 215–236, Springer, Cham, 2019.
48. Kekuda, T.P., Lavanya, D., Pooja, R., Lichens as promising resources of enzyme inhibitors: A review. *J. Drug Delivery Therapeut.*, 9, 2-s, 665–676, 2019.

2

Comparative Perspectives on Extraction Methods for Organic Metabolites and Pollutants from Lichens

Patricia Forbes*, **Leandri van der Wat and Johanna Strumpher**

Department of Chemistry, Faculty of Natural and Agricultural Science, University of Pretoria, Pretoria, South Africa

Abstract

Lichens have been used extensively as biomonitors, due to their uptake of both organic and inorganic air pollutants. These organic air pollutants include polycyclic aromatic hydrocarbons, phenols, polychlorinated dibenzo-p-dioxins (PCDDs), polychlorinated dibenzofurans (PCDFs), and polychlorinated biphenyls (PCBs). In addition, lichens produce a range of organic secondary metabolites (aliphatic acids, esters, terpenes, quinones, etc.) which have found various uses such as in pharmacology and the dye industry. A number of methods are available to extract organic compounds from the lichen matrix and range from traditional Soxhlet and ultrasonic extraction methods, to microwave-assisted extraction and most recently the QuEChERS technique. For each extraction option, there are parameters which require optimization to enable efficient extraction, while tuning thereof may adjust the efficiencies of extraction of different organic compound classes from the lichen. Experimental parameters of relevance include choice of solvent, extraction time, solvent volume, and temperature. Here, we discuss the different reported approaches for extraction of organic lichen-derived compounds and compare advantages and disadvantages thereof. We also highlight extract cleanup procedures and the impact these may have on the final lichen extract composition and analysis thereof.

Keywords: Extraction, secondary metabolites, organic pollutants, Soxhlet, ultrasonic extraction, microwave-assisted extraction, QuEChERS

**Corresponding author*: patricia.forbes@up.ac.za

Mohd Yusuf (ed.) Lichen-Derived Products: Extraction and Applications, (27–74) © 2020 Scrivener Publishing LLC

2.1 Introduction

Many plants and other organisms are being investigated as potential sources of useful products for numerous applications, including the quest for medicinal and pharmaceutical products which can reduce disease prevalence and enhance life expectancy. This wide realm of "bioprospecting" also applies to lichens, in which many potentially useful compounds have already been identified, as discussed briefly in this chapter and more broadly in this book.

In order to effectively identify potentially useful compounds in lichens, it is necessary to first determine what compounds are present in the thalli of a particular lichen species of interest (based on the lichen species which is prevalent in the area where the study is being conducted, for instance). The first step, post harvesting of the lichen, is to efficiently extract possible target compounds from the collected material. There are numerous parameters which can be evaluated and optimized during this process; from extraction temperature and extraction time, to solvent employed and extraction technique.

Sample preparation introduces the most uncertainty in the final analytical outcome and, as a result, is considered the most crucial step of any analysis [1]. In order to optimize sample preparation for analytes of interest, these should be selectively isolated and the noise and potential interference from the matrix should be minimized. Sample preparation steps also often serve as pre-concentration measures for analytes that are present at trace levels, in order to improve the sensitivity of analysis [2].

After extraction, it may be necessary to first cleanup the extract before analysis, depending on the analytical method employed. Inlet liners and columns of gas chromatographs can be very severely negatively affected by residual chlorophyll in sample extracts, while liquid chromatography (LC) is more robust in this regard due to the sample dilution effects of the liquid mobile phase. Different approaches to lichen extract cleanup are thus discussed in Section 2.5 of this chapter, while analytical approaches employed in the analysis of lichen extracts are covered in Section 2.6.

This information is also important in the wider context of utilizing useful compounds present in lichen extracts, as these will usually require extensive purification prior to further testing (for example, for biological efficacy or biocompatibility, or even dye permanency). These purification steps may thus be informed by the analytical sample extraction processes developed during the determination of the chemical fingerprint of the lichen species, in terms of the compounds found to be present in the extract via chromatographic analysis with, for example, mass spectrometric (MS) detection.

This chapter therefore covers the extraction methods (with related parameters) which have been employed in the use of lichens as biomonitors of organic air pollutants, as well those extraction methods relating to lichen metabolites. The methods and procedures reported in the literature are compared and advantages and disadvantages of each are discussed. Other options are also highlighted. Means to clean up the extracts prior to analysis and typical analytical methods are included, as it is expected that this information will inform future studies on lichen-derived products and open the door to the discovery of novel compounds and applications.

2.1.1 Organic Lichen-Derived Compounds

A wide range of organic compounds have been extracted from lichens. These may be broadly categorized into organic air pollutants and lichen metabolites. The former arises from the uptake of these organic pollutants from the environment, specifically from the atmosphere, by the lichen thallus. In contrast, lichen metabolites are produced by the lichen itself. These two categories of organic lichen-derived compounds are discussed briefly in the following sections before details on the options for their extraction are explored.

2.1.1.1 Organic Air Pollutants

Lichens have been used extensively for many years as biomonitors of air pollution, due to their direct uptake of air pollutants. Initially, lichens were used as biomonitors of inorganic air pollutants such as metals associated with particles, but more recently, the potential of lichens to serve as biomonitors of organic air pollution was realized [3, 4]. The organic air pollutants which have been detected in lichens include polycyclic aromatic hydrocarbons (PAHs), phenols, polychlorinated dibenzo-p-dioxins (PCDDs), polychlorinated dibenzofurans (PCDFs), polychlorinated biphenyls (PCBs), and organochlorine pesticides (OCPs).

2.1.1.2 Lichen Metabolites

Lichens produce metabolic compounds as either primary or secondary metabolites. Primary metabolites encompass amino acids, proteins, carotenoids, polyols, polysaccharides, and vitamins [5]. These compounds play a direct role in metabolic activities, such as respiration or reproduction. Primary metabolites are found intracellularly, synthesized by either the fungal or algal component of the lichen, or through a combination of both. Conversely, secondary metabolites are exclusively synthesized by the fungal component of the lichen. These metabolites are found extracellularly, on the exterior of the hyphae [6], and do not play a direct role in the lichen's metabolic processes. However, secondary lichen metabolites may serve a number of biological roles. Some secondary metabolites, such as vulpinic acid for example, may protect the lichen against consumption by herbivores [7]. Another widely studied secondary metabolite, usnic acid, has been shown to exhibit antimicrobial and larvicidal effects, as well as possibly playing a role in protecting the lichen from ultraviolet (UV) radiation [8].

Primary metabolites are usually not unique to lichens and can be found in other fungi, algae, or higher plants. Most, but not all, of the known secondary metabolites identified in lichens are specific to these organisms [9], and some are characteristically found in certain lichen species, as is evident from Figure 2.1 which shows the different amino acid profiles of some Indian lichen species. In 1971, Asahina and Shibata described and classified about 80 lichen metabolite structures. However, improvements in instrumentation and analytical techniques led to the rapid discovery and structure elucidation of many more lichen substances. A catalogue has been published describing known lichen substances, with the second edition published in 1993 comprising 605 lichen substances, while the third edition released in 2014 contains 854 lichen substances [10].

Classification of secondary metabolites is based on biosynthetic pathways, namely, the acetyl-polymalonyl, mevalonic, or shikimic acid pathways [5]. The acetyl-polymalonyl pathway accounts for the synthesis of most secondary metabolites, with the majority existing as phenolic compounds. Table 2.1 shows the classification of lichen secondary metabolites, according to the classification system proposed by Culberson and Elix [9], with examples of each class of secondary metabolite. The di- and tri-aryl derivatives of simple phenolic units (such as depsides, depsidones, depsones, and dibenzofurans) are classified as such, since these compounds are formed through the linkage of simple orcinol-type phenolic units shown in Figure 2.2.

The mevalonic-acid pathway leads to the synthesis of terpenoids. Simple monoterpenes consist of two isoprene units (C_5H_8) and are characterized by a backbone of ten carbons, whereas diterpenes consist of four isoprene units and are characterized by a backbone of twenty carbons. Similarly, triterpenes consist of six isoprene units and are characterized by a backbone of thirty carbons. Sesterterpenes are more rare terpenoids, consisting of five isoprene units, and thus characterized by a backbone of 25 carbon atoms.

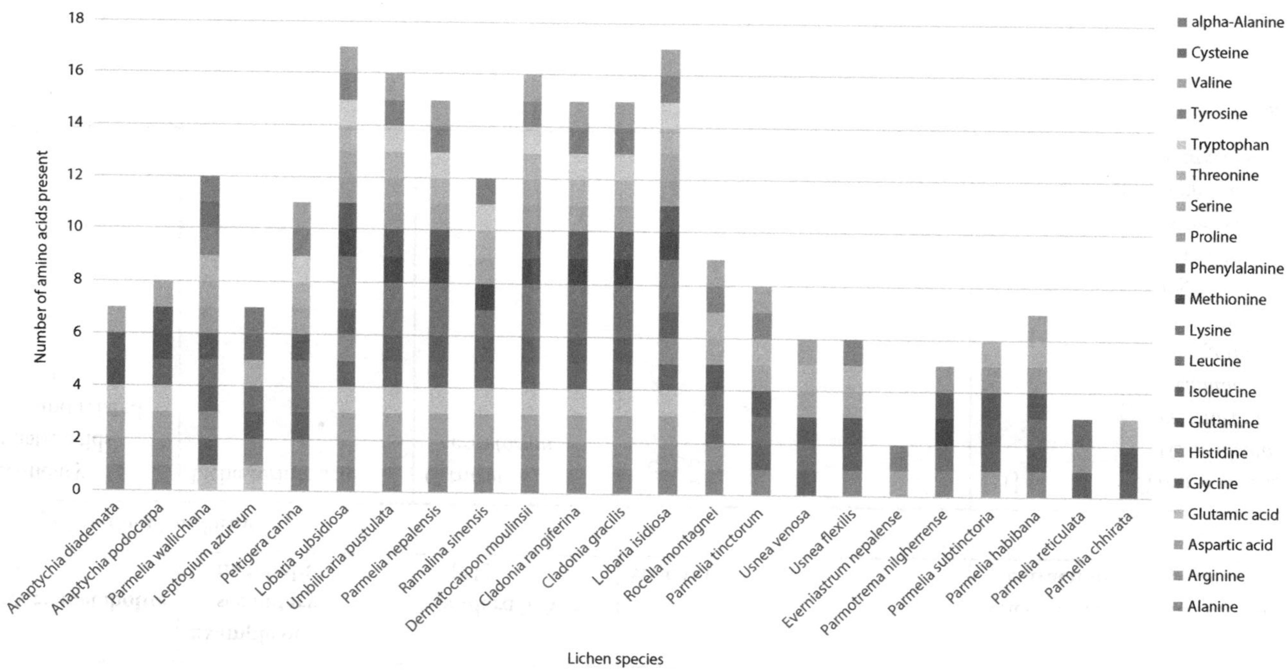

Figure 2.1 Amino acid profiles of some Indian lichen species, adapted from [6]. The amino acids were separated from extracts by paper chromatography.

Table 2.1 Classification of secondary metabolites as synthesized by lichens.

Secondary metabolite class	Example of secondary metabolite	Standard lichen [31]	Structure	Reference of extraction	Referenced lichen species
1. ***Acetyl-polymalonyl pathway***					
1.1 Secondary aliphatic acids, esters and related derivatives	Lichesterinic acid	*Cetraria ericetorum*		[11]	*Cetraria lichens (C. aculeata, ericetorum, icelandica, laevigata, muricata, nigricans, obtusata, odontella, sepincola, xizangensis)*

(*Continued*)

Table 2.1 Classification of secondary metabolites as synthesized by lichens. (*Continued*)

Secondary metabolite class	**Example of secondary metabolite**	**Standard lichen [31]**	**Structure**	**Reference of extraction**	**Referenced lichen species**
1.2 Polyketide derived aromatic compounds					
1.2.1 Mononuclear phenolic compounds	Orsellinic acid	*Pseudocyphellaria crocata*	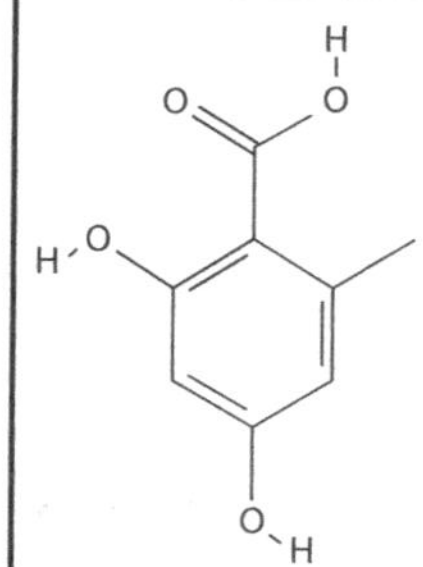	[12]	*Parmotrema lichens* (*P. tinctorum, andinum, austrosinense*)

(*Continued*)

Table 2.1 Classification of secondary metabolites as synthesized by lichens. (*Continued*)

Secondary metabolite class	Example of secondary metabolite	Standard lichen [31]	Structure	Reference of extraction	Referenced lichen species
1.2.2 Di- and tri-aryl derivatives of simple phenolic units					
1.2.2.1 Depsides, tridepsides, and benzyl esters	Atranorin (depside)	*Parmotrema tinctorum*	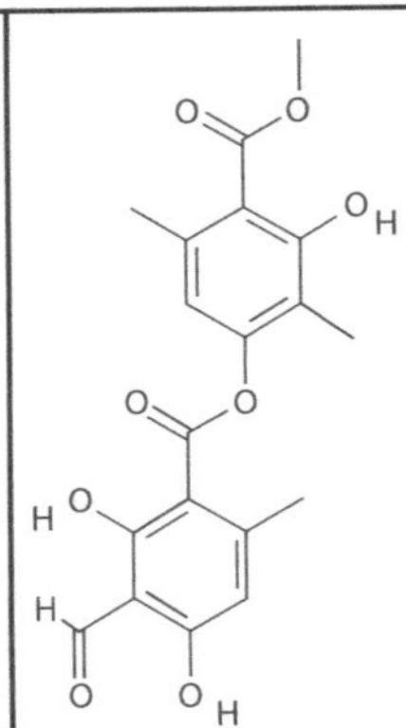	[12]	*Parmotrema lichens (P. tinctorum, andinum, praesorediosum, grayanum, austrosinense)*

(*Continued*)

Table 2.1 Classification of secondary metabolites as synthesized by lichens. (*Continued*)

Secondary metabolite class	Example of secondary metabolite	Standard lichen [31]	Structure	Reference of extraction	Referenced lichen species
	Crustinic acid (tridepside)	*Umbilicaria crustulosa*	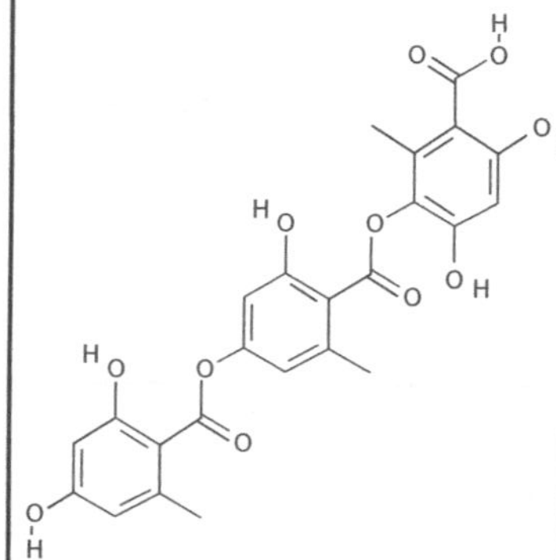	[13]	*Umbilicaria crustulosa*
	Barbatolic acid (benzyl ester)	*Bryoria capillaris*	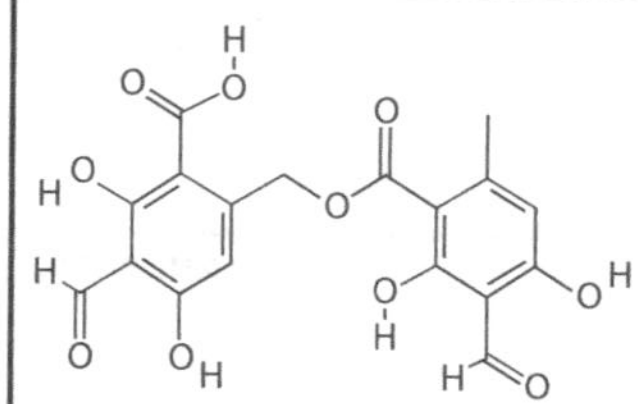	[14]	*Bryoria capillaris*

(*Continued*)

Table 2.1 Classification of secondary metabolites as synthesized by lichens. (*Continued*)

Secondary metabolite class	Example of secondary metabolite	Standard lichen [31]	Structure	Reference of extraction	Referenced lichen species
1.2.2.2 Depsidones and diphenyl ethers	Protocetraric acid (depsidone)	*Flavoparmelia caperata, Ramalina farinacea*	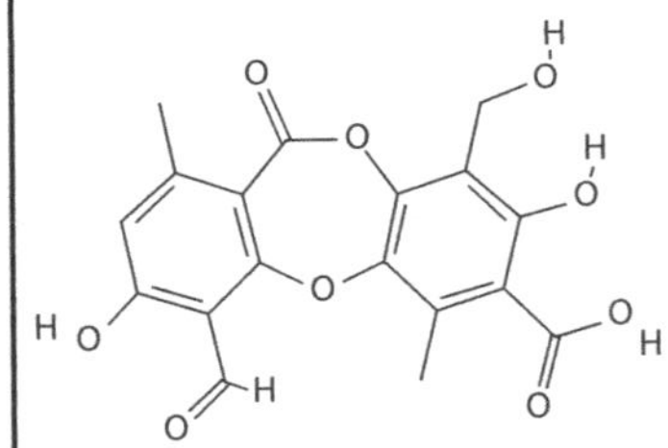	[11]	*Cetreria icelandica, Cetreria xizangensis*
	Alectoronic acid (diphenyl ether)	*Asahinea chrysantha, Parmelia birulae*	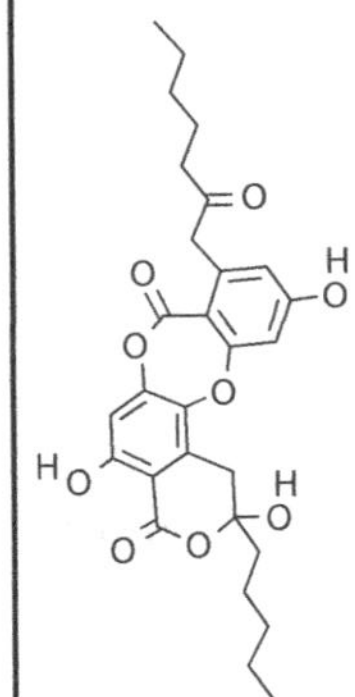	[12]	*Parmotrema tinctorum*

(*Continued*)

Table 2.1 Classification of secondary metabolites as synthesized by lichens. (*Continued*)

Secondary metabolite class	Example of secondary metabolite	Standard lichen [31]	Structure	Reference of extraction	Referenced lichen species
1.2.2.3 Depsones	Picrolichenic acid	*Pertusaria amara*	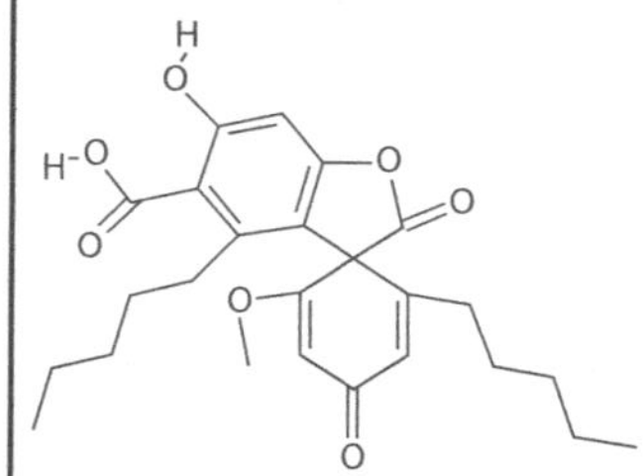	[15]	*Pertusaria truncata*
1.2.2.4 Dibenzofurans, usnic acids and derivatives	Strepsilin (Dibenzofuran)	*Cladonia strepsilis*	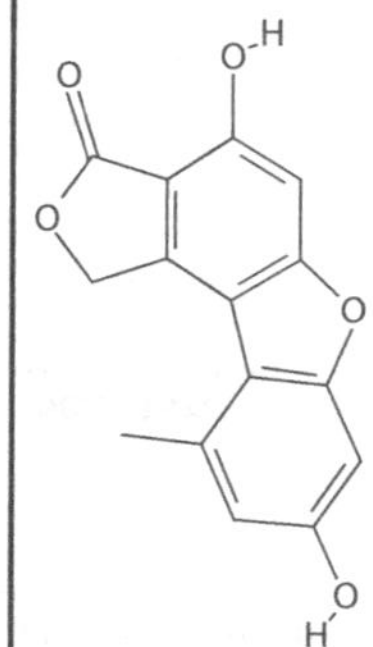	[16]	*Stereocaulon azoreum*

(*Continued*)

Table 2.1 Classification of secondary metabolites as synthesized by lichens. (*Continued*)

Secondary metabolite class	Example of secondary metabolite	Standard lichen [31]	Structure	Reference of extraction	Referenced lichen species
	Usnic acid	*Evernia prunastri, Cladonia stellaris*	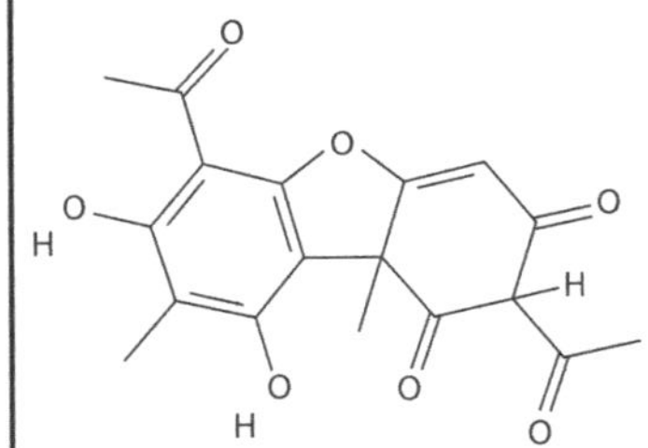	[17]	*Usnea Florida, Flavoparmelia caperata*
1.2.3 Anthraquinones and biogenetically related xanthones	Parietin	*Xanthoria parietina*	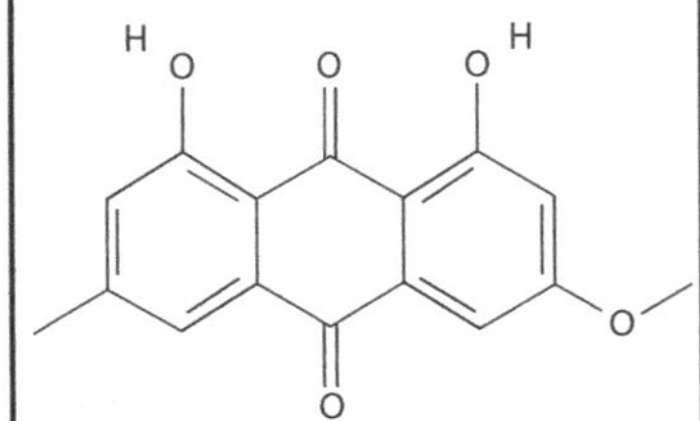	[18]	*Xanthoria parietina*

(*Continued*)

Table 2.1 Classification of secondary metabolites as synthesized by lichens. (*Continued*)

Secondary metabolite class	Example of secondary metabolite	Standard lichen [31]	Structure	Reference of extraction	Referenced lichen species
1.2.4 Naphthaquinones	Trypethelone	*Trypethelium eluteriae*	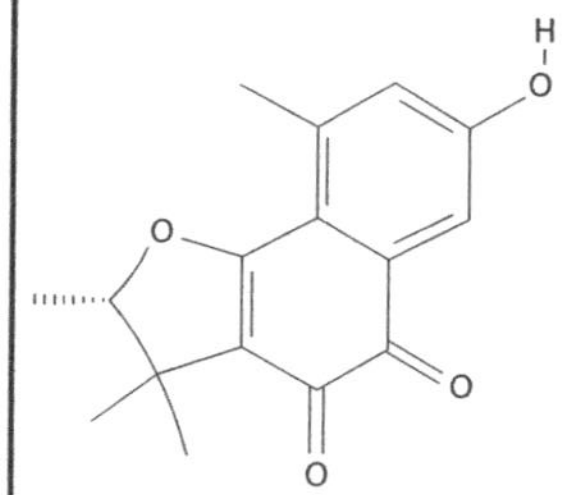	[19]	*Trypethelium eluteriae*
1.2.5 Xanthones	Lichexanthone	*Hypotrachyna formosana, Lecidella stigmatea*	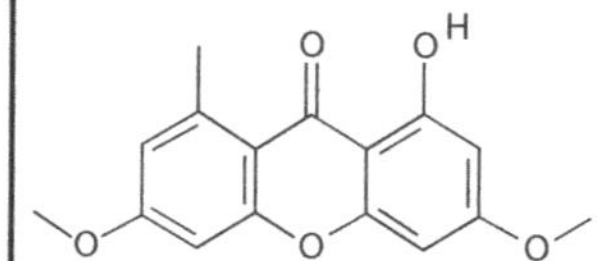	[20]	*Pyxine consocians, Heterodermia leucomelos*
2. *Mevalonic acid pathway*					
2.1 Mono, di-, sester-, and triterpenes	Limonene (Monoterpene)	*Evernia prunastri*	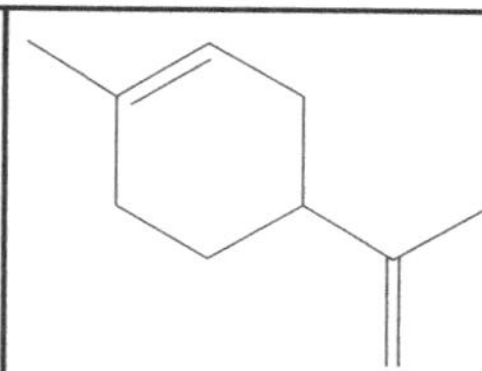	[21]	*Parmelia perlata*

(*Continued*)

Table 2.1 Classification of secondary metabolites as synthesized by lichens. (*Continued*)

Secondary metabolite class	Example of secondary metabolite	Standard lichen [31]	Structure	Reference of extraction	Referenced lichen species
	(-)-Sandaracopimaric acid (Diterpene)	*Ramalina hierrensis*	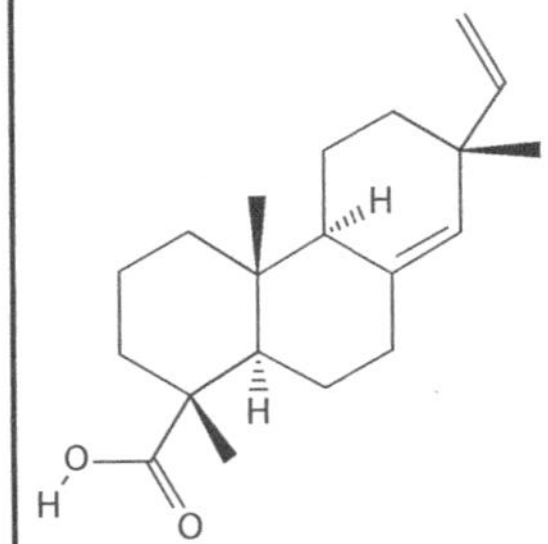	[22]	*Ramalina hierrensis*
	Retigeranic acid A (sesterterpene)	*Lobaria isidiosa*	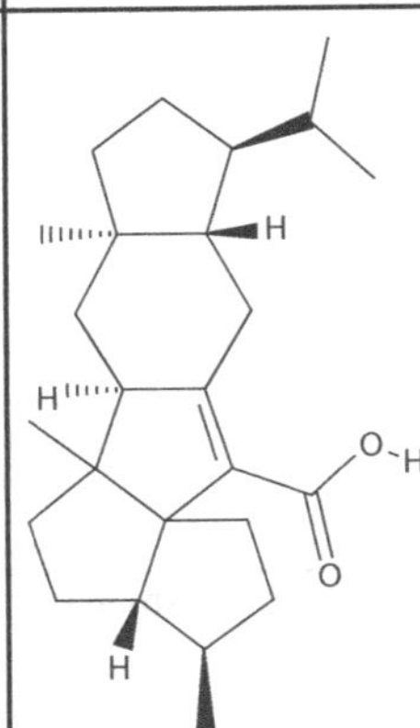	[23]	*Lobaria retigera*

(*Continued*)

Table 2.1 Classification of secondary metabolites as synthesized by lichens. (*Continued*)

Secondary metabolite class	Example of secondary metabolite	Standard lichen [31]	Structure	Reference of extraction	Referenced lichen species
	Lupeol (triterpene)	*Flavocetraria nivalis*	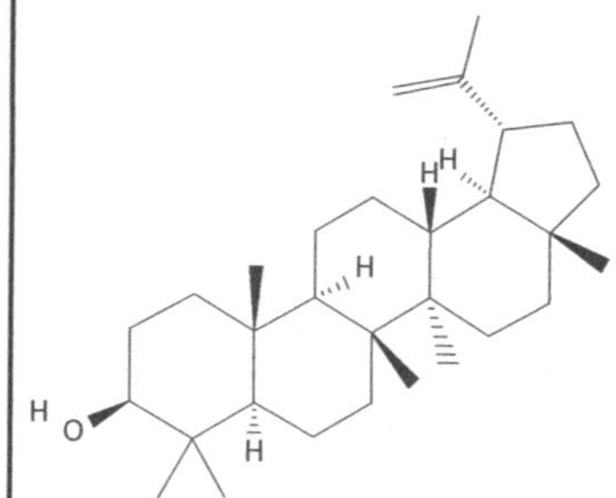	[16]	*Stereocaulon azoreum*
2.2 Steroids	Ergosterol	*Cladonia rangiferina, Xanthoria parietina*	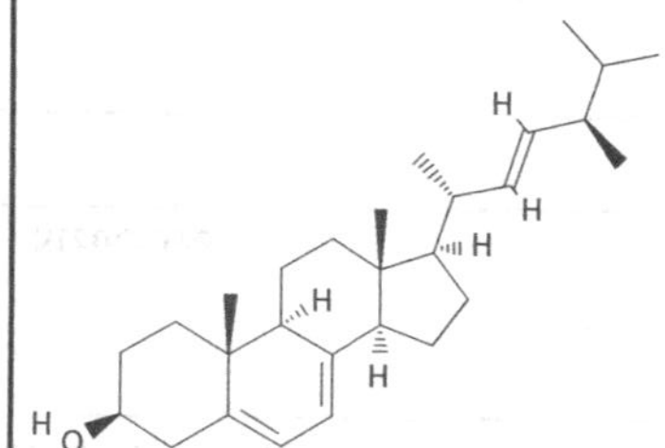	[24]	*Lobaria pulmonaria, Lobaria scrobiculata, Usnea longissima*

(*Continued*)

Table 2.1 Classification of secondary metabolites as synthesized by lichens. (*Continued*)

Secondary metabolite class	Example of secondary metabolite	Standard lichen [31]	Structure	Reference of extraction	Referenced lichen species
3. Shikimic acid pathway					
3.1 Terphenylquinones	Polyporic acid	*Pseudocyphellaria coronata*	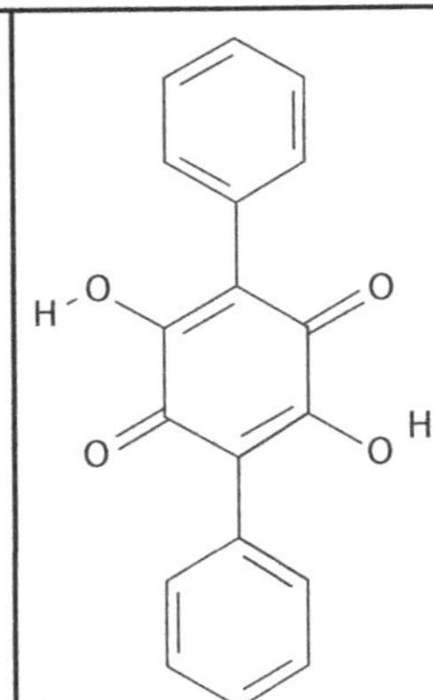	[25]	*Sticta mougeotiana*

(*Continued*)

Table 2.1 Classification of secondary metabolites as synthesized by lichens. (*Continued*)

Secondary metabolite class	Example of secondary metabolite	Standard lichen [31]	Structure	Reference of extraction	Referenced lichen species
3.2 Pulvinic acid derivatives	Vulpinic acid	*Letharia vulpina*	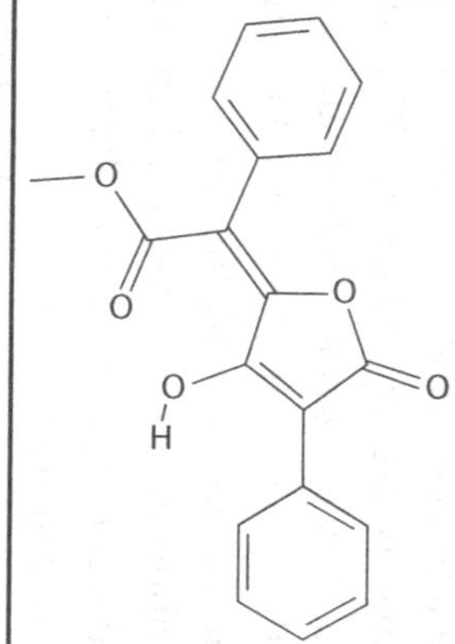	[26]	*Vulpicida canadensis*

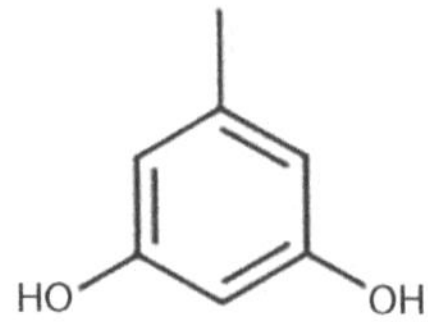

Figure 2.2 The molecular structure of orcinol.

2.2 Sample Preparation and Solvent Extraction

Prior to extraction, the fresh lichen samples can be oven dried to remove excess moisture and achieve stable mass. An oven temperature of 30°C, held for 4–5 days, is usually sufficient [27]. Alternatively, freeze drying can also be used. Any exogenous matter, such as bark or leaves, should be removed from the lichen using tweezers. Lichens are ground to a powder prior to extraction to provide a higher surface area. Additionally, secondary metabolite distribution varies across the lichen body, with some metabolites located in the cortex, and other in the medulla [28], thus grinding the dried lichen homogenizes the sample.

Three common means can be employed to grind the lichen, namely, a) mortar and pestle, b) ball mill grinder, and c) rotating blade blender. There is evidence that the method of grinding can have an effect on extraction efficiency or even be used to selectively increase extraction efficiency of a certain secondary metabolite. Atranorin, a prominent metabolite in *Pseudevernia furfuracea*, is mostly found at the lichen's surface, in the cortex layer. Using a blender to grind this lichen will selectively grind the cortex into a fine powder, which can be recovered from the larger medulla pieces, to yield a high extraction efficiency of this compound. Conversely, a ball mill or mortar and pestle technique will reduce the whole lichen to a powder, leading to lower extraction efficiencies of atranorin [29]. Similarly, it has been shown that a mortar grinding technique can selectively improve erythrin yields from *Rocella fuciformis* lichen, whereas a ball mill grinding technique can improve lepraric acid extraction efficiency from the same lichen [30].

When extracting organic compounds from lichens, another important consideration is the selection of an extraction solvent. Different solvents will show varying solubility of organic pollutants or secondary metabolites, due to the differences in polarity across the range of organic compounds potentially present. Most lichen secondary metabolites are polar and will therefore be soluble in a polar solvent. Huneck [31], for example, recommended acetone (polarity index 5.1) as an appropriate extraction solvent. A more recent comparative study between n-hexane, diethyl ether, dichloromethane, tetrahydrofuran (THF), ethyl acetate, acetone, acetonitrile, n-butanol, ethanol, methanol, and water found acetonitrile (polarity index 5.8) to be the optimal solvent, with the highest total extraction efficiency [12]. In some cases, the stability of the desired secondary metabolite in a specific solvent must also be considered. Atranorin has been found to destabilize over time when in solution, but less so in acetonitrile [32].

The method of extraction can also play a role when selecting an extraction solvent. For example, if Soxhlet extraction apparatus is used, the boiling point of the extraction solvent should be considered, as solvents with high boiling points will require more energy to heat and may be potentially less environmentally friendly than another comparative solvent. A study by Bonny *et al.* [33] focused on the extraction of norstictic acid from *Pertusaria pseudocorallina*. Results showed that microwave-assisted extraction (MAE) of this secondary metabolite improved as the ratio of THF increased in a THF:acetone mixture, even though acetone is traditionally used in the extraction of norstictic acid [33]. Furthermore, removal of the extraction solvent is sometimes necessary, usually in order to concentrate the sample for subsequent analysis steps or to reconstitute the sample in another solvent. In these cases, it may also be beneficial to consider how easily the selected extraction solvent can be removed through rotary evaporation or through evaporation under an inert gas flow.

2.3 Extraction of Organic Air Pollutants

There are a number of options in terms of extraction methods which can be employed for the extraction of organic air pollutants in lichens, as discussed in this section.

2.3.1 Soxhlet Extraction

Since the development of Soxhlet as an extraction method over a century ago by Von Soxhlet [34], it has been used to leach analytes of interest from their solid matrices into a solvent of choice [35]. The principle behind Soxhlet extraction is the continual dripping of condensed extracting solvent onto the sample (in this case, powdered lichen) held in a porous thimble, which encourages extraction of the analytes into the solvent [36, 37]. The thimbles are usually made of glass microfibers and ensure that extraction can take place without any solid materials transferring to the solvent flask. When an over-flow volume is exceeded in the thimble, the liquid is poured back into the distillation flask, allowing the transfer equilibrium to be continually disturbed over a period of hours, improving the extraction efficiency when compared to other static leaching techniques [37]. Despite this advantage, Soxhlet extractions still require a high amount of solvent compared to more modern extraction methods. Soxhlet extractions also employ long extraction times and a high energy input to keep the solvent heated throughout the run, and therefore, Soxhlet is arguably not an environmentally friendly extraction method [27].

Historically, Soxhlet extraction has been the extraction method of choice when studying organic analytes extracted from lichens [38] and still continues to be used for the extraction of PAHs, PCDDs, PCDFs, organochlorine pesticides and insecticides (OCPs), and polybrominated diphenyl ethers (PBDE) from the lichen matrix [39–44]. Soxhlet extraction of a lichen sample using hexane has also been employed to determine PCB concentrations [45]. Table 2.2 shows some Soxhlet solvent schemes mentioned in the literature where PCDD/Fs, OCPs, PBDEs, and PAHs were studied in lichens.

2.3.2 Ultrasonic Extraction

The principle of cavitation, where tiny bubbles form and pop in liquids and the subsequent erosion of solid materials within the liquids, forms the basis of ultrasound-assisted solvent extraction (USAE) [55]. When ultrasound energy, typically at frequencies between 16 and 500 kHz, is applied to the matrix in a solvent, an interface between solids and liquids is created by increasing the frictional contact between the extraction solvent and the solid from which analytes are to be extracted [56]. As sound waves translate through the matrix, they are converted to mechanical energy. Pressure variations induced by the sound waves cause cavities to form and collapse, which disrupts cellular structures including cell membranes [57]. Disruption of the cellular structures allows enhanced contact between the solvent and the desired compound. This, in turn, leads to higher extraction efficiencies. Cavitation, friction, and the absorption of ultrasonic energy all generate heat within the vessel, increasing the temperature of extraction and thus improving extraction efficiency [58], as long as the heat is maintained at a temperature below which the analytes of interest are susceptible to degradation (as an example, below 35°C for PAHs [59]).

An ultrasonic water bath is commonly used for extraction. However, when using this apparatus, it must be remembered that reproducibility and

Table 2.2 Different studies using Soxhlet to extract organic pollutants from lichens with related extraction parameters.

Analytes	Sample mass (g)	Solvent scheme	Solvent volume (ml)	Extraction time (h)	References
PCDD/Fs	10	Methylene chloride: hexane (1:1)	Not reported	18–24	[38, 46]
PAHs	2	Dichloromethane	100	16	[41, 42, 47, 48]
PAHs	2	Acetonitrile	200	24	[39, 46, 49]
PAHs	0.2	Dichloromethane	250	6	[50]
PAHs	0.6 – 0.8	Hexane: acetone (1:1)	150	2	[51]
PCDD/Fs	Not reported	Toluene	200	24	[46, 52]
Chlorinated hydrocarbons	Not reported	Hexane	Not reported	8	[53]
PCBs, PCDD/Fs	Not reported	Toluene	Not reported	Overnight	[44]
PAHs	Not reported	Acetone	Not reported	Overnight	
PBDEs	Not reported	Hexane: acetone	Not reported	Overnight	
PAHs	2.0	Acetonitrile	300	24	[54]
PAHs	0.2	Dichloromethane	100	6	[27]
	0.2	Acetonitrile	100	6	

repeatability are negatively influenced both as a result of the insufficient uniformity of the distribution of ultrasound energy throughout the bath as well as a loss of power with time [60]. The recent use of a probe, as opposed to a bath, improves upon these limitations and usually allows for a faster sonication time. By optimizing parameters such as the sonication amplitude and the sonication time, higher extraction efficiencies can also be obtained. An increase in sonication time will usually lead to an increase in extraction efficiency, but after a certain time, the relationship between these variables will plateau [61]. Regarding the sonication amplitude, an increase in this variable will lead to increased extraction efficiencies, to an extent. At very high sonication amplitudes, the conversion of sound waves to mechanical energy can lead to the formation of many cavities in the matrix, which may have a dampening effect, thereby restricting the translation of sound waves through the medium [62]. Thus, too high sonication amplitude can actually be detrimental to extraction efficiency.

The advantages of using USAE over other techniques are given in Table 2.3 and include reduced sample size and solvent volumes compared to traditional solvent extraction techniques as well as preparation time [50, 63]. Various solvent systems and sample sizes can be used to optimize extraction, whereas Guidotti *et al.* [63, 64] used cyclohexane as the solvent for ultrasound assisted extraction of PAHs from lichens (sample mass of 2 g), combining only two extractions of 30 ml each, Domeño *et al.* [50] extracted a 0.2-g lichen sample using portions of 15-ml dichloromethane in each of four subsequent extractions.

A novel dynamic ultrasonic-assisted extraction technique (DSASE) for lichens has been developed by Domeño *et al.* [50], where the new DSASE technique using hexane extracted PAHs more efficiently than traditional USAE and has been popular in numerous studies since [59, 65, 66]. As a result of the quick preparation times that require lower solvent and energy consumption, the method has grown in popularity.

However, uncertainties can be introduced during USAE by temperature fluctuations as well as possible degradation of analytes by sonication. A study by Manariotis *et al.* [67] showed that three PAHs: phenanthrene, naphthalene, and pyrene, were degraded at frequencies above 582 kHz, particularly when the extractions were performed for periods of time longer than 30 minutes. This should therefore be carefully considered when planning the experimental procedure. Since temperature can impact extraction efficiencies and solvent evaporation, the generation of heat can impact repeatability, and so, temperature control is important when performing USAE.

Table 2.3 Examples of studies using USAE to extract organic pollutants from lichens.

Analytes	Sample mass (g)	Solvent scheme	Solvent volume (ml)	Extraction time	References
PAHs	0.2	Dichloromethane, hexane: acetone (1:1)	12 ml × 3 (36 ml total)	15 min × 3 (45 min total)	[27]
PAHs + PACs (alkyl PAHs and dibenzothiophenes)	0.2	Cyclohexane	2 ml × 3 (6 ml total)	30 min × 3 (90 min total)	[68]
PAHs	2	Cyclohexane	30 ml × 2 (60 ml total)		[63]
PAHs	0.2	Dichloromethane	15 ml × 5 (75 ml total)		[50]
PAHs	4.0	Acetonitrile	15 ml	2 h	[69]

2.3.3 Microwave-Assisted Extraction

In MAE, analytes of interest adsorbed or absorbed by the substrate are partitioned into the extraction solvent by the application of microwave energy which is transferred to the solvent resulting in heating thereof. Transfer of microwave energy takes place through the simultaneous mechanisms of ionic conduction and dipole rotation. During ionic conduction, the electromagnetic field induces an ionic current in the solvent. Friction between the ions and solvent molecules heat up the solution. During dipole rotation, molecules rapidly realign with the electromagnetic field. As a result, the solution heats up due to these quick molecule movements [70]. Rapid solvent heating through microwaves has the advantage of reducing extraction time and solvent consumption, compared to more conventional heating methods [71].

Typical parameters which require optimization with respect to MAE are extraction time, irradiation power, and volume, and type of solvent employed. MAE methods are generally advantageous in cases where the target molecules to be extracted are thermally stable, as they tend to require significantly less solvent volume and extraction time compared to conventional reflux extraction methods. The choice of extraction solvent heavily impacts extraction efficiency, since the solvent should be able to readily absorb microwave energy [72]. Therefore, unless used in conjunction with solvents with large dipole moments, the extraction efficiency of a non-polar solvent system would be poor [73]. Typically, the experimental design involves the absorption of microwave heat by the solvent and not by the sample within the solvent by selecting a solvent with a high dielectric constant.

Several modifications of the MAE process have been developed. During nitrogen-protected MAE, oxidation of compounds is prevented by displacing oxygen through an inert gas, such as nitrogen or argon. For compounds prone to thermal degradation, vacuum MAE can be conducted under low pressure and low temperature conditions. Ultrasonic MAE combines microwave energy with ultrasound to lower extraction times, as the additional ultrasound energy aids in disrupting cellular structures [74].

MAE is not widely used for organic analyte extraction in lichens [27] but has been used for organic analyte extraction in pine needles [75]. Combination of stir bar microextraction (SBME) with MAE has been reported for the extraction of PAHs from soil samples, which demonstrated improved extraction efficiencies compared to traditional MAE [72]. The application of this technique to lichens might be challenging due to the fact that many pigments could co-extract with the organic analytes of interest (for example, PAHs) and thus nullify the apparent advantage that SBME offers regarding minimal required cleanup.

In general, when compared to traditional Soxhlet extraction, MAE offers a reduction in extraction time (between 5 and 20 minutes compared to >2 to 24 hours for Soxhlet), reduced solvent consumption (10 and 50 ml for MAE compared to 250 to 500 ml for Soxhlet) as well as enhanced reproducibility and reduced solvent wastage [73, 76].

2.3.4 QuEChERS

The quick, easy, cheap, effective, rugged, and safe extraction technique, known as QuEChERS, was developed by Anastassiades *et al.* [77] and quickly found popularity as an effective sample preparation technique for pesticide analysis in foodstuffs [78]. In this method, the number of steps, total solvent consumption, and extraction time are reduced and it serves as an alternative to traditional liquid-liquid extraction and solid phase extraction techniques [79]. Commercialized kits are available containing extraction salts, centrifuge tubes, and cleanup salts and can be customized according to the analytes of interest and the extraction matrix. The steps involved in this technique are shown in Figure 2.3.

The first step is the extraction of homogenized samples using an organic solvent of choice with a solution of extraction salts, often made up of magnesium sulphate (absorbs water and increases ionic strength), sodium chloride (improves partitioning to organic phase), trisodium citrate dihydrate (acts as a buffer), and disodium hydrogen citrate sesquihydrate (acts as a buffer), followed by a centrifugation step. The addition of the extraction salts (Na_2SO_4 or $MgSO_4$) is an exothermic process and therefore increases the temperature of the reaction vessel, improving the extraction efficiency [77]. It has been noted that water should be added to the extract, so that the extraction efficiency is improved as a result of the partitioning of polar compounds between the organic and aqueous phases [80, 81]. The selectivity of the extraction salt can be varied through the addition of sodium chloride, as this compound can control the polarity range of extraction. For the purpose of pesticide analysis, addition of sodium chloride can avoid extraction of some interfering compounds, such as fructose [79].

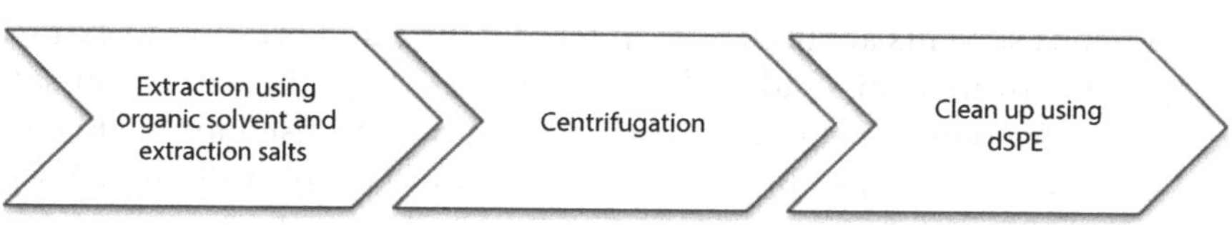

Figure 2.3 A simplified flow diagram showing the QuEChERS extraction process.

The second step then involves the cleanup of the supernatant using a dispersive solid phase extraction (dSPE) which differs from SPE since it does not use a fixed solid bed. The dSPE step involves the use of a drying salt (magnesium sulphate) as well as SPE sorbent packing materials such as primary secondary amine (PSA), often supplemented with graphitized carbon black (GCB) or C_{18} material to clean the extracts and remove moisture. The dSPE step also removes sugars, fats, organic acids, proteins, pigments, and sterols [82], thereby simplifying what would be many consecutive cleanup steps into one cleanup procedure.

Seafood was the first sample tested for PAHs using the QuEChERS technique [83, 84]. Perhaps, closer to our samples of interest, the lichen matrix, Sadowska-Rociek and co-workers [85] evaluated the QuEChERS methodology for the application of PAHs in fresh herbs, cleaned up using 0.15-g PSA, 0.05-g GCB, and 0.9-g $MgSO_4$, followed by GC—single ion monitoring mass spectrometry (SIM-MS) analysis. Acetonitrile as well as ethyl acetate were used, and it was found that acetonitrile was more efficient at extracting PAHs from the herbs with recoveries between 71.6 and 116.9%. Included in the methodology was a 15-minute shaking step, which deviates from the original QuEChERS methodology of extractions between 1 and 5 minutes [80] but remains, however, a considerably faster extraction technique compared to Soxhlet, USAE, and some MAE methods.

The first study using QuEChERS to extract PAHs from the lichen *Parmotrema austrosinense* by Van der Wat and Forbes [27] found that QuEChERS outperformed USAE, Soxhlet, and MAE in terms of extraction efficiency and recoveries. A short cooling step was introduced to both minimize degradation of analytes and solvent volatilization. The optimized QuEChERS extraction procedure for lichens is shown in Figure 2.4. Acetonitrile is typically considered the standard extraction solvent for QuEChERS extractions. However, solvent mixtures, such as 1:1 hexane:acetone (v/v), have also been used successfully [27].

2.3.5 Other Extraction Methods

Accelerated solvent extraction (ASE) is a technique in which organic acids and aqueous solvents are used at high temperature and pressure to extract compounds from semi-solid and solid matrices [86]. The extraction solvents used are typically the same as those traditionally used for Soxhlet or USAE. Although larger sample sizes are typically used for ASE, the solvent volumes and extraction times are reduced with this method compared to traditional methods such as Soxhlet and USAE.

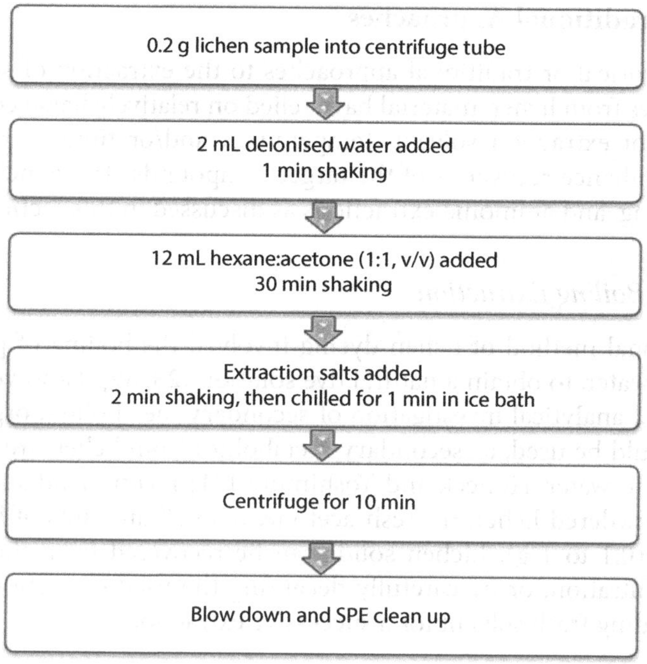

Figure 2.4 Flow diagram of the optimized lichen QuEChERS method (adapted from [27]).

In a study on the occurrence of dechloranes, which are highly chlorinated flame retardants, in lichens growing on the Tibetan plateau, Yang *et al.* [87] used ASE to prepare a 2-g lichen sample using a solvent combination of hexane: dichloromethane (1:1, v/v) at a temperature of 150°C and a pressure of 1,500 psi. The extractions were performed for three cycles, with a 7-minute thermal equilibration time. Similarly, Kodnik *et al.* [88] studied PAHs in lichens in Italy using ASE, with a 600-mg lichen sample, extraction salts (600-mg Na_2SO_4 and 600-mg florisil), and dichloromethane: acetone (1:1, v/v) as solvent, at 140°C and 100 atm. ASE employing the same dichloromethane:acetone (1:1, v/v) solvent mixture and using three consecutive extractions of 5 minutes each at 120 bar and 100°C was used to extract PCDD/Fs from lichen samples [45].

2.4 Extraction of Lichen Metabolites

In this section, examples of reported methods for the extraction of secondary metabolites from lichens are briefly discussed. The operational principles of these techniques have been explained in Section 2.3.

2.4.1 Traditional Approaches

Many historical or traditional approaches to the extraction of secondary metabolites from lichen material have relied on relatively harsh conditions in terms of extraction solvent, temperature, and/or time employed, in order to enhance recoveries of the target compounds. These include boiling, shaking, and ammonia extractions as discussed in this section.

2.4.1.1 Boiling Extraction

A traditional method of lichen dyeing involved the boiling of powdered lichen in water, to obtain a natural dye solution [28, 89]. However, for the purpose of analytical investigation of secondary metabolites, organic solvents should be used, as secondary metabolites from lichen are generally insoluble in water. Huneck and Yoshimura [31] recommend successively boiling powdered lichen in fresh acetone, if small amounts of lichen are analyzed (0.1 to 1 g). Lichen solids can be recovered from the mixture through filtration, or by carefully decanting the used extraction solvent, before adding fresh solvent for a successive extraction.

2.4.1.2 Shaking

Vu *et al.* [90] tested five different lipid extraction methods for lichen samples and found that shaking at 80°C with chloroform:methanol (2:1, v/v) was the most efficient and reproducible for fatty acid extraction.

2.4.1.3 Ammonia Extraction

Dye baths which utilize lichens often employ ammonia extraction; for example, 30 % ammonia with the addition of a small amount of K_2CO_3 for 20 days with daily mechanical stirring, after which air flow was employed to assist in ammonia removal [91].

2.4.2 Alternative Extraction Approaches for Lichen Metabolites

2.4.2.1 Soxhlet Extraction

For lichen amounts ranging between 5 and 100 g, extraction can be achieved through the use of a Soxhlet apparatus [31]. Soxhlet extraction is considered an efficient extraction method and is therefore still a popular means of obtaining a lichen extract [92–94]. Reflux extraction of lichen metabolites for 1.5 to 2 hours using acetone or acetone:ethyl acetate was employed prior to LC with tandem MS (LC-MS/MS) analysis by Parrot *et al.* [95], for example.

When comparing the extraction efficiency of Soxhlet extraction to that of ultrasonic assisted extraction (refer to Sections 2.3.2 and 2.4.2.2), Soxhlet extraction was found to be more efficient at extracting terpenoids and steroids using alcohol [96]. However, another study pertaining to the extraction of usnic acid (a dibenzofuran) from *Cladonia* lichen species found that Soxhlet extraction was a less efficient method compared to supercritical fluid extraction (SFE) [97].

2.4.2.2 Ultrasonic Extraction

Ultrasonic extraction has been shown to efficiently extract secondary metabolites from lichens in around 30 minutes, where Musharraf *et al.* [98] used methanol:dichloromethane (1:1, v/v) for extraction of secondary metabolites from ground lichens, employing a vortex followed by 30 minutes sonication prior to centrifugation and drying under vacuum. In another study, a 30-minute acetone ultrasonic extraction of ground lichens was used for quantitative usnic acid, perlatolic acid, and atranorin determinations, which was then allowed to stand for 23.5 hours prior to analysis by capillary electrophoresis (CE) (refer to Section 2.6.3) [99].

2.4.2.3 Microwave-Assisted Extraction

Bonny *et al.* [100] developed a method for extraction of secondary metabolites, specifically the depsidone norstictic acid, from crustose lichens using MAE with THF as the extraction solvent.

2.4.2.4 QuEChERS

Even though results pertaining to the extraction of secondary metabolites from lichens using the QuEChERS method have not yet been published, preliminary outcomes suggest that the QuEChERS method can indeed be used to successfully extract a number of metabolites from this matrix (Strumpher and Forbes, unpublished). GC-MS analysis showed repeated extraction of terpenoids such as phytol, ergosterol, lichesterol, campesterol, and stigmasterol from *Parmotrema Austrosinense*, using a 1:1 hexane:acetone solvent scheme. The compounds orcinol and n-heptadecane were also detected in numerous extracts. Both palmitic acid and stearic acid were found in a single extract. LC results of extracts obtained using the QuEChERS method with acetonitrile as an extraction solvent, showed repeated, efficient extraction of the depside lecanoric acid—a secondary metabolite which has been noted for its strong anti-oxidant activity [101]. QuEChERS extraction with acetonitrile also extracted palmitic acid, stearic acid, orcinol, 4,5-dimethyl resorcinol, and atraric acid, as well as the depside atranorin.

2.4.2.5 *Solid Phase Microextraction*

During solid phase microextraction (SPME), a fiber coated with an extraction phase (such as polydimethylsiloxane, carbowax/divinylbenzene, or poly(acrylate) [102] is placed in a liquid or gas sample, allowing compounds to partition between the sample phase and the extraction phase. Thereafter, the fiber is thermally desorbed and analyzed by GC-MS [103]. Different extraction phase coatings have different polarities, which render it more or less selective toward a particular analyte or group of analytes. Other parameters which can influence extraction efficiency are the sample temperature, extraction time, and desorption conditions employed [104].

Since SPME only requires a small quantity of material, it can be readily employed when large lichen samples are not available or when only small amounts can be feasibly removed due to sensitivity of the substrate. De Angelis *et al.* [105] determined secondary metabolites present in lichens growing on the surface of monuments as potential markers of environmental stress. The use of SPME (utilizing a polydimethylsiloxane fiber) required the removal of only 20 mg of lichen, thereby preserving the stone monument substrate [105].

2.4.2.6 *Other Extraction Methods*

SFE has been applied in the extraction of usnic acid from lichens, where use of supercritical carbon dioxide was found to provide higher extraction efficiencies than with 24-hour Soxhlet extraction employing acetone, ethanol, or petroleum ether [97]. The SFE process took 40 minutes and the addition of co-solvents (acetone, ethanol, methylene chloride) to the CO_2 was found to slightly increase extraction yields. SFE is considered to be particularly advantageous in the extraction of biologically active secondary metabolites from lichens for pharmaceutical applications, as removal of potentially toxic solvents is not needed.

An interesting alternative method for extraction of norstictic acid, a common depsidone with anti-oxidant properties isolated from crustose lichens growing on seaside rocks, involves the use of ionic liquids (ILs) with MAE as shown in Figure 2.5 [100]. ILs have received considerable

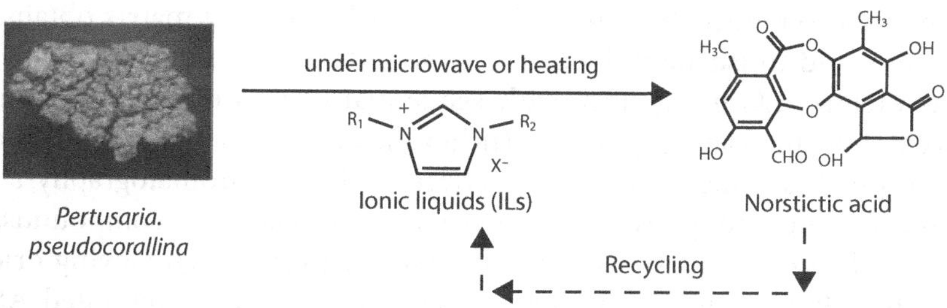

Figure 2.5 Extraction of lichen compounds isolated with imidazolium derived ionic liquid (ILs) using microwave-assisted extraction or heat-extraction [100]. Reprinted with permission from Elsevier.

research attention as a "green solvent". Here, different ILs were investigated and extraction efficiencies were compared when using them in conjunction with traditional or conventional heat-reflux (boiling) approaches to that of microwave-assisted methods. The IL extraction results were also compared to heat-reflux and MAE utilizing THF as a solvent (which prevents hydrolysis which occurs in methanol or water). The heat-reflux methods required a 2-hour extraction, while the IL-MAE process only required 5 minutes. The authors found that the alkyl chain of the IL and the anion influenced the extraction recoveries, with sulphate-based ILs being optimal. MAE increased recoveries by more than 1.5 times relative to heat-reflux methods. A drawback of this extraction method is the lengthy multi-step process required to recover the metabolites of interest from the extract, which involve precipitation, liquid-liquid extraction, and washing steps employing a range of additional solvents. This also resulted in lower recoveries for IL-based extractions than with THF, although ILs provide more selective extraction of target metabolites, they can be regenerated and have no flammability. They provide a softer extraction method required for depsides which are readily hydrolyzed [100].

2.5 Extract Cleanup

In some cases, particularly with respect to LC based analyses, it is merely necessary to filter sample extracts prior to analysis, such as with 0.45-μm filters [106]. For example, dried lichen extracts were re-dissolved and simply filtered through 0.45-μm syringe filters prior to LC-MS/MS analysis for metabolite determinations [98]. Filtration was also the only extract cleanup required prior to micellar electrokinetic chromatographic analysis [99].

In order to reduce the complexity of the lichen extract matrix obtained by reflux and to isolate individual compounds present, Parrot *et al.* [95] used vacuum LC using silica with sequential elution of fractions using solvents of increasing polarity (n-hexane:dichloromethane and then dichloromethane:methanol). Alternatively, column chromatography and flash chromatography (using C_{18}) were used to isolate other compounds.

Kodnik *et al.* [88] also used column chromatography employing dried activated alumina and dichloromethane to cleanup a concentrated ASE lichen extract prior to GC-MS analysis of PAHs, while silica gel cleanup after ultrasonic extraction of PAHs and polycyclic aromatic compounds (PACs) prior to GC-MS analysis was employed by Studabaker *et al.* [68]. A more extensive cleanup was required prior to analysis of lichen extracts for trace PCDD/Fs using florisil 3% with toluene elution followed by a second purification using carbon-celite 18% [45].

Cyclohexane sonication extracts from lichens were cleaned up using silica gel solid phase extraction (SPE) cartridges to prevent interferences during subsequent GC-MS analysis for PAHs [107]. The SPE cartridges were made of glass and had glass fiber frits to avoid potential interferences from or losses to plastic.

Dispersive solid phase extraction (dSPE) is applied in QuEChERS methods, as mentioned under Section 2.3.4, where sorbent mixtures typically consist of differing ratios of primary amine exchange material (PSA), C_{18}, and/or GCB. Bearing in mind that these sorbents were developed to provide optimal cleanup of samples for pesticide analysis, PSA removes organic acids and sugars which may act as interfering substances in such analyses. GCB primarily removes pigments, another important consideration, as excess chlorophyll can result in matrix effects such as peak enhancement or suppression, particularly in GC based analyses [27]. GCB can also remove sterols and other nonpolar interferences. Finally, C_{18} primarily removes lipids. The ratios of C_{18} and GCB typically decrease relative to PSA, as sample pigments decrease. In the case of general fruit and vegetable analyses, C_{18} and GCB may be absent from the sorbent mixture altogether. Figure 2.6 shows three lichen extracts before and after cleanup using dSPE sorbents with decreasing C_{18} and GCB sorbent ratios from left to right. The sample extract which was cleaned using a dSPE sorbent containing only PSA had a vibrant green color after sample cleanup, due to residual chlorophyll (vial on the right in Figure 2.6b).

When considering the application of QuEChERS dSPE sorbents in the cleanup of lichen extracts intended for metabolite analyses, sorbents containing a high ratio of C_{18} and GCB (intended for the cleanup of highly pigmented samples) may additionally remove target metabolites. Preliminary

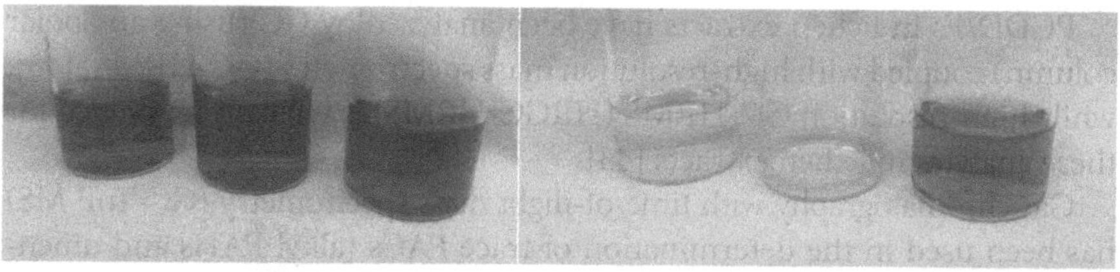

(a) (b)

Figure 2.6 Lichen extracts before dSPE cleanup (a), and lichen extracts after dSPE cleanup (b), with decreasing sorbent ratios from left to right.

findings pertaining to the comparison of *Parmotrema austrosinense* extracts (obtained using QuEChERS salts, with acetonitrile as the extracting solvent) showed that lecanoric acid was present in the sample if cleaned using a low GCB and C_{18} ratio dSPE sorbent, while it was absent when the same lichen sample extract was cleaned up using a high GCB and C_{18} ratio dSPE sorbent (Strumpher and Forbes, unpublished).

2.6 Analysis of Lichen Extracts

The choice of analytical method to employ depends primarily on the target organic lichen-derived compounds. Options in this regard are discussed in this section.

2.6.1 GC

Some of the organic compounds present in lichens are amenable to gas chromatographic (GC) analysis, including PAHs, PCDD/Fs, and PCB air pollutants. Indeed, Herzig *et al.* [44] used GC with MS detection for the analysis of 94 individual and 27 sum parameters of persistent organic pollutants (POPs) and PAHs in lichen extracts. The POPs investigated included PCDD/Fs, organochlorine pesticides, PCBs, and PBDEs.

Gas chromatography with flame ionization detection (GC-FID) and GC-MS have been used to compare the fatty acid composition of different lichen species, where total, saturated, unsaturated, and branched-chain fatty acids were considered [90].

PCBs have been analyzed by both GC-electron capture detection (GC-ECD) and GC-MS [45] and dechloranes (which are highly chlorinated flame retardants) were determined in lichen extracts from the Tibetan plateau by means of GC-MS using negative chemical ionization employing methane [1].

PCDD/Fs in lichen extracts have been analyzed by GC (using an apolar column) coupled with high-resolution mass spectrometry (GC-HRMS) [45], while high-resolution GC-HRMS (HRGC-HRMS) is typically employed for these analytes in lichen extracts [43].

Gas chromatography with time-of-flight mass spectrometry (GC-ToF-MS) has been used in the determination of trace PACs (alkyl PAHs and dibenzothiophenes) in lichen extracts, to allow for improved characterization and quantification [68]. The authors noted the potential future benefits of utilizing more advanced instrumental methods such as GCxGC-MS and HRMS to enhance compound resolution and mass resolution respectively of complex lichen extracts.

2.6.2 LC

The less volatile organic lichen-derived compounds may be analyzed by LC systems. In some cases, analytes can be analyzed by either GC or LC. PAHs, for example, can be analyzed by GC or by high performance LC (HPLC) with UV or fluorescence detectors due to the fact that they fluoresce. In the case of PAHs, a C_{18}LC column is typically employed [45].

LC has been used extensively in the analysis of bioactive secondary metabolites in lichens as it provides sensitivity, and multiple reaction monitoring (MRM) assists with selective quantification. Typically, upon MS/MS analysis of pure standards of the compounds of interest, the transitions that provide the two highest intensities of fragment ions are selected as quantifier and qualifier transitions, respectively. The MRM mode can then be used to collect mass spectral data of precursor and product ion transitions for quantification purposes. In order to maximize sensitivity in terms of the instrument response for each of the target analyte transitions, MRM parameters such as collision energy and voltages need to be optimized. Transitions specific to individual target compounds should be chosen to overcome the possible chromatographic overlap of compounds. Electrospray tandem mass spectrometry (ESI-MS/MS) has proved to be a robust technique for the identification and characterization of secondary metabolites in crude lichen extracts. In order to use this technique for structure elucidation, however, it is necessary to have a knowledge and understanding of the fragmentation patterns of the specific compounds of interest [108].

For example, ultra HPLC with triple quadrupole mass spectrometry (UHPLC-QqQ-MS) employing MRM has been utilized in the determination of depsides, depsidones, dibenzofuran, and monocyclic phenols [98]. The optimized method had the advantage of short run times (6 minutes) in which nine metabolites could be determined. A reverse-phase C_{18} column was used with a mobile phase consisting of 0.5% ammonium acetate in MilliQ water at pH 4 and 0.5% ammonium acetate in methanol at pH 4, respectively.

LC-MS/MS employing negative-ion electrospray ionization has been used to compare salazinic, norstictic, and usnic acid concentrations in acetone ultrasonic extracts of lichens found in areas where animals had experienced toxic effects. Here, an acetonitrile-water mobile phase gradient was employed with a C_{18} LC column [109].

HPLC with a diode array detector (DAD) allowed for the analysis of norstictic acid, a common depsidone, again using a C_{18} column and a mobile phase consisting of 0.6% acetic acid aqueous solution and methanol (60:40) [100]. Similarly, HPLC-DAD (detection at 245 nm, with a mixture of methanol and phosphate buffer (pH 7.4) (70:30, (v/v) as mobile phase) was used to determine usnic acid concentrations extracted with acetone from a range of lichens growing in Turkey, as this compound has reported antimicrobial properties [106].

The LC-ESI-MS/MS fragmentation patterns of metabolites from the lichen *Ramalina siliquosa* were determined by Parrot *et al.* [95], which included 10 main acidic compounds and 22 minor compounds. The aim of this study was to develop chemical fingerprints for the different chemotypes of this lichen to assist with identification. The metabolites were characterized based on their retention times, UV-Vis absorption spectra (280 nm), and mass spectra using a LC-UV-Vis-MS approach, while LC with photodiode array detection (LC-PDA) was also used in some cases. A gradient system was applied for the HPLC analyses using 0.1% formic acid in water and 0.1% formic acid in acetonitrile as mobile phases. The major compounds were further characterized by their LC-MS/MS and HRMS spectra, by analysis of authentic standards, and by consultation of the literature.

From this, it is evident that the full characterization of organic lichen-derived compounds in extracts is analytically very challenging and requires a multi-faceted approach. This is particularly the case for secondary metabolite lichen extracts, for which alternative analytical approaches have been explored as discussed in the following section.

2.6.3 Other Analytical Methods

Thin layer chromatography (TLC) with UV light detection is commonly used in the initial analysis of metabolites extracted from lichens, typically employing a deuterium lamp (270 nm) as a light source, where supplementary ^{1}H and ^{13}C nuclear magnetic resonance (NMR) spectra may provide useful information on isolated compounds [100]. Both ^{13}C NMR spectroscopy and optical rotation were used by Parrot *et al.* [95], for example, to assist with the identification of isolated lichen metabolites.

Quantitative analysis of secondary metabolites extracted from lichens is possible with TLC when using a spectrophotodensitometer [33]. Interestingly, surface enhanced Raman spectroscopy (SERS) directly on a TLC plate after deposition of silver colloid on the TLC spots (Figure 2.7) has found recent application in the elucidation of differences in chemical composition of orcein dyed yarns from different lichens [91].

CE has been applied in the analysis of usnic acid contained in ethyl acetate extracts from pastilles containing lichen material, using reverse polarity capillary zone electrophoresis (CZE) of the extracts reconstituted in methanol. It was found that a high methanol content was required in the CZE buffer due to the low water solubility of usnic acid, necessitating the reversed polarity configuration with the anode on the detector side of the capillary [110].

Micellar electrokinetic chromatography (MEKC) is a common mode of CE separation in which charged micelles act as a pseudostationary phase. MEKC has been employed to quantitatively determine secondary metabolites from lichens. Specifically, usnic acid, perlatolic acid, and atranorin have been determined in acetone lichen extracts, with a diode array UV detector providing detection limits at 214 nm in the low μg/ml range and an analysis time of less

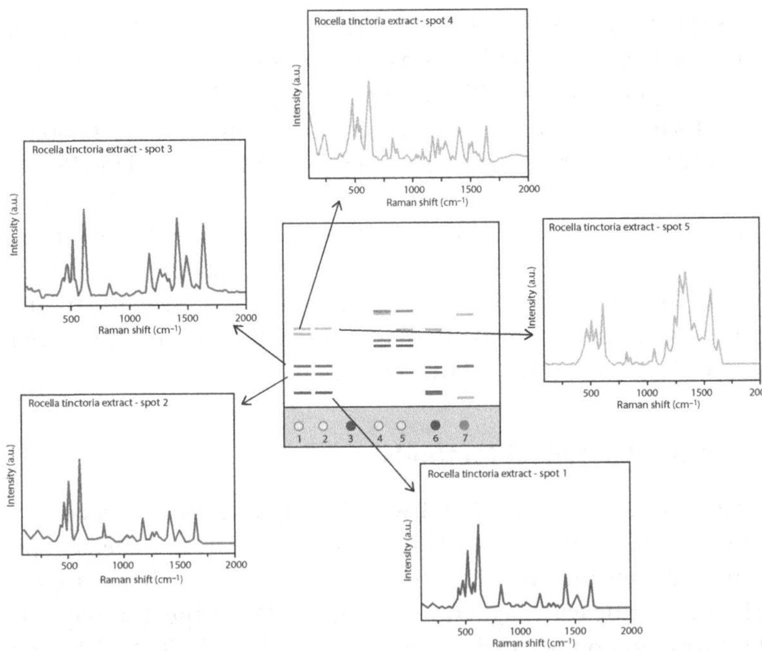

Figure 2.7 TLC-SER spectra of the principal dyes found in *R. tinctoria* lichen extract and in ammonia extracts of *R. tinctoria* dyed yarns [91]. Reprinted with permission from Elsevier.

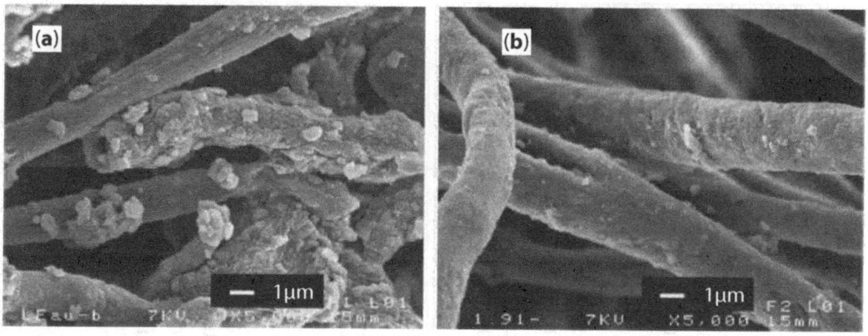

Figure 2.8 SEM images of a lichen surface after (a) water extraction and (b) successive extractions with n-hexane, diethyl ether, acetone, and then tetrahydrofuran [29]. Reprinted with permission from Elsevier.

than 10 minutes [99]. The background electrolyte employed in this case was a high-pH borate buffer containing micelle-forming sodium docecyl sulfate and β-cyclodextrin, which decreased migration time and thus peak width.

Scanning electron microscopy (SEM) has been used to visually evaluate differences in lichen extraction efficiencies when different solvents are employed. Komaty *et al.* [29] found that crystals of secondary metabolites were present on the lichen surface after water extraction but were dissolved when various organic solvents were successively employed (Figure 2.8). In this study, ToF secondary ionization mass spectrometry (ToF-SIMS) was used as a surface analytical technique to determine variations in the composition of the raw surface of the lichen with respect to atranorin content and that of sublayers after etching of the surface via an argon ion beam. This confirmed that the atranorin was primarily present on the surface of the lichen thallus, which was shown to be removed after acetone washing.

2.7 Conclusion

Extraction of organic lichen-derived compounds is challenging due to the range of analytes involved which have large variations in physicochemical properties. In addition, the lichen matrix itself is complex.

Fortunately, the ongoing development of analytical extraction methods allows researchers to meet these challenges, facilitating the discovery of new organic compounds of interest in lichen extracts, from both biomonitoring and secondary metabolite perspectives. Table 2.4 compares the main (most used) extraction methods discussed in this chapter and highlights the advantages and disadvantages thereof.

Table 2.4 A comparison between different sample extraction methods for organic analytes in lichens.

Extraction technique	Advantages	Disadvantages	References
Soxhlet	Robust Easy No further filtration required	Time consuming (2–24 h) Consumes large volumes of solvents and energy Solvent waste has environmental concerns Low sample throughput Loss of thermo-labile species	[37, 46, 51]
USAE	Lower sample mass required Faster than Soxhlet Requires fewer preparation steps	Reproducibility and repeatability can be poor when using an ultrasonic bath Analytes may be degraded by sonication and heat hotspots	[50, 60, 63, 67]
MAE	Lower solvent usage than USAE and Soxhlet Quick extraction Less energy and solvent wastage High reproducibility	Expensive equipment required Not widely reported in the literature	[73, 76]
QuEChERS	No heat energy required Low solvent usage No pre-concentration or further cleanup required Cheap Improved recoveries	Cleanup with dSPE should be carefully considered to minimize interference from chlorophyll or loss of analytes of interest	[27]

After extraction, cleanup of the sample extract may be required to enhance analytical sensitivity and avoid inaccuracies. The powerful analytical technologies currently available allow for the chemical characterization of lichen extracts in a comprehensive manner, especially when a range of analytical techniques are simultaneously employed.

Ultimately, choice of extraction (and analytical) methodologies depends on a range of factors, including instrument availability, cost, environmental impact, efficiency, sensitivity, selectivity, and robustness. In summary, the methods must be fit for purpose in order to gain optimal insight into the organic compounds which are contained in lichens.

References

1. Yang, C., Wang, J., Li, D., Microextraction techniques for the determination of volatile and semivolatile organic compounds from plants: A review. *Anal. Chim. Acta*, 799, 8–22, 2013.
2. Romanik, G., Gilgenast, E., Przyjazny, A., Kaminski, M., Techniques of preparing plant material for chromatographic separation and analysis. *J. Biochem. Biophys. Methods*, 70, 253–261, 2007.
3. Van der Wat, L. and Forbes, P.B.C., Lichens as biomonitors for organic air pollutants. *TrAC, Trends Anal. Chem.*, 64, 165–172, 2015.
4. Forbes, P.B.C., van der Wat, L., Kroukamp, E.M., Chapter 3: Biomonitors, in Comprehensive Analytical Chemistry vol. 70, in: *Monitoring of Air Pollutants: Sampling, Sample Preparation and Analytical Techniques*, P. Forbes (Ed.), pp. 53–108, Elsevier, Netherlands, 2015.
5. Nash, T.H., *Lichen biology*, 2nd ed., p. 106, Cambridge University Press, Cambridge, 2010.
6. Shukla, V., Upreti, D.K., Baipai, R., *Lichens to biomonitor the environment*, pp. 22–33, Springer India, Delhi, 2014.
7. Nimis, P.L. and Skert, N., Lichen chemistry and selective grazing by the coleopteran *Lasioderma serricorne. Environ. Exp. Bot.*, 55, 1, 175–182, 2006.
8. Molnár, K. and Farkas, E., Current results on biological activities of lichen secondary metabolites: A review. *Z. Naturforsch., C: Biosci.*, 65, 3–4, 157–173, 2010.
9. Culberson, C.F. and Elix, J.A., 15 - Lichen Substances, in: *Methods in Plant Biochemistry*, J.B. Harborne (Ed.), p. 511, Academic Press, Cambridge, 1989.
10. Elix, J.A. and Ernst-Russell, K.D., *A catalogue of standardized thin layer chromatographic data and biosynthetic relationships for lichen substances*, 2014, Published by the author, Canberra. *http://www.anbg.gov.au/abrs/lichenlist/Chem%20Cat%203.pdf.*
11. Xu, M., Heidmarsson, S., Olafsdottir, E.S., Buonfiglio, R., Kogej, T., Omarsdottir, S., Secondary metabolites from cetrarioid lichens:

Chemotaxonomy, biological activities and pharmaceutical potential. *Phytomedicine*, 23, 5, 441–459, 2016.

12. Kumar, K., Siva, B., Sarma, V.U.M., Mohabe, S., Reddy, A.M., Boustie, J., Tiwari, A.K., Rao, N.R., Babu, K.S., UPLC–MS/MS quantitative analysis and structural fragmentation study of five *Parmotrema* lichens from the Eastern Ghats. *J. Pharm. Biomed. Anal.*, 156, 45–57, 2018.
13. Huneck, S., Porzel, A., Schmidt, J., Feige, G.B., Posner, B., Crustinic acid, a tridepside from *Umbilicaria crustulosa*. *Phytochemistry*, 32, 2, 475–477, 1993.
14. Varol, M., Anti-breast cancer and anti-angiogenic potential of a lichen-derived small-molecule: Barbatolic acid. *Cytotechnology*, 70, 6, 1565–1573, 2018.
15. Wei, X., Schmitt, I., Hodkinson, B., Flakus, A., Kukwa, M., Divakar, P.K., Kirika, P.M., Otte, J., Meiser, A., Lumbsch, H.T., Circumscription of the genus Lepra, a recently resurrected genus to accommodate the "*Variolaria*"-group of *Pertusaria sensu lato (Pertusariales, Ascomycota)*. *PLoS ONE*, 12, 7, e0180284, 2017.
16. González, A.G., Pérez, E.M.R.G., Barrera, J.B., Chemical constituents of the lichen *Stereocaulon azoreum*. *Z. Naturforsch., C: Biosci.*, 47, 7–8, 503–507, 1992.
17. Dieu, A., Mambu, L., Champavier, Y., Chaleix, V., Sol, V., Gloaguen, V., Millot, M., Antibacterial activity of the lichens *Usnea Florida* and *Flavoparmelia caperata (Parmeliaceae)*. *Nat. Prod. Res.*, 1–5, 2019.
18. Vannini, A., Paoli, L., Ceccarelli, S., Sorbo, S., Basile, A., Carginale, V., Nali, C., Lorenzini, G., Pica, M., Loppi, S., Physiological and ultrastructural effects of acute ozone fumigation in the lichen *Xanthoria parietina*: The role of parietin and hydration state. *Environ. Sci. Pollut. Res.*, 25, 9, 8104–8112, 2018.
19. Srinivasan, M., Shanmugam, K., Kedike, B., Narayanan, S., Shanmugam, S., Gopalasamudram Neelakantan, H., Trypethelone and phenalenone derivatives isolated from the mycobiont culture of *Trypethelium eluteriae Spreng.* and their anti-mycobacterial properties. *Nat. Prod. Res.*, 1–8, 2019.
20. Kathirgamanathar, S., Ratnasooriya, W.D., Baekstrom, P., Andersen, R.J., Karunaratne, V., Chemistry and bioactivity of physciaceae lichens *Pyxine consocians* and *Heterodermia leucomelos*. *Pharm. Biol.*, 44, 3, 217–220, 2006.
21. Kyslychenko, O.A., Protska, V.V., Zhuravel, I.O., Hutsol, V.V., The study of volatile compounds of *Parmelia perlata* thalluso. *Pharmacia*, 65, 3, 11–16, 2018.
22. Gonzalez, A.G., Bermejo Barrera, J., Rodriguez Perez, E.M., Hernandez Padron, C.E., Chemical constituents of the lichen *Ramalina hierrensis*. *Planta Med.*, 58, 2, 214–218, 1992.
23. Kaneda, M., Takahashi, R., Iitaka, Y., Shibata, S., Retigeranic acid, a novel sesterterpene isolated from the lichens of *Lobaria retigera* group. *Tetrahedron Lett.*, 13, 45, 4609–4611, 1972.

24. Safe, S., Safe, L.M., Maass, W.S.G., Sterols of three lichen species: *Lobaria pulmonaria, Lobaria Scrobiculata* and *Usnea Longissima*. *Phytochemistry*, 14, 8, 1821–1823, 1975.
25. Corbett, R.E. and Cumming, S.D., Lichens and fungi. Part VII. Extractives from the lichen *Sticta mougeotiana* var. dissecta del. *J. Chem. Soc. C*, 955–960, 1971.
26. Fernández-Moriano, C., Divakar, P.K., Crespo, A., Gómez-Serranillos, M.P., Neuroprotective activity and cytotoxic potential of two *Parmeliaceae* lichens: Identification of active compounds. *Phytomedicine*, 22, 9, 847–855, 2015.
27. Van der Wat, L. and Forbes, P.B.C., Comparison of extraction techniques for polycyclic aromatic hydrocarbons from lichen biomonitors. *Environ. Sci. Pollut. Res.*, 26, 11, 11179–11190, 2019.
28. Shukla, P., Upreti, D.K., Nayaka, S., Tiwari, P., Natural dyes from Himalayan lichens. *Indian J. Trad. Know.*, 13, 1, 195–201, 2014.
29. Komaty, S., Letertre, M., Dang, H.D., Jungnickel, H., Laux, P., Luch, A., Carrié, D., Merdrignac-Conanec, O., Bazureau, J.-P., Gauffre, F., Tomasi, S., Paquin, L., Sample preparation for an optimized extraction of localized metabolites in lichens: Application to *Pseudevernia furfuracea*. *Talanta*, 150, 525–530, 2016.
30. Parrot, D., Peresse, T., Hitti, E., Carrie, D., Grube, M., Tomasi, S., Qualitative and spatial metabolite profiling of lichens by a LC-MS approach combined with optimised extraction. *Phytochem. Anal.*, 26, 1, 23–33, 2015.
31. Huneck, S. and Yoshimura, I., *Identification of Lichen Substances*, pp. 11–12, Springer, New York City, 1996.
32. Vos, C., McKinney, P., Pearson, C., Heiny, E., Gunawardena, G., Holt, E.A., The optimal extraction and stability of atranorin from lichens, in relation to solvent and pH. *Lichenologist*, 50, 4, 499–512, 2018.
33. Bonny, S., Hitti, E., Boustie, J., Bernard, A., Tomasi, S., Optimization of a microwave-assisted extraction of secondary metabolites from crustose lichens with quantitative spectrophotodensitometry analysis. *J. Chromatogr. A*, 1216, 7651–7656, 2009.
34. Von Soxhlet, F., Die Gewichtsanalytische Bestimmung Des Milchfettes. *Dingler's Polytech. J.*, 232, 461, 1879.
35. Jensen, W.B., The Origin of the Soxhlet Extractor. *J. Chem. Educ.*, 84, 12, 1913–1914, 2007.
36. De Castro, M.L. and Garcıa-Ayuso, L.E., Soxhlet extraction of solid materials: An outdated technique with a promising innovative future. *Anal. Chim. Acta*, 369, 1–2, 1–10, 1998.
37. De Castro, M.L. and Priego-Capote, F., Soxhlet extraction: Past and present panacea. *J. Chromatogr. A*, 1217, 16, 2383–2389, 2010.
38. Augusto, S., Pinho, P., Branquinho, C., Pereira, M.J., Soares, A., Catarino, F., Atmospheric dioxin and furan deposition in relation to land-use and other pollutants: A survey with lichens. *J. Atmos. Chem.*, 49, 53–65, 2004.

39. Augusto, S., Máguas, C., Matos, J., Pereira, M.J., Branquinho, C., Lichens as an integrating tool for monitoring PAH atmospheric deposition: A comparison with soil, air and pine needles. *Environ. Pollut.*, 158, 2, 483–489, 2010.
40. Augusto, S., Pereira, M.J., Máguas, C., Soares, A., Branquinho, C., Assessing human exposure to polycyclic aromatic hydrocarbons (PAH) in a petrochemical region utilizing data from environmental biomonitors. *J. Toxicol. Environ. Health, Part A*, 75, 13–15, 819–830, 2012.
41. Shukla, V. and Upreti, D.K., Polycyclic aromatic hydrocarbon (PAH) accumulation in lichen, *Phaeophyscia hispidula* of DehraDun City, Garhwal Himalayas. *Environ. Monit. Assess.*, 149, 1–4, 1–7, 2009.
42. Shukla, V., Patel, D.K., Upreti, D.K., Yunus, M., Lichens to distinguish urban from industrial PAHs. *Environ. Chem. Lett.*, 10, 2, 159–164, 2012.
43. Augusto, S., Pinho, P., Santos, A., Botelho, M.J., Palma-Oliveira, J., Branquinho, C., Tracking the spatial fate of PCDD/F emissions from a cement plant by using lichens as environmental biomonitors. *Environ. Sci. Technol.*, 50, 2434–2441, 2016.
44. Herzig, R., Lohmann, N., Meier, R., Temporal change of the accumulation of persistent organic pollutants (POPs) and polycyclic aromatic hydrocarbons (PAHs) in lichens in Switzerland between 1995 and 2014. *Environ. Sci. Pollut. R.*, 26, 10562–10575, 2019.
45. Ratier, A., Dron, J., Revenko, G. *et al.*, Characterization of atmospheric emission sources in lichen from metal and organic contaminant patterns. *Environ. Sci. Pollut. Res.*, 25, 8364–8376, 2018.
46. Augusto, S., Máguas, C., Branquinho, C., Understanding the performance of different lichen species as biomonitors of atmospheric dioxins and furans: Potential for intercalibration. *Ecotoxicology*, 18, 8, 1036–1042, 2009.
47. Shukla, V., Upreti, D.K., Patel, D.K., Yunus, M., Lichens reveal air PAH fractionation in the Himalaya. *Environ. Chem. Lett.*, 11, 1, 19–23, 2013.
48. Bajpai, R., Karakoti, N., Upreti, D.K., Performance of a naturally growing Parmelioid lichen *Remototrachyna awasthii* against organic and inorganic pollutants. *Environ. Sci.*, 20, 5577–5592, 2013.
49. Schauer, C., Niessner, R., Pöschl, U., Polycyclic aromatic hydrocarbons in urban air particulate matter: Decadal and seasonal trends, chemical degradation, and sampling artifacts. *Environ. Sci. Technol.*, 37, 2861–2868, 2003.
50. Domeño, C., Blasco, M., Sánchez, C., Nerín, C., A fast extraction technique for extracting polycyclic aromatic hydrocarbons (PAHs) from lichen samples used as biomonitors of air pollution: Dynamic sonication versus other methods. *Anal. Chim. Acta*, 569, 103–112, 2006.
51. Nascimbene, J., Tretiach, M., Corana, F., Schiavo, F.L., Kodnik, D., Dainese, M., Mannucci, B., Patterns of traffic polycyclic aromatic hydrocarbon pollution in mountain areas can be revealed by lichen biomonitoring: A case study in the Dolomites (Eastern Italian Alps). *Sci. Total Environ.*, 475, 90–96, 2014.
52. Augusto, S., Pereira, M.J., Soares, A., Branquinho, C., The contribution of environmental biomonitoring with lichens to assess human exposure to dioxins. *Int. J. of Hyg. Envir. Heal.*, 210, 3–4, 433–438, 2007.

53. Villeneuve, J.P. and Holm, E., Atmospheric background of chlorinated hydrocarbons studied in Swedish lichens. *Chemosphere*, 13, 10, 1133–1138, 1984.
54. Domínguez-Morueco, N., Augusto, S., Trabalón, L., Pocurull, E., Borrull, F., Schuhmacher, M., Domingo, J.L., Nadal, M., Monitoring PAHs in the petrochemical area of Tarragona County, Spain: Comparing passive air samplers with lichen transplants. *Environ. Sci. Pollut. R.*, 24, 11890–11900, 2017.
55. Szreniawa-Sztajnert, A., Zabiegała, B., Namieśnik, J., Developments in ultrasound-assisted microextraction techniques for isolation and preconcentration of organic analytes from aqueous samples. *TrAC, Trends Anal. Chem.*, 49, 45–54, 2013.
56. Capote, F.P. and De Castro, M.L., *Analytical applications of ultrasound*, vol. 26, Elsevier, Amsterdam, 2007.
57. Garcia, V., Rajauria, M., O'Doherty, G., Sweeney, J.V., Polysaccharides from macroalgae: Recent advances, innovative technologies and challenges in extraction and purification. *Food Res. Int.*, 3, 99, Part 3, 1011–1020, 2017.
58. Babić, S., Petrović, M., Kaštelan-Macan, M., Ultrasonic solvent extraction of pesticides from soil. *J. Chromatogr. A*, 823, 1–2, 3–9, 1999.
59. Blasco, M., Domeño, C., Nerín, C., Use of lichens as pollution biomonitors in remote areas: Comparison of PAHs extracted from lichens and atmospheric particles sampled in and around the Somport tunnel (Pyrenees). *Environ. Sci. Technol.*, 40, 6384–6391, 2006.
60. Namieśnik, J. and Szefer, P., Preparing samples for analysis-the key to analytical success. *Ecol. Chem. Eng. S.*, 15, 2, 167–244, 2008.
61. Krishna, M.V.B. and Arunachalam, J., Ultrasound-assisted extraction procedure for the fast estimation of major, minor and trace elements in lichen and mussel samples by ICP-MS and ICP-AES. *Anal. Chim. Acta*, 522, 2, 179–187, 2004.
62. Filgueiras, A.V., Capelo, J.L., Lavilla, I., Bendicho, C., Comparison of ultrasound-assisted extraction and microwave-assisted digestion for determination of magnesium, manganese and zinc in plant samples by flame atomic absorption spectrometry. *Talanta*, 53, 2, 433–441, 2000.
63. Guidotti, M., Stella, D., Owczarek, M., De Marco, A., De Simone, C., Lichens as polycyclic aromatic hydrocarbon bioaccumulators used in atmospheric pollution studies. *J. Chromatogr. A.*, 985, 1–2, 185–190, 2003.
64. Guidotti, M., Stella, D., Dominici, C., Blasi, G., Owczarek, M., Vitali, M., Protano, C., Monitoring of Traffic-Related Pollution in a Province of Central Italy with Transplanted Lichen *Pseudoverniafurfuracea*. *Bull. Environ. Contam. Toxicol.*, 83, 852–858, 2009.
65. Blasco, M., Domeño, C., Bentayeb, K., Nerín, C., Solid-phase extraction clean-up procedure for the analysis of PAHs in lichens. *Int. J. Environ. Anal. Chem.*, 87, 833–846, 2007.
66. Blasco, M., Domeño, C., Lopez, P., Nerin, C., Behaviour of different lichen species as biomonitors of air pollution by PAHs in natural ecosystems. *Environ. Monit. Assess.*, 13, 2588–2596, 2011.

67. Manariotis, I.D., Karapanagioti, H.K., Chrysikopoulos, C.V., Degradation of PAHs by highfrequency ultrasound. *Water Res.*, 45, 8, 2587–2594, 2011.
68. Studabaker, W.B., Puckett, K.J., Percy, K.E., Landis, M.S., Determination of polycyclic aromatic hydrocarbons, dibenzothiophene, and alkylated homologs in the lichen *Hypogymnia physodes* by gas chromatography using single quadrupole mass spectrometry and time-of-flight mass spectrometry. *J. Chromatogr. A.*, 1492, 106–116, 2017.
69. Puy-Alquiza, M.J., Reyes, V., Wrobel, K., Wrobel, K., Elguera, J.C.T., Miranda-Aviles, R., Polycyclic aromatic hydrocarbons in urban tunnels of Guanajuato city (Mexico) measured in deposited dust particles and in transplanted lichen *Xanthoparmelia mexicana* (Gyeln.) Hale. *Environ. Sci. Pollut. R.*, 23, 12, 11947–11956, 2016.
70. Sparr Eskilsson, C. and Björklund, E., Analytical-scale microwave-assisted extraction. *J. Chromatogr. A*, 902, 1, 227–250, 2000.
71. Llompart, M., Garcia-Jares, C., Celeiro, M., Dagnac, T., Extraction - Microwave-Assisted Extraction, in: *Encyclopedia of Analytical Science (Third Edition)*, P. Worsfold, C. Poole, A. Townshend, M. Miró (Eds.), pp. 67–77, Academic Press, Oxford, 2019.
72. Guo, L. and Lee, H.K., Microwave assisted extraction combined with solvent bar microextraction for one-step solvent-minimized extraction, cleanup and preconcentration of polycyclic aromatic hydrocarbons in soil samples. *J. Chromatogr. A.*, 1286, 9–15, 2013.
73. Camel, V., Microwave-assisted solvent extraction of environmental samples. *TrAC*, 19, 4, 229–248, 2000.
74. Chan, C.H., Yusoff, R., Ngoh, G.C., Kung, F.W.L., Microwave-assisted extractions of active ingredients from plants. *J. Chromatogr. A*, 1218, 37, 6213–6225, 2011.
75. Ratola, N., Lacorte, S., Barceló, D., Alves, A., Microwave-assisted extraction and ultrasonic extraction to determine polycyclic aromatic hydrocarbons in needles and bark of *Pinus pinaster Ait.* and *Pinus pinea L.* by GC–MS. *Talanta*, 77, 1120–1128, 2009.
76. Baumard, P., Budzinski, H., Garrigues, P.H., Sorbe, J.C., Burgeot, T., Bellocq, J., Concentrations of PAHs (polycyclic aromatic hydrocarbons) in various marine organisms in relation to those in sediments and to trophic level. *Mar. Pollut. Bull.*, 36, 12, 951–960, 1998.
77. Anastassiades, M., Lehotay, S.J., Štajnbaher, D., Schenck, F.J., Fast and easy multiresidue method employing acetonitrile extraction/partitioning and "dispersive solid-phase extraction" for the determination of pesticide residues in produce. *J. AOAC Int.*, 86, 2, 412–431, 2003.
78. Santana-Mayor, Á., Socas-Rodríguez, B., Herrera-Herrera, A.V., Rodríguez-Delgado, M.Á., Current trends in QuEChERS method. A versatile procedure for food, environmental and biological analysis TrAC-Special Issue "Green Extraction Techniques. *TrAC, Trends Anal. Chem.*, 116, 214–235, 2019.

79. Wilkowska, A. and Biziuk, M., Determination of pesticide residues in food matrices using the QuEChERS methodology. *Food Chem.*, 125, 3, 803–812, 2011.
80. Albinet, A., Tomaz, S., Lestremau, F., A really quick easy cheap effective rugged and safe (QuEChERS) extraction procedure for the analysis of particle-bound PAHs in ambient air and emission samples. *Sci. Total Environ.*, 450, 31–38, 2013.
81. Zhao, J., He, Q., Yao, B.H., Kong, X.H., Li, J.H., Rapid determination of polycyclic aromatic hydrocarbons in farmland soils by GC-MS coupled with modified QuEChERS. *Chinese J. Anal. Lab.*, 31, 2, 33–37, 2012.
82. Restek Corporation, Fast, Simple QuEChERS Extraction and Cleanup of Pesticide Residue Samples, 2009, [Online] Available at: http://www.restek.com/pdfs/GNAN1097A.pdf (Accessed: 2 June 2020).
83. João Ramalhosa, M., Paíga, P., Morais, S., DelerueMatos, C., Prior Pinto Oliveira, M.B., Analysis of polycyclic aromatic hydrocarbons in fish: Evaluation of a quick, easy, cheap, effective, rugged, and safe extraction method. *J. Sep. Sci.*, 32, 20, 3529–3538, 2009.
84. Kalachova, K., Pulkrabova, J., Drabova, L., Cajka, T., Kocourek, V., Hajslova, J., Simplified and rapid determination of polychlorinated biphenyls, polybrominated diphenyl ethers, and polycyclic aromatic hydrocarbons in fish and shrimps integrated into a single method. *Anal.Chim.Acta*, 707, 1–2, 84–91, 2011.
85. Sadowska-Rociek, A., Surma, M., Cieślik, E., Application of QuEChERS method for simultaneous determination of pesticide residues and PAHs in fresh herbs. *Bull. Environ. Contam. Toxicol.*, 90, 4, 508–513, 2013.
86. Richter, B.E., Jones, B.A., Ezzell, J.L., Porter, N.L., Avdalovic, N., Pohl, C., Accelerated solvent extraction: A technique for sample preparation. *Anal. Chem.*, 68, 6, 1033–1039, 1996.
87. Yang, R., Zhang, S., Li, X., Luo, D., Jing, C., Dechloranes in lichens from the southeast Tibetan Plateau: Evidence of long-range atmospheric transport. *Chemosphere*, 144, 446–451, 2016.
88. Kodnik, D., Carniel, F.C., Licen, S., Tolloi, A., Barbieri, P., Tretiach, M., Seasonal variations of PAHs content and distribution patterns in a mixed land use area: A case study in NE Italy with the transplanted lichen *Pseudevernia furfuracea*. *Atmos. Environ.*, 113, 255–263, 2015.
89. Brough, S.G., Navajo Lichen Dyes. *Lichenologist*, 20, 03, 279–290, 1988.
90. Vu, T., Catheline, D., Delmail, D., Boustie, J., Legrand, P., Lohézic-le Dévéhat, F., Gas chromatographic analysis to compare the fatty acid composition of fifteen lichen species, with a focus on *Stereocaulon*. *The Lichenologist*, 48, 4, 323–337, 2016.
91. Serafini, I., Lombardi, L., Reverberi, M., Ciccola, A., Calà, E., Sciubba, F., Guiso, M., Postorino, P., Aceto, M., Bianco, A., New advanced extraction and analytical methods applied to discrimination of different lichen species used for orcein dyed yarns: Preliminary results. *Microchem J*, 138, 447–456, 2018.

92. Bézivin, C., Tomasi, S., Lohezic-Le Devehat, F., Boustie, J., Cytotoxic activity of some lichen extracts on murine and human cancer cell lines. *Phytomedicine*, 10, 6–7, 499–503, 2003.
93. Bilgin Sökmen, B., Kinalioğlu, K., Aydin, S., Evaluation of Antifungal Potentials and Antioxidant Capacities of Some Foliose Lichen Species. *Süleyman Demirel Üniversitesi Fen Bilimleri Enstitüsü Dergisi*, 22, 1, 148–152, 2018.
94. Chauhan, R. and Abraham, J., *In vitro* antimicrobial potential of the lichen *Parmotrema sp.* extracts against various pathogens. *Iran. J. Basic Med. Sci.*, 16, 7, 881–885, 2013.
95. Parrot, D., Saleem, S., Baert, N. *et al.*, Comparative metabolite profiling and chemical study of *Ramalina siliquosa* complex using LC–ESI-MS/MS approach. *Phytochemistry*, 89, 114–124, 2013.
96. Shen, J. and Shao, X., A comparison of accelerated solvent extraction, Soxhlet extraction, and ultrasonic-assisted extraction for analysis of terpenoids and sterols in tobacco. *Anal. Bioanal. Chem.*, 383, 6, 1003–1008, 2005.
97. Brovko, O.S., Ivakhnov, A.D., Palamarchuk, I.A., Boitsova, T.A., Supercritical fluid extraction of usnic acid from lichen of *Cladonia* Genus. *Russ. J. Phys. Chem. B*, 11, 8, 1306–1311, 2017.
98. Musharraf, S.G., Siddiqi, F., Ali, A., Thadhani, V.M., Sensitive analysis of bioactive secondary metabolites in lichen species using liquid chromatography–mass spectrometry. *J. Pharm. Biomed. Anal.*, 146, 279–284, 2017.
99. Falk, A., Green, T.K., Barboza, P., Quantitative determination of secondary metabolites in *Cladina stellaris* and other lichens by micellar electrokinetic chromatography. *J. Chromatogr. A*, 118, 141–144, 2008.
100. Bonny, S., Paquin, L., Carrié, D., Boustie, J., Tomasi, S., Ionic liquids based microwave-assisted extraction of lichen compounds with quantitative spectrophotodensitometry analysis. *Anal. Chim. Acta*, 707, 69–75, 2011.
101. Luo, H., Yamamoto, Y., Kim, J.A., Jung, J.S., Koh, Y.J., Hur, J.S., Lecanoric acid, a secondary lichen substance with antioxidant properties from *Umbilicaria antarc*tica in maritime Antarctica (King George Island). *Polar Biol.*, 32, 7, 1033–1040, 2009.
102. Kumar, A., Gaurav, K.A., Malik, A.K., Tewary, D.K., Singh, B., A review on development of solid phase microextraction fibers by sol-gel methods and their applications. *Anal. Chim. Acta*, 610, 1, 1–14, 2008.
103. Piri-Moghadam, H., Ahmadi, F., Pawliszyn, J., A critical review of solid phase microextraction for analysis of water samples. *TrAC, Trends Anal. Chem.*, 85, 133–143, 2016.
104. Souza-Silva, T.A., Gionfriddo, E., Pawliszyn, J., A critical review of the state of the art of solid-phase microextraction of complex matrices II. Food analysis. *TrAC, Trends Anal. Chem.*, 71, 236–248, 2015.
105. De Angelis, F., Ceci, R., Quaresima, R., Reale, S., Di Tullio, A., Investigation by solid-phase microextraction and gas chromatography/mass spectrometry of secondary metabolites in lichens deposited on stone monuments. *Rapid Commun. Mass Spectrom.*, 17, 6, 526–531, 2003.

106. Cansaran, D., Kahya, D., Yurdakulola, E., Atakol, O., Identification and quantitation of usnic acid from the lichen *Usnea* species of Anatolia and antimicrobial activity. *Zeitschrift fur Naturforschung C*, 61, 773–776, 2006.
107. Graney, J.R., Landis, M.S., Puckett, K.J., Studabaker, W.B., Edgerton, E.S., Legge, A.H., Percy, K.E., Differential accumulation of PAHs, elements, and Pb isotopes by fivelichen species from the Athabasca Oil Sands Region in Alberta, Canada. *Chemosphere*, 184, 700–710, 2017.
108. Musharraf, S.G., Kanwal, N., Thadhani, V.M., Choudhary, M.I., Rapid identification of lichen compounds based on the structure-fragmentation relationship using ESI–MS/MS analysis. *Anal. Methods*, 7, 6066–6076, 2015.
109. Dailey, R., Siemion, R., Raisbeck, M., Jesse, C., Analysis of salazinic, norstictic, and usnic acids in *Xanthoparmelia chlorochroa* by ultra-performance liquid chromatography/tandem mass spectrometry. *J. AOAC Int.*, 93, 1137–1142, 2010.
110. Kreft, S. and Štrukelj, B., Reversed-polarity capillary zone electrophoretic analysis of usnic acid. *Electrophoresis*, 22, 2755–2757, 2001.

3

Review on Advanced Methods for Extraction and Identification of Nature-Derived Bioactive and Economic Products

Sudarshana Deepa Vijaykumar

Department of Biotechnology, National Institute of Technology, Tadepalligudem, West Godavari District, Andhra Pradesh, India

Abstract

In recent years, the interest in the development of procedures for extraction of bioactive compounds from natural sources with more liability gains increased attention globally, due to their potential applications in many sectors such as food and beverages, paper, textile, chemical, and pharmaceutical industries. Problems like consumption of energy and time and solvent quantity are associated with conventional extraction procedures led to demand for alternative techniques for extraction of nature-derived products. So, various novel extraction techniques have been developed and refined in last few decades because of their potential to efficiently extract a great variety of valuable bioactive compounds. This review intends to discuss different advanced extraction techniques along with new identification methods for bioactive and economic nature-derived products.

Keywords: Lichens, extraction, solvent, ultrasound, microwave

3.1 Introduction

Nature-derived products have been used as the source of food, traditional medicines, dyes, and cosmetics from ancient times. Undisputedly, nature has always had upper hand over the synthetic products, due to their bio-compatible and native advantages that gained them the priority over synthetic products [1]. Most compounds bear at least one aromatic group

Email: sudarshanadeepa@nitandhra.ac.in

Mohd Yusuf (ed.) Lichen-Derived Products: Extraction and Applications, (75–100) © 2020 Scrivener Publishing LLC

with one or more chromophoric groups and their structures may range from a simple phenolic molecule to a complex high-molecular polymer [2]. The application of nature-extracted compounds in various commercial/industrial sectors such as food, pharmaceutical, and chemical industries signifies the need for the standardization of isolation, identification, and characterization of these active components. In this regard, numerous new methods have been established along with conventional methods. The efficiencies of the extraction methods generally depend on the critical input parameters, understanding the nature of plant matrix, and the chemical composition of bioactive compounds.

Conventional natural compounds extraction methods such as thermal agitation, boiling, or refluxing have been used to predominantly extract the natural phenolic/other compounds from the biomass. But, these procedures lead to the loss of molecule-skeletal moiety, possibly due to ionization, hydrolysis, physicochemical aberrations of functional groups during the extraction, as well as the extended extraction [3]. The classical/traditional extraction techniques practiced for the extraction of natural phenolic compounds are maceration, infusion, decoction, hydrodistillation, and soxhlet extraction [4].

Recently, various advanced extraction techniques have been developed for the extraction of natural products from natural sources like plants, animals, insects, bacteria, fungi, and lichens. Some of the techniques are the accelerated solvent extraction, ultrasound-assisted extraction, microwave-assisted extraction (MAE), supercritical fluid extraction (SFE), high pressure extraction, high hydrostatic pressure extraction, enzyme supported extraction, etc. This chapter deals with the updated overview on advanced extraction techniques of lucrative nature-derived products. The chapter also discusses the analytical techniques employed for the identification of nature-derived compounds responsible for the specific applications claimed in the traditional usage.

3.2 Accelerated Solvent Extraction

The accelerated solvent extraction as explained by Richter *et al.* (1996) is the extraction procedure that combines elevated temperatures and pressures with liquid solvents. The various operational parameters that influence the yield and effectiveness of the extraction procedure are temperature, pressure, and the solvent volume. The experimental procedure of accelerated solvent extraction can be briefly deliberated as to enclose the biomass in a sample cartridge filled with the extraction fluid and statically extract the sample under elevated temperature (50–200°C) and pressure (500–3,000 psi) for the short duration (5–10 min). Sample is purged from the cell into a collection vessel using the compressed gas [5].

The two major phenomena that improve the extraction efficiency of the accelerated solvent extraction technique are as follows.

3.2.1 Effect of Solubility and Mass Transfer

The solubility efficiency of solvent is increased with the increasing temperatures while extracting from the sample. In many researches to extract compounds, the elevated temperature has been practiced for better yield. In one of the researches, the temperature dependency of solubility of an ideal solution, the solubility of anthracene was found to have increased more than 12 folds while increasing the temperature from 50 to 150°C. The solubility of hydrocarbons such as *n*-eicosane in the solvents has been reported to increase several hundred folds while increasing the temperature by 100°C [6]. Further, the miscibility of water with the organic solvents increases with the increasing temperature because of the molecular agitation. This would be helpful in the cases in which solvents at low temperatures and pressures are excluded from water-sealed pores that contain analytes. The increased miscibility of the water with the organic solvents driven by the elevated temperatures facilitates the availability of the sealed pores and the analytes contained therein [7]. The diffusion rates of the compounds from cells into the extraction solvent increases with increasing the temperature of the extraction. It would be difficult to predict and simulate the exact relationship for the effect of temperature on diffusion rates, especially with finite concentrations and a multicomponent system. But the diffusion rates of the compounds were found to increase approximately 2–10 folds while increasing the temperature from 25 to 150°C through the illustrations [8].

According to the Fick's first law of diffusion, the mass transfer rate or flux is directly proportional the concentration gradient. Hence, the fresh solvent is added during a static extraction step to enhance the mass transfer thereby increasing the extraction rate. The fresh solvent is added to the extracting system to intensify the concentration gradient between the solutions present in the cell and the surface of the sample matrix [5]. This concentration gradient accelerates the mass transfer.

3.2.2 Lowering the Surface Equilibrium

The solute-matrix interactions formed by weak bonds such as Van der Waals forces, hydrogen bonding, and dipole attractions between the solute molecules, and active sites on the matrix are broken by the increasing temperature. Thermal energy decreases the activation energy to trounce the cohesive (solute-solute) and adhesive (solute-matrix) interactions, leading to desorption of the extracted solutes from the solution [9]. Another surface phenomenon of the liquids that determine the penetration of the solutes through the membranes is the viscosity of the extract solvents. The viscosity is decreased by increasing the temperature of the system, thus allowing better penetration of matrix particles and enhancing extraction. It further decreases the surface tension of the solvents, solutes, and matrix, allowing the solvent to moisten the sample matrix effectively. These changes together will facilitate better contact of the compounds with the extraction liquid and enhance the diffusion of compounds from the biomass. The solvent cavities are formed more easily by decreasing the solvent surface tension [10].

The extraction temperature can be increased during the extractions more than the threshold boiling point by applying pressure sufficiently on the solvent. The applied pressure also facilitates the extraction of the analytes that are trapped in matrix pores during the extraction. The applied pressure drives the solvent into regions of the matrices that would not normally be reached by the solvents under the atmospheric conditions. The analytes trapped in the pores resist the water to enter the pores, thus the organic solvents may not be able to contact these analytes and extract them. The use of elevated pressures along with elevated temperatures remove the analytes in the pores and the reduced solvent surface tensions as mentioned earlier forces the solvent into the pore and contact the analytes [5].

These principles are best used by scientists for extraction from lichens where the extraction was accelerated by a brief mechanical agitation such as vortexing with the coolants to control the raise in temperature of the extracting system [11].

3.3 Ultrasound/Ultrasonic Extraction

The bioactive compound extraction methods from lichens involve high energy techniques like ultrasound-assisted extraction. One of the methods of ultrasonic-assisted extraction is using ultrasonic baths. A portion of the pulverized lichen sample is placed into vial and spiked with required amount of surrogate standard before the extraction solvents are added. The contents are sonicated or ultrasonicated based on the required energy for 15 min in an ultrasonic bath with water temperature maintained at 30°C. The sonicated contents are then centrifuged at high speed for 10 min and the pellet is collected. The processed sample is then concentrated under nitrogen atmosphere to minimal quantity. The extracts are then cleaned up by the sample extract cleanup procedure [11]. Unlike the ultrasonic baths, in the probe sonication process, the high frequency ultrasonic probes that can produce high energy sonic waves drastically accelerate the extraction of compounds from solid samples [12].

The sample extract cleanup procedure is usually followed as per the following protocol. The filters like Strata–NH_2 solid-phase extraction (SPE) cartridges are used to filter the extract from the ultrasonicated solution that has the biomass. The extract obtained is isolated into different compounds using stationary phase adsorption silica beds (florisil) along with alkalies like Na_2SO_4. The proper ratio of polar/non-polar organic solvent is used as the elution solvent. The eluent is then evaporated to obtain the concentrated extract fractions and washed with inorganic solvents to remove the traces of the organic solvents and the yield is usually stored at −18°C until analysis [13].

The sonication time required for the extraction using ultrasonic probe is less than the ultrasonic bath extraction. From the literature, the process duration of ≤5 min is usually sufficient for the probe ultrasonication for solid liquid extraction [14, 15]. The reduced time is advantageous in the aspect of heat generation which can affect the nature of the compounds and solvents.

In a work by Krishna and Arunachalam, the ultrasonic probe was used to extract the contents from lichens. To a 0.1 g biomass of lichen, 5 ml of acidic solvent (HNO_3/HCl) solution had been added. Then the extraction mixture was ultrasonicated at room temperature by varying the parameters such as duration and amplitude of sonication. After the extraction using ultra sonic waves, the supernatant was separated from the solution by centrifugation and then transferred the known volume of clear supernatant to another pre-cleaned centrifuge tube for analysis [16].

Thus from the understandings, it is evident that the factors to be considered in ultrasonication process are amplitude of sonication [17, 18], sonication time [12], and solvent concentration [19].

3.4 Microwave Extraction

MAE is another high energy extraction procedure that is receiving high attention as an alternative to classical solid-liquid extraction methods. This technique is advantageous over the traditional extraction techniques which require long processing time and produce poor yield. This technique uses the microwave energy to heat the solvents containing the biomass sample in order to dissociate the required active compounds from the sample matrix into the solvent. The MAE method was first introduced by Ganzler and Szinai by using a domestic microwave oven [20]. The process is carried out in the closed environment for the extraction to be performed at elevated temperatures accelerating the mass transfer of target compounds from the sample matrix. The general labscale setup for the extraction from lichens in this method requires small solvent volumes in the range of 10–30 ml and takes 15–30 min of microwave irradiation. The conventional methods require at least 10 folds higher solvent volumes compared to MAE method. Further, the sample throughput is enhanced as several biomasses can be extracted simultaneously. The product recovery and reproducibility are more in MAE than the conventional techniques [21].

The MAE technique uses the phenomena of dipole rotation and ionic conduction of molecules under microwave irradiated environment. These two molecular vibrations take place simultaneously and are used in many applications. In the ionic conduction, the ions are electrophoreticaly migrated in an electromagnetic field. This electronic migration is resisted by the solution leading to the friction thereby increasing the temperature of the solution. In the dipole rotation, the dipoles of the solvent are realignment with respect to the applied field. At the commercial microwave frequency of 2,450 MHz, a 4.9×10^9 times per second dipole randomization and alignment is observed [22]. This randomization and alignment mobilizes the molecules developing a friction thereby resulting in the heating [23–25].

When the solvents are heated above the boiling point, the extraction rate and efficiency are improved. This can be achieved in closed vessels. The heating imparted to the surrounding media will vary with the chemical substances used because the different chemical substances absorb microwave energy to different extents. Hence, for samples with non-homogeneous structural characteristics, or that contain various chemical species with different dielectric properties dispersed into a homogeneous environment, it is possible to produce a selective heating of some areas, or components of the sample in microwave heating. This phenomenon is sometimes called superheating [21].

The parameter optimization for the MAE technique has been reported in several findings. The concentration and composition of the extraction liquid, microwave exposure time, temperature, and the matrix compositions such as the water content are the most considered factors to obtain the highest yield. A correct choice of solvent is fundamental for obtaining an optimal extraction process.

3.4.1 Solvent Composition

Solvents have to be chosen based on their microwave-absorbing properties, their efficiency to interact with the extraction matrix, and the degree of compound solubility in the solvent. The selectivity of the solvent is majorly influenced by the polarity of the compound to be extracted. The analytical procedure for the identification of the compounds also determines the solvent of interest. These optimizations are complicated in conventional techniques because of the longer duration and involvement of more parameter variables. There will be no heating if the solvent molecules do not absorb microwave energy. Different solvents have been studied and used for different analytes previously. Atrazine has been extracted in aqueous solution [26]. Polar compounds like taxanes are extracted with pure ethanol (95%) [27]. Low molecular oligomers are extracted most effectively using organochloride solvent Dichloromethane (DCM) [28], while the stable organic solvent like pure tetrahydrofuran (THF) has been used for the extraction of organochlorine pesticides [29]. The aqueous buffers (pH 10) have been used as the best extractant for imidazolinone herbicides [30], and the non-buffered aqueous solution of polyoxy-ethylene 10 lauryl ether has been used to extract polycyclic aromatic hydrocarbons from marine sediments [31].

3.4.2 Solvent Volume

The lesser the solvent required, the more economical the process is. The volume of solvent required for a single biomass sample is generally in the range of 10–30 ml. The efficiency of the extraction is determined by the volume of the extraction liquid. The solvent should be sufficient to completely immerse the biomass, and a little more when extracting from a matrix that swells during the extraction process. The fraction of the solvent in the extraction solution should be more than 60% and less than 75% (v/w) for the optimum yield. From various research findings on the MAE extraction, it was found that the recovery of the yield is reduced with higher concentration of the solvent unlike in the conventional extraction techniques [21]. Hence, the fraction of the solvent should not exceed 75% in MAE technique.

3.4.3 Temperature

In any extraction technique, the most vital parameter to be optimized for the best extraction efficiency is the temperature as it influences the extent of the recovery. In the extraction techniques that take place in a closed environment, the temperature raises during the process due to the increasing pressure. The MAE technique is conducted in closed vessels, and thus, the temperature of the extraction system would rise more than the required boiling point of the solvent. Hence, maintaining the temperature using temperature sensors and coolants are basic necessity of the extraction process. The temperature rise is used to desorb the compounds from the biomass matrix thereby enhancing the efficiency of the extraction. Besides, the solubility of the solvents is increased with the increasing temperatures, and other physical parameters such as surface tension and solvent viscosity are inversely proportional to the temperature. These reduced surface tension and solvent viscosity which improves the sample wetting and matrix penetration, respectively.

3.4.4 Extraction Time

The duration of MAE technique is very brief when compared to the conventional techniques. Since the microwave energy produces heat along with the elevated temperature, rate of effusion from the lichens is amplified. This reduces the extraction time than the conventional methods.

3.4.5 Biomass Characteristics and Water Content

The physicochemical properties of the biomass (matrix) in which the compounds of interest are bound have a substantial influence on the yield efficiency. This phenomenon has been examined by spiking experiments, where solid samples were spiked with the compounds of interest and compared with extractions from the native samples. The results proved that the highest extraction was obtained from the spiked samples, reiterating the impact of the binding of compounds to the biomass matrix [32]. Aging of matrices leads to decreased recoveries. This has been proved from the experiments varying the sample preparation techniques such as SFE and interpreted that the native analytes strongly bind to the biomass matrix than the spiked samples owing to the increased contact time [21]. It is thus critical to monitor the aging kinetics and its effect when simulating the protocol so as to achieve the suitable techniques for exhaustive extractions.

3.5 Ultrasound-Microwave Mixed

Both the ultrasonication- and microwave-assisted extraction techniques are widely identified as the most effective techniques for the rapid and effective extraction, with highest quality. The sonic waves in the ultrasonication mechanically disrupt the biomass cells and release the soluble metabolites, accelerating the mass transfer, and facilitate the infusion of the solvent into the cell core contact the intracellular contents. This phenomenon is inversely proportional to the frequency. The micro wave vibrates the molecules and raises the temperature of the whole sample content simultaneously. This breaks the weak bonding such as hydrogen bonds and Van der Waal's force, through the molecular dipole rotations, an effect that is resisted by the viscosity of the solvent. Consequently, these movements of dissolved ions of the compounds increase solvent penetration into the biomass matrix and thus facilitate compound solvation.

Cravotto *et al.* along with Danacamerini sas, Italy, have designed an equipment for combined ultrasonic-microwave extraction functioning at different frequencies (19, 25, 40, and 300 kHz). The equipment contains a high-power immersion titanium horn probe working at 25 and 40 kHz and a titanium cup-horn working at 300 kHz. The transducer can work without overheating at high power (up to 500 W) thanks to an efficient cooling system that employs a chiller and an oil circulation circuit; a thin hollow titanium cylinder whose base is fixed on a booster placed in a high-power cavitating tube, an innovative cup-horn-like system. A small hydrophone was used to evaluate the efficiency of the transfer of acoustic energy from the bottom and the wall to the liquid. A coolant fluid was run through the outer polytetrafluoroethylene (PTFE) jacket. Two transducers were operated spontaneously at different frequencies in this immersion horn technique, a microwave probe made of PEEK® (polyether ether ketone) was immersed for the simultaneous production of microwaves along with ultrasound (US) waves produced by the titanium probes immersed in the vessel. The yield from combined ultrasonic and microwave (MW) irradiation in a closed chamber at 120°C was found to be higher compared to the conventional methods. In the case of bioactive compounds extraction from the marine weeds, the result was even higher as the rougher cell wall of the microalga was effectively ruptured by the combined US/MW irradiation. The duration of the process was reduced up to 10 folds and the yield was increased by 50–500% in comparison with the traditional techniques [33].

3.6 High Hydrostatic Pressure Extraction

High Hydrostatic Pressure Extraction (HHPE) is an emerging technique used for the extraction of the phytochemical compounds from the natural resources [34]. It is one of the technologies that have instigated to show its latent to assure both the end user and research needs and an effective alternate for the thermal processing in many industrial applications. HHPE has been reported to achieve pure bioactive compound yield and enhance the shelf-life of the compounds as it maintains its physicochemical, nutritional features, and bioactive profiles [35, 36]. HHPE is a cold isostatic super high hydraulic pressure that ranges from 100 to 800 MPa or more. The pressure can cause some structural modifications in fragile foods, such as membrane damage, protein denaturation, cell deformation, etc. The solubility is high as the pressure which is based on the phase behavior theory [37, 38]. Pressurized cells show elevated permeability according to the mass transfer theory [39]. The high hydrostatic pressure causes more solvent enter into the cell and as a result high number of compounds permeate into the cell membrane. The pressure gradient across the cell membrane is high which leads to rapid permeation under high hydrostatic pressure. The flow of ions leads to the concentration gradient across the cell membranes to attain equilibrium within a short period. HHPE can be used for the extraction of various compounds using different solvents based on polarity. It is operated without heating process at room temperature with the exception of increase in temperature ensuing from compression. There are numerous principal measures followed during HHPE. Initially, the raw herb is mixed with solvent and then exposed to HHP. Finally, the extraction solution is filtered to get rid of the biomass sludge. The resulting filterate contains the active compounds of the herb. The extractants of the herb be used instead of chemicals in cosmetic, pharmaceutical, and even in food industries as drugs or additives which has attracted the attention of researchers. The resulting solution during extraction is considerably influenced by the parameters such solvent volume with respect to the biomass, pressure, temperature, and the duration [40].

While comparing with other conventional methods of extraction, the HHPE method offer increased yield and selectivity. This type of extraction have many advantages such as very less processing time, low power consumption, and higher extraction yields [41, 42]. Additionally, this method shows minimal deterioration of the activity and physicochemical characteristics of the bioactive compounds due to heat while compared to other conventional solvent extraction methods. This technology is particularly beneficial for thermo sensitive products. It functions at low temperatures and high pressures to extract the compounds rapidly, involving fewer amounts of organic solvents. The temperature of the process throughout pressure treatment ranges between 0 and 100°C, and the duration of extraction ranges from few seconds to 20 min. It has been established to inactivate a number of microorganisms and enzymes. However, temperature and pressure do not have much impact on low molecular weight components such as vitamins, pigments, flavoring agents, and other nutritional compounds. The nature friendly and rapid extraction favors the food and pharmaceutical industries than the other methods of extraction.

HHPE apart from augmenting the quantity of the compounds of interest in the extract but also retain their functionality and native quality since the heating is less. This technique creates an extensive divergence in accumulating value to the food by-product and will consequently diminish the expenditure of the finishing product. Consumer observation of product quality depends on microbial quality, enzymatic and biochemical reactions, and the physicochemical fluctuations. Topical studies have exposed that HHPE can condense processing time and offer increased extraction efficiency as it has fewer negative effects on the antioxidant activity and structure of bioactive constituents.

In this technique, the organic wastes are used as the source of bioactive compounds. This makes HPPE economically feasible for the farmers and entrepreneurs and decisively a substitute to waste management. HPPE is pertinent for the consumption to a wide range of food products or by-products such as fruit and vegetables, oil seeds, cereal bran and husk, sea food, meat and poultry industry waste, or herbs and medicinal plants. Attempts are taken by the researchers to scale up the process to accomplish the market demands of high quality bio-active compounds or extracts. A valiant effort to analyze and recognize the geo-economic benefits such as energy and water conservation aspects of HPPE along with the safety and pure attributes improved the consumer perception of non-thermal technologies. This led to the adaption of HPPE in industrial scale globally.

3.7 Enzyme Supported Extraction

Enzymes are highly competent and specific [43, 44]. Enzyme supported compound extraction can potentially replace the conventional techniques as it is eco-friendly [45]. This method is a high bioactive compound yielding technology where the compounds of interest are released by which cell walls are broken down. It also helps in removing the unwanted biochemicals from the cell walls. It aids the remuneration of increased catalytic effectiveness and conserves the unique value of the natural products [46].

3.7.1 The Types and Characteristics of Enzyme

Enzymes are derived from microbes including bacteria and fungi, and animals or plants. They are functionally classified into seven subsequent types, namely, hydrolyzing enzymes, oxidation-reduction enzymes, ligases, group trasnsfer enzymes, desmolases, isomerizing enzymes, and carboxylation enzymes. A particular enzyme acts on its specific complementary substrate based on their catalytic property of catalyzing definite reactions. Enzymes also degrade the cell walls and membranes as they act as ideal catalysts that can aid in the extraction of complex bioactive compounds. As a result, extraction of bioactive compounds is increased as they have high cell wall permeability [47].

3.7.2 The Breakdown of Cell Walls by Enzyme

The cell wall is the toughest organelle in the biomass to break. They resist the bio-active compounds from diffusing out of the cells during the extraction process. Hence, it is important to degrade the cell wall to extract the various bioactive compounds within the cell walls. Enzyme supported extraction is based on the hydrolysis of cell walls to disrupt their structural integrity by the enzymes under mild process conditions, to permit the proficient extraction and discharge of the bioactive compounds [48, 49]. When enzymes and substrates are complexed, the shape of the enzyme can alter by itself into an optimal fit for the interaction between enzyme and substrate, which could cause stress and strain on the substrate molecule that results in breakage of bond thereby uphold the reaction. When the concentration of substrate increases, the accumulation of enzymes can promote the reaction rate, until the volume of substrate turns insignificant. Until the enzyme concentration becomes restrictive the rate of reaction is directly proportional to the substrate concentration [50].

The operational systems such as reaction temperature, duration of the process, pH of the extraction system, enzyme concentration, and particle size of substrate are assisting the enzyme supported extraction. The enzyme supported extraction reduces the solvent volume and extraction time in addition to the quality of product and increased yield.

The elements of the bioactive products in natural sources are complex, they combine with pectin, starch, protein, and cellulose in cells. This contamination lowers the extraction yields of products and also diminishes the transparency of system. Therefore, it becomes difficult to improve the quality of products. The addition of suitable enzyme into the system of extraction helps in disrupting the complex bond found between pectin and proteins in the cells. Thus, the transparency of system and the quality of product could enhance significantly. Zhang treated blackberry juice with the mixture of pectinase and multiplexed peitinase for a study. It showed the optimal enzymatic treatment condition which is as followed: 0.13 %

enzyme was included and the portion of two enzymes was 1:1. The maximum extract was 79.5 %, the anthocyanins content of the extract was 676.24 mg·L^{-1} [51]. Some natural source consists of very less bioactive molecules but shows their chemical structure might be similar to that of target products. Therefore, particular enzyme can be added to the extraction system could modify their chemical structure into target structure. This trait of enzyme shows improvement in products' eminence. Enzymes are also used in chemical applications to cleavage the ester bonds, to gain unique optically active compounds through the partition of racemic mixtures and to break the double bonds. Ginsenosides are the principle components accountable for the pharmaceutical activities of ginseng. Some minor ginsenosides displayed a range of biological and pharmacological activities that includes anti-tumor, immune-modulatory, anti-inflammatory, and anti-aging effects. These ginsenosides subsisted in very small amount or did not even naturally present in ginseng. Jin treated some high-content ginsenosides which had parallel structure as Rh2 with glycoside hydrolases. After it has been treated with enzyme, the yield of conversion from high-content ginsenosides to Rh2 approached 60 %. This result is 500–700 times high than that of extraction from ginseng without enzyme.

Enzyme supported extraction of natural product facilitates the shortened extraction time. The requirement of organic solvent is reduced, increased yield and purity of the product. However, enzyme supported extraction has latent commercial and scientific limitations [47]. 1) Enzymes are comparatively costly for large industrial production. 2) Available enzyme cannot break down the cell walls entirely. 3) Enzyme supported extraction is impossible to be practical in industrial scale applications as the enzymes are require stringently ideal environment for their activity. The above-mentioned restrictions were required to be solved in order to effectively achieve the enzymes' fullest application value in the utilization and development of natural products.

3.8 Pulsed Electric Field Extraction

Pulsed electric field (PEF) extraction technology is yet another promising extraction technique in the league of microwave extraction and ultrasonication for extracting from the natural source [52]. Moreover, PEF is less expensive, and also, the production of high value sensitive compounds is possible as it does not generate heat during the extraction process. The application of electric fields for a short duration creates pores through the electroporation on the cell membrane, and consequently, the cell membrane become permeable for cellular components [53]. Pore formation is a dynamic process that depends on the intensity of the PEF treatment. The pore size depends direct proportionally on the intensity, strength (E), and the duration (t) of the application of electric field. With the increased magnitude of these parameters, the reversible permeabilization would become irreversible disruption of the cell membrane, which is used in some of the respective applications that require irreversible permeabilization. The most important factor determining the electroporation efficiency is the electric field strength (E). Cell disruption, which leads to enhanced mass transfer of cellular components and high membrane permeability, is the chief significance of this non-thermal technology [54]. Besides, the yield amount and purity are increased, while the duration, concentration of the solvent, and heat generation for the extraction are shortened. The reduced heating is optimum for the extraction of heat-sensitive compounds [55].

At the ideal temperature, the PEF facilitates the electroporation on the cell membranes that consequently enhances the diffusion of the valuable bioactive compounds. This type of "cold" extraction method supported by PEF does not degrade the cell wall, but just creates pores to extract the cellular compounds into the fluid. The disadvantage of undesirable chemical reactions among the extracted compounds at high temperatures is eliminated in the PEF technology, and thus, it is suitable for the extraction of pure heat sensitive compounds [56]. The conductive biomass cell placed between the electrodes is exposed to intense electric pulses for electroporation. The external electric field induces permeabilization of cells membranes in untreated matrices which supports the production processes. The mechanism of modification of transmembrane permeability has also gained increased interest in many industries as it has possible application in extraction of bioactive components and also especially in maceration stages of red wine process. As the compounds responsible for the coloration of red wine is present in the peripheral layer of the grape berries, the mass transfer mechanism and increased permeability of cytoplasmic membrane are used for accomplishing efficient extraction.

El-Belghiti and Vorobiev [57] investigated the purposes of PEF to sugar beet slices and the consequent diffusion (2 h) at appropriate temperature led to a revitalization of 93% in solute yield while comparing to the 40% acquired in untreated samples. The purity of treated slices was found higher than that of untreated slices after thermal diffusion at 70°C. PEF, in the past, was used for the extraction of inulin from chicory tissues [58], anthocyanins from red cabbage [59], podophyllotoxin from Podophyllum peltatum [60], polyphenols from fresh tea leaves [61], chitosan from shrimp shells [62], calcium from fishbone [63], and crocin, safranal, and picrocrocin from Crocus sativus [64]. It was investigated the impact of PEF treatment on the yield and quality parameters of the extract obtained paprika. PEF has also been used for the aroma compounds extraction from toasted wood in order to flavor wines [65].

3.9 Supercritical Fluid Extraction

The components present in the natural sources are conjugated to one another that are to be isolated. The process of isolating compounds using the supercritical fluid (SCF) as the solvent is called supercritical fluid extraction (SFE). The most preferred SCF is carbon dioxide (CO_2), modified in few cases using the polar co-solvents such as ethanol or methanol, which can also be used. The conditions for supercritical CO_2 extractions are higher than the critical temperature of 31°C and critical pressure of 74 bar. The high value products from natural sources can be extracted by means carbon dioxide at high pressure. Furthermore, the CO_2 is inert, non-toxic, nonflammable, odorless, tasteless, and inexpensive. Carbon dioxide can be completely adapted for many industrial purposes as it has critical temperature 31°C.

It is moderately fast since the SCFs have low viscosity and better flow and diffusivity. There will be selective extraction process thereby controlling the density of the medium and by depressurizing the material which is extracted can be removed and the SCF is allowed to evaporate completely without leaving residues in the medium.

In SFE, by altering the pressure or the temperature of the system within crossing phase boundaries, the physicochemical properties such as density, diffusivity, dielectric constant, and viscosity can be controlled [66]. The typical single component phase diagram is illustrated in Figure 3.1. The dissolving power of a SCF depends on its density, whereas that of other liquid solvents is extremely bendable by changing few parameters such as pressure or temperature. Mainly, the isothermal raise in the pressure enhances the density of the fluid while the effect reduces with increase in distance from critical values. Therefore, it is easy to maintain the solvent strength than that of liquid solvent. The SCF has higher standard diffusion coefficient and lower surface tension and viscosity than other liquid solvents, which causes possible mass transfer [67]. There is a high requirement for advanced analytical techniques in environmental and food sciences which strongly demand of information n quality control, contamination, processing, and also in food regulations. The chemists and other regulatory agencies requires faster, powerful, and cheaper protocols to be followed in laboratories as their requirements can be accomplished by SFE.

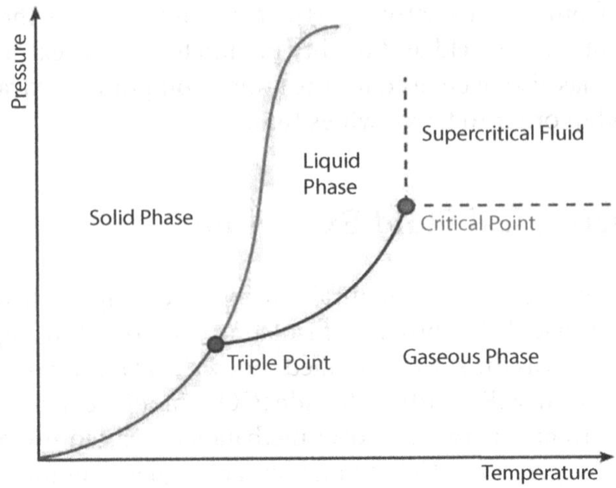

Figure 3.1 Single component phase diagram.

3.9.1 Supercritical Fluid (SCF)

Any substance above the critical point of its temperature and pressure is termed as SCF. It can able to dissolve substances like a liquid and diffuse through solid materials like gas. If there is small changes happened in temperature or pressure, it results in varied density and permits several properties of a SCF to be tuned finely. The SCF can be used as an alternative for other organic solvents in both laboratory and industrial processes. The widely used SCFs are CO_2 for decaffeination and water for power generation. The extraction characteristics can be specifically manipulated with fine changes in pressure and temperature.

3.9.2 Properties of SCF

The general properties of the SCFs are:

(i) Comprised of extremely compressed gases, which mix the characteristic of both gases and liquids;
(ii) Can aid the reactions that are generally impossible using the conventional solvents;
(iii) The solubility of SCF is as effective as the light hydrocarbons for most of the solutes;
(iv) Solubility augments with increasing density and pressure;
(v) SCFs have the properties intermediate between gas and liquid.

3.10 Modern Identification Methods Based on Spectroscopy Approach

In recent era, various spectral methods are being used in structure elucidation and characterization of phenolic and other chemical compounds from natural sources. Some spectral methods frequently employed such as mass spectroscopy, NMR spectroscopy, and infrared spectroscopy given herein.

3.10.1 Mass Spectroscopy (MS)

Mass spectrometry/spectroscopy is a powerful spectral analysis method for the quantification of known materials, identification of unknown compounds in a sample, and for the elucidation of the structure and chemical structures of biomolecules such as polyphenols, peptides, lipids, and other biochemical moieties. The process involves, principally, vaporizing the samples into gaseous ions (parental ions), with or without fragmentation, and the characterization of the ionic gases by their mass-to-charge ratios (m/z) and relative abundances [68]. The results are illustrated as the spectrum of mass plotted against the function of mass-to-charge ratio. Thus, the MS is used in various fields for the identification of compounds based on molar mass. Currently, in the analysis of phenolic and other constituent compounds, the methods used are: electrospray ionization (ESI), fast atom bombardment (FAB), thermospray analysis (TSP) (the thermal application makes it unfit for the analysis of oligomers and polymers), direct flow injection electrospray ionization (DFIESI) (used to establish polyphenol fingerprints of complex extracts), and atmospheric pressure ionization (API) that involved two variations: atmospheric pressure chemical ionization (APCI) and atmospheric pressure photo-ionization (APPI).

Consequently, matrix-assisted laser desorption/ionization (MALDI) was invented that generate ions with minimal fragmentation from sample molecules using a laser energy absorbing matrix and analyze the biomolecules (biopolymers, i.e., sugars, DNA, proteins, and peptides) and large organic molecules [69, 70]. The high molecular compounds are identified with better accuracy using the matrix-assisted laser desorption ionization time-of-flight (MALDI-TOF) technique [71].

Off late, the coupled chromatography-MS systems are very popular as the efficiency of each technique is enhanced mutually. Liquid chromatography–mass spectrometry (LC-MS) and gas chromatography–mass spectrometry (GC-MS) techniques are the best analytical approaches to study polyphenols in natural samples for characterization [72]. Currently, there are many physical and spectroscopic methods that have been developed for the isolation and identification of lichen substances. For isolation, broadly accepted methods are micro and semi-micro crystallization, derivatization, thin layer chromatography (TLC), high performance liquid chromatography (HPLC), gas-liquid chromatography (GLC), and gas chromatography integrated with mass spectrometry (GCMS). Identification of organic substance was characterized by chiefly melting-boiling points, microcrystallization, spectra of UV-vis, IR, FTIR, ^{1}H-NMR, ^{13}C-NMR, and mass and chromatography-assisted spectroscopy [73].

3.10.2 NMR Spectroscopy

Nuclear magnetic resonance (NMR) spectroscopy or magnetic resonance spectroscopy (MRS) is a technique that predicts the nature of C and H moieties, and besides identification, it provides detailed information about the structure, dynamics, reaction state, and chemical environment of molecules in organic samples by observing local magnetic fields around atomic nuclei. The fingerprint of the electronic structure of a molecule and its individual functional groups are obtained by the resonating intra-molecular magnetic field around an atom in a molecule with the applied magnetic field.

The two major types of NMR are proton (PMR) and carbon-13 NMR (CMR). NMR spectra are unique in nature, well-resolved, and highly predictable for small organic molecules. Different functional groups with identical groups but with varying neighboring substituent give distinguishable signals/chemical shifts [74]. Standard ^{1}H, ^{13}C and high resolution/magic angle spinning (HR/MAS) NMR spectra generate the necessary chemical information on liquid and even semi-solid organic samples. 2D-NMR or correlation spectroscopy, [75] best known by its acronym, COSY, was proposed to get more accurate and highly fragrant observations. Other examples of 2D-NMR include J-spectroscopy (JSY), exchange spectroscopy (EXSY), nuclear overhauser effect spectroscopy (NOESY), total correlation spectroscopy (TOCSY), and heteronuclear correlation experiments, such as Heteronuclear Single Quantum Coherence (HSQC), Heteronuclear Multiple Quantum Coherence (HMQC), and Heteronuclear Multiple Bond Correlation (HMBC).

3.10.3 UV-Visible Spectroscopy

In the electromagnetic spectrum, 200 to 400 nm constitute the ultraviolet (UV) region and 400 to 800 nm constitute the visible portion. The atoms and molecules in the rest (ground) energy gets excited to higher energy level by absorbing the ultraviolet and visible radiation. The absorption wavelength of each atom or molecule is different since the energy levels of matter are quantized. The possible electronic transitions that light might cause are: $\sigma \rightarrow \sigma^*$, $n \rightarrow \sigma^*$, $n \rightarrow \pi^*$, and $\pi \rightarrow \pi^*$. The maximum wavelength (λ_{max}) and the intensity of the absorption bands are indicative of conjugation/unsaturation and basic skeletal system in the organic molecule, and if the wavelength is larger, the conjugation is greater. Also, UV-visible spectroscopy is a useful technique for the identification of various vitamins (organic molecules) qualitatively such as riboflavin, cyanocobalmin, etc., and catrotenoids, flavonoids, and many more [76, 77].

3.10.4 Infrared Spectroscopy

Nature-derived products are pertaining to the certain classes of compounds such as alkaloids, flavonoids, carotenoids, terpenes, essential oils, and others. However, several spectrometric methods such as HPLC, LC, GC, and MS are being utilized for elucidation, qualitative, and quantitative analysis of individual compounds from different natural sources [78]. Infrared region occurs between 2,500 and 16,000 nm and give most useful data regarding functional groups present in the organic molecules. FTIR spectroscopy is the advanced form of IR with Fourier Transform mathematical approach and has superior clarity as well as accuracy [76].

Consequently, the introduction of near-infrared (NIR) spectroscopy (NIR region 750–2500 nm) by William Herschel was in 18th century but neglected for many years. Nowadays, it has now become one of the most promising techniques for molecular spectroscopy and is applied for the simultaneous analysis of the content of free amino acids, caffeine, total polyphenols, and amylose in green tea. The principle of NIR reflectance spectroscopy is that light of wavelengths between 1,100 and 2,500 nm, reflected off powdered solids, contains compositional information which can be unravelled by a computer to report multiple analyses almost instantaneously [79–81]. This method can provide simultaneous, rapid, and non-destructive quantitation of major components in many organic substances such as fatty acid, food grade ingredients, etc. In order to get accurate distinguishing or to classify among a large organic moieties: alcohols, acetone, acetonitrile, hydrocarbons, carboxylic acids, aldehydes, ketones, esters, carboxylic acids, and non-carboxylic acids, NIRS was found as a very powerful analytical tool [82, 83]. NIRS integrated with chemometric technique has provided forwarded applications and become fast and precise method for quality control in the petrochemical as well food industry

[83]. Recently, Malaspina *et al.* (2018) studied spectroscopic techniques integrated with chemometrics for the interpretation of lichen biomonitoring of air pollution. In this study, lichen transplanted thalli was analyzed through monitoring techniques, including Near-Infrared Spectroscopy (NIRS), Front-Face Fluorescence Spectroscopy (FFFS), and Plant Efficiency Analyzer (PEA) [84].

3.11 Conclusion

Traditional methods for extraction of bioactive and economic nature-derived products need alternative techniques so that their maleficent associates such as heavy use of toxic and hazardous solvents and consumption of energy and time could be minimize. In this regard, scientists have developed novel extraction and identification methodologies in last few decades because of their potential to efficiently extract a great variety of valuable bioactive compounds and ease of characterization steps. Broadly, all the modern methods for extraction and identification for bioactive and economic nature-derived products are based on spectroscopic considerations/approaches. More and more researches are being on track that may enable further R&D progress in the extraction, separation, and identification of organic compounds.

References

1. Yusuf, M., Shabbir, M., Mohammad, F., Natural colorants: Historical, processing and sustainable prospects. *Nat. Prod. Bioprospect.*, 7, 1, 123–145, 2017.
2. Balasundram, N., Sundram, K., Samman, S., Phenolic compounds in plants and agri-industrial by-products: Antioxidant activity, occurrence, and potential uses. *Food Chem.*, 99, 191–203, 2006.
3. Hui, L., Bo, C., Shouzhuo, Y., Application of ultrasonic technique for extracting chlorogenic acid from *Eucommia ulmodies* Oliv. (E. ulmodies). *Ultrason. Sonochem.*, 12, 295–300, 2005.
4. Shahid, M., Yusuf, M., Mohammad, F., Plant Phenolics: A Review on Modern Extraction Techniques in: Analytical and Processing Techniques, in: *Recent Progress in Medicinal Plants*, vol. 41, M. Pathak (Ed.), J.N. Govil (Series Ed.) pp. 1–23, Studium Press LLC, USA, 2016.
5. Richter, B.E., Jones, B.A., Ezzell, J.L., Porter, N.L., Accelerated Solvent Extraction: A Technique for Sample Preparation. *Anal. Chem.*, 68, 1033–39, 1996.
6. Pitzer, K.S. and Brewer, L., *Thermodynamics*, 2nd ed, McGraw-Hill, New York, 1961, Chapter 18.

7. Sekine, T. and Hasegawa, Y., *Solvent Extraction Chemistry*, Marcel Dekker, New York, 1977.
8. Perry, R.H., Green, D.W., Maloney, J.O., *Perry's Chemical Engineers' Handbook*, 6th ed, McGraw-Hill, New York, 1984, Chapter 3.
9. Pimentel, G.C. and McClellan, A.L., *The Hydrogen Bond*, Freeman, San Francisco, 1960.
10. Mockel, H.J., Welter, G., Melzer, H., Correlation between reversed-phase retention and solute molecular surface type and area: I. Theoretical outlines and retention of various hydrocarbon classes. *J. Chromatogr.*, 388, 255–266, 1987.
11. Leandri, V.W. and Patricia, B.C., Forbes. Comparison of extraction techniques for polycyclic aromatic hydrocarbons from lichen biomonitors. *Env. Sci. Poll. Res.*, 26, 11179–90, 2019.
12. Mamba, S. and Kratochvil, B., Application of Ultrasound to Dissolution of Environmental Samples for Elemental Analysis. *Int. J. Environ. Anal. Chem.*, 60, 295, 1995.
13. Blasco, M., Domeno, C., Bentayeb, K., Nerín, C., Solid-phase extraction clean-up procedure for the analysis of PAHs in lichens. *Int. J. Environ. An. Chem.*, 87, 833, 2007.
14. Lima, E.C., Barbosa, F., Krug, F.J., Silva, M.M., Vale, M.G.R., Comparison of ultrasound-assisted extraction, slurry sampling and microwave-assisted digestion for cadmium, copper and lead determination in biological and sediment samples by electrothermal atomic absorption spectrometry. *J. Anal. Atom. Spectrom.*, 15, 995–1000, 2000.
15. Santos, C., Alava-Moreno, F., Lavilla, I., Bendicho, C., Total As in seafood as determined by transverse heated electrothermal atomic absorption spectrometry-longitudinal Zeeman background correction: An evaluation of automated ultrasonic slurry sampling, ultrasound-assisted extraction and microwave-assisted digestion methods. *J. Anal. Atom. Spectrom.*, 15, 987–994, 2000.
16. Krishna, M.V.B. and Arunachalam, J., Ultrasound-assisted extraction procedure for the fast estimation of major, minor and trace elements in lichen and mussel samples by ICP-MS and ICP-AES. *Anal. Chimica Acta*, 522, 179–187, 2004.
17. Capelo, J.L., Maduro, C., Vilhena, C., Discussion of parameters associated with the ultrasonic solid–liquid extraction for elemental analysis (total content) by electrothermal atomic absorption spectrometry. An overview. *Ultrason. Sonochem.*, 12, 225–232, 2005.
18. Capelo, J.L., Filgueiras, A.V., Lavilla, I., Bendicho, C., Solid-liquid extraction of copper from slurried samples using high intensity probe sonication for electrothermal atomic absorption spectrometry. *Talanta*, 50, 905–911, 1999.
19. Ashley, K., Andrews, R.N., Cavazos, L., Demange, M., Ultrasonic extraction as a sample preparation technique for elemental analysis by atomic spectrometry. *J. Anal. Atom. Spectrom.*, 16, 1147–1153, 2001.

20. Ganzler, K. and Szinai, I., Microwave extraction. A novel sample preparation method for chromatography. *J. Chromatogr.*, 271, 299, 1986.
21. Eskilsson, C.S. and Bjorklund, E., Analytical-scale microwave-assisted extraction. *J. Chromatogr. A.*, 902, 227–50, 2000.
22. Zuloaga, O., Etxebarria, N., Fernandez, L.A., Madariaga, J.M., Optimization and comparison of microwave assisted extraction and Soxhlet extraction for the determination of polychlorinated biphenyls in soil samples using an experimental design approach. *Talanta*, 50, 345–57, 1999.
23. Jassie, L., Revesz, R., Kierstead, T., Hasty, E., Metz, S., *Microwave-Enhanced Chemistry*, H.M. Kingston and S.J. Haswell (Eds.), p. 569, American Chemical Society, Washington, DC, 1997.
24. Kingston, H.M. and Jassie, L.B. (Eds.), *Introduction to Micro-wave Sample Preparation*, American Chemical Society, Washington, DC, 1988.
25. Thuery, J., *Microwaves: Industrial, Scientific and Medical Applications*, E.H. Grant (Ed.), Artech House, Boston, MA, 1992.
26. Xiong, G., Liang, J., Zou, S., Zhang, Z., Microwave-assisted extraction of atrazine from soil followed by rapid detection using commercial ELISA kit. *Anal. Chim. Acta*, 371, 97, 1998.
27. Mattina, M.J.I., Berger, W.A.I., Denson, C.L., Microwave assisted extraction of taxanes from Taxus biomass. *J. Sci. Food Agri.*, 45, 4691, 1997.
28. Costley, C.T., Dean, J.R., Newton, I., Carroll, J., Extraction of Oligomers From Poly(ethylene terephthalate) by Microwave-assisted Extraction. *Anal. Commun.*, 34, 89, 1997.
29. Silgoner, I., Krska, R., Lombas, E., Gans, O., Rosenberg, E., Grasserbauer, M., Microwave assisted extraction of organochlorine pesticides from sediments and its application to contaminated sediment samples. *Frensenius J. Anal. Chem.*, 362, 120, 1998.
30. Stout, S.J., daCunha, A.R., Allardice, D.G., Microwave Assisted Extraction Coupled with Gas Chromatography/Electron Capture Negative Chemical Ionization Mass Spectrometry for the Simplified Determination of Imidazolinone Herbicides in Soil at the ppb Level. *Anal. Chem.*, 68, 653–658, 1996.
31. Pino, V., Ayala, J.H., Afonso, A.M., Gonzalez, V., Determination of polycyclic aromatic hydrocarbons in marine sediments by high-performance liquid chromatography after microwave-assisted extraction with micellar media. *J. Chromatogr. A*, 869, 515, 2000.
32. Barnabas, I.J., Dean, J.R., Fowlis, I.A., Owen, S.P., Extraction of polycyclic aromatic hydrocarbons from highly contaminated soils using microwave energy. *Analyst.*, 120, 1897, 1995.
33. Cravotto, G., Boffa, L., Mantegna, S., Perego, P., Avogadro, M., Cintas, P., Improved extraction of vegetable oils under high-intensity ultrasound and/or microwaves. *Ultrason. Chem.*, 15, 898–902, 2008.
34. US Food and Drug Administration Center for Food Safety and Applied Nutrition, Kinetics of Microbial Inactivation for Alternative Food Processing Technologies—High Pressure Processing, 2000.

35. Peter, B.B., Ivan, D., Robert, E.M., *High Pressure Biology and Medicine*, University of Rochester Press, New York, 1998.
36. Paul, P.L. and Morita, R.Y., Effects of hydrostatic pressure and temperature on the uptake and respiration of amino acids by a facultively psychrophilic marine bacterium. *J. Bacteriol.*, 108, 835–843, 1971.
37. Richard, J.S., *High Pressure Phase Behaviour of Multicomponent Fluid Mixtures*, Elsevier, Amsterdam, 1992.
38. Noble, W.J. le, *Organic High Pressure Chemistry*, Elsevier, Amsterdam, 1988.
39. Yan, H., *Separation Engineering*, China Petrochemical Press, Beijing, 2002.
40. Chen, R., Meng, F., Zhang, S., Liu, Z., Effects of ultrahigh pressure extraction conditions on yields and antioxidant activity of ginsenoside from ginseng. *Sep. Purif. Technol.*, 66, 2, 340–346, 2009.
41. Corrales, M., García, A.F., Butz, P., Tauscher, B., Extraction of anthocyanins from grape skins assisted by high hydrostatic pressure. *J. Food Eng.*, 90, 4, 415–421, 2009.
42. Jun, X., Caffeine extraction from green tea leaves assisted by high pressure processing. *J. Food Eng.*, 94, 1, 105–109, 2009.
43. Linnell, J.C., Mackenzie, H.M., Wilson, J., Matthews, D.M., Patterns of plasma cobalamins in control subjects and in cases of vitamin B_{12} deficiency. *J. Clin. Pathol.*, 22, 545–550, 1969.
44. Jeon, Y.J., Wijesinghe, W., Kim, S.K., Handbook of Marine Macroalgae. *Biotechnol. Appl. Phycol.*, 221–228, 2011.
45. Fleurence, J., Massiani, L., Guyader, O., Mabeau, S., Use of enzymatic cell wall degradation for improvement of protein extraction from Chondrus crispus, *Gracilaria verucosa* and *Palmaria palmate. J. Appl. Phycol.*, 7, 393–397, 1995.
46. Shen, L., Wang, X., Wang, Z., Wu, Y., Chen, J., Studies on tea protein extraction using alkaline and enzyme methods. *Food Chem.*, 107, 929–938, 2008.
47. Puri, M., Sharma, D., Barrow, C.J., Enzyme-assisted extraction of bioactives from plants. *Trends Biotechnol.*, 30, 37–44, 2012.
48. Gardossi, L., Poulsen, P.B., Ballesteros, A., Hult, K., Švedas, V.K., Vasić-Rački, D., Carrea, G., Magnusson, A., Schmid, A., Wohlgemuth, R., Guidelines for reporting of biocatalytic reactions. *Trends Biotechnol.*, 28, 171–180, 2010.
49. Pinelo, M., Arnous, A., Meyer, A.S., Upgrading of Grape Skins: Significance of Plant Cell-Wall Structural Components and Extraction Techniques for Phenol Release. *Trends Food Sci. Technol.*, 17, 579–590, 2006.
50. Sowbhagya, H. and Chitra, V., Enzyme-Assisted Extraction of Flavorings and Colorants from Plant Materials. *Crit. Rev. Food Sci. Nutr.*, 146–161, 2010.
51. Zhang, L., Zhou, J., Liu, H., Huang, K., Gu, Z., Process optimization of preparation of clarified blackberry juice using two commercial enzymes. *Trans. Chinese Soc. Agric. Eng.*, 26, 372–376, 2010.
52. Asavasanti, S., Ristenpart, W., Stroeve, P., Barret, M., Permeabilization of plant tissue by monopolar pulsed electric fields: effect of frequency. *J. Food Science*, 76, 1, 98–111, 2011.

53. Huang, K., Hongping, T., Gai, L., Wang, J., A review of kinetic models for inactivating microorganisms and enzymes by pulsed electric field. *J. Food Eng.*, 111, 191–207, 2012.
54. Angersbach, A., Heinz, V., Knorr, D., Effects of pulsed electric fields on cell membranes in real food systems. *Innovative Food Sci. Emerging Technol.*, 1, 135–149, 2000.
55. Vorobiev, E. and Lebovka, N.I., *Electrotechnologies for Extraction from Food Plants and Biomaterials*, 1st ed, pp. 39–82, Springer, New York, NY, USA, 2008.
56. Puertolas, E., Luengo, E., Alvarez, I., Raso, J., Improving mass transfer to soften tissues by pulsed electric fields: Fundamentals and applications. *Annu. Rev. Food Sci. Technol.*, 3, 263–282, 2012.
57. El-Belghiti, K. and Vorobiev, E., Modelling of solute aqueous extraction from carrots subjected to a pulsed electric field pre-treatment. *Biosyst. Eng.*, 90, 289–294, 2005.
58. Loginova, K., Shynkaryk, M.V., Lebovka, N.I., Vorobiev, E., Acceleration of soluble matter extraction from chicory with pulsed electric fields. *J. Food Eng.*, 96, 374–379, 2010.
59. Gachovska, T.K., Cassada, D., Subbiah, J., Hanna, M., Thippareddi, H., Snow, D., Enhanced anthocyanin extraction from red cabbage using pulsed electric field processing. *J. Food Sci.*, 75, E323–E329, 2010.
60. Abdullah, S.H., Zhao, S., Mittal, G.S., Baik, O.-D., Extraction of podophyllotoxin from *Podophyllum peltatum* using pulsed electric field treatment. *Sep. Purif. Technol.*, 93, 92–97, 2012.
61. Zderic, A., Zondervan, E., Meuldijk, J., Breakage of cellular tissue by pulsed electric field: Extraction of polyphenols from fresh tea leaves. *Chem. Eng. Trans.*, 32, 1795–1800, 2013.
62. He, G., Yin, Y., Yan, L., Yu, Q., Fast extraction of chitosan from shrimp shell by high intensity pulsed electric fields. *Nongye Gongcheng Xuebao/Trans. Chin. Soc. Agric. Eng.*, 27, 344–348, 2011.
63. Zhou, Y., Sui, S., Huang, H., He, G., Wang, S., Yin, Y., Ma, Z., Process optimization for extraction of fishbone calcium assisted by high intensity pulsed electric fields. *Nongye Gongcheng Xuebao/Trans. Chin. Soc. Agric. Eng.*, 28, 265–270, 2012.
64. Pourzaki, A., Mirzaee, H., Hemmati Kakhki, A., Using pulsed electric field for improvement of components extraction of saffron (*Crocus sativus*) stigma and its pomace. *J. Food Process. Preserv.*, 37, 1008–1013, 2013.
65. Drosou, F., Yang, E., Marinea, M., Dourtoglou, E.G., Chatzilazarou, A., Dourtoglou, V.G., An assessment of potential applications with pulsed electric field in wines, in: *Proceeding of the 40th World Congress of Vine and Wine*, Sofia, Bulgaria, 2017.
66. Bravi, E., Pperretti, G., Motanari, L., Favati, F., Fantozzi, P., Supercritical fluid extraction for quality control in beer industry. *J. Supercrit. Fluids*, 42, 342–346, 2007.

67. José, A.M., Miguel, H., Alejandro, C., Elena, I., Use of compressed fluids for sample preparation. *Food Appli. J. Chromatog.*, 1152, 234–246, 2007.
68. Ignat, I., Volf, I., Popa, V.I., A critical review of methods for characterisation of polyphenolic compounds in fruits and vegetables. *Food Chem.*, 126, 4, 1821–1835, 2011.
69. Hillenkamp, F., Karas, M., Beavis, R.C., Chait, B.T., Matrix-assisted laser desorption/ionization mass spectrometry of biopolymers. *Anal. Chem.*, 63, 24, 1193A–1203A, 1991.
70. Karas, M. and Krüger, R., Ion formation in MALDI: the cluster ionization mechanism. *Chem. Rev.*, 103, 2, 427–440, 2003.
71. Fulcrand, H., Mané, C., Preys, S., Mazerolles, G., Bouchut, C., Mazauric, J.-P. *et al.*, Direct mass spectrometry approaches to characterize polyphenol composition of complex samples. *Phytochemistry*, 69, 3131–3138, 2008.
72. Dunn, W.B., Broadhurst, D., Begley, P., Zelena, E., Francis-McIntyre, S., Anderson, N., Brown, M., Knowles, J.D., Halsall, A., Haselden, J.N., Nicholls, A.W., Procedures for large-scale metabolic profiling of serum and plasma using gas chromatography and liquid chromatography coupled to mass spectrometry. *Nature Protoc.*, 6, 7, 1060–1083, 2011.
73. Huneck, S. and Yoshimura, I., Identification of lichen substances, in: *Identification of lichen substances*, pp. 11–123, Springer, Berlin, Heidelberg, 1996.
74. Hore, P.J., *Nuclear magnetic resonance*, Oxford University Press, USA, 2015.
75. Eshuis, N., Aspers, R.L., van Weerdenburg, B.J., Feiters, M.C., Rutjes, F.P., Wijmenga, S.S., Tessari, M., 2D NMR trace analysis by continuous hyperpolarization at high magnetic field. *Angew. Chem. Int. Ed.*, 54, 48, 14527–14530, 2015.
76. Pretsch, E., Buehlmann, P., Affolter, C., Pretsch, E., Bhuhlmann, P., Affolter, C., *Structure determination of organic compounds*, p. 108, Springer-Verlag, Berlin, 2000.
77. Rodríguez, J., Crews, P., Jaspars, M., *Contemporary strategies in natural products structure elucidation*, pp. 423–517, Handbook of marine natural products, 2012.
78. Gudi, G., Krähmer, A., Kr̈ger, H., Hennig, L., Schulz, H., Discrimination of fennel chemotypes applying IR and Raman spectroscopy: Discovery of a new γ-asarone chemotype. *J. Agric. Food Chem.*, 62, 16, 3537–3547, 2014.
79. Miller, ChE., Chemical principles of near infrared technology, in: *Near infrared technology in the agricultural and food industries*, P.C. Williams and K.H. Norris (Eds.), pp. 9–29, American Association of Cereal Chemist;, St. Paul, 2001.
80. Osborne, B.G., Fearn, T., Hindle, P.H., *Practical near infrared spectroscopy with applications in food and beverage analysis*, 2nd edition, p. 227, Longman Scientific and Technical;, Harlow, 1993.
81. McClure, F.W., 204 years of near infrared technology: 1800–2003. *J. Near Infrared Spectrosce.*, 11, 487–488, 2004.

82. Xie, L., Ye, X., Liu, D., Ying, Y., Quantification of glucose, fructose and sucrose in bayberry juice by NIR and PLS. *Food Chem.*, 114, 3, 1135–1140, 2009.
83. Reboucas, M.V., Brandão, D.S., Trindade, A., Pimentel, M.F., Teixeira, L.S.G., Chemical composition determination of complex organic-aqueous mixtures of alcohols, acetone, acetonitrile, hydrocarbons and water by near-infrared spectroscopy. *Vib. Spectrosc.*, 55, 2, 172–182, 2011.
84. Malaspina, P., Casale, M., Malegori, C., Hooshyari, M., Di Carro, M., Magi, E., Giordani, P., Combining spectroscopic techniques and chemometrics for the interpretation of lichen biomonitoring of air pollution. *Chemosphere*, 198, 417–424, 2018.

4

Lichens and Lichenology: Historical and Economic Prospects

Munesh Sharma[1]* and Aash Mohammad[2]

[1]*Department of Chemistry, Faculty of Science, J.V. College, C.C.S. University, Baraut, Baghpat, Uttar Pradesh, India*
[2]*Department of Bioinformatics, Faculty of Science, J.V. College, C.C.S. University, Baraut, Baghpat, Uttar Pradesh, India*

Abstract
Lichens are a type of whole organism that is interconnected by symbiosis. In this symbiotic relationship, one organism is algae while the other organism is a fungus, which live together. They grow in various places such as on the walls, roofs, bark of tree, gravestones, rocks, and soil etc. They produce a number of phytochemicals and crude extracts of medicinal potential as well as economic values. Wide medicinal impacts include antibiotic, antibacterial, antifungal, antiviral, insecticidal, anthelmintic, antiproliferative, antipyretic, anti-inflammatory, etc. In present scenario of eco-preservation, the interest in the development of methods for extractions and applications of secondary compounds of economically viable from natural resources with more liability gains increased attention worldwide, due to their explored applicability in various applied sectors. This chapter deals with the description on lichen's historical background and their economic viabilities of lichens in several vistas.

Keywords: Lichens, lichenology, lichenometry, antimicrobial activity, biological activity

4.1 Historical Prospects of Lichens

The history of lichenology, a branch of science deals with the study of Lichens, does not date very far back due to their little economic value all over the world.

**Corresponding author*: munesh004@gmail.com

Mohd Yusuf (ed.) Lichen-Derived Products: Extraction and Applications, (101–118) © 2020 Scrivener Publishing LLC

From ancient times, lichens had hardly been studied. Linnaeus, the father of binomial system gives them "*rustici pauperrimi*" of vegetation, which has been considered latterly to "*poor trash*" of vegetation. Therefore, lichens did not attract much plant's scientific attention nor they have any peculiar class until the beginning of the 18th century in comparison to their biological compounds isolated. Lichens somewhat remarked between old age society due to their medicinal properties to treat various diseases and to dye their articles of clothing but were limited to litmus, cudbear, and orchil. *Lobaria pulmonaria, Peltigera aphthosa*, and *Xanthoria parietina* have been used to treat the diseases of the lungs, thrush, and jaundice, respectively [1, 2].

As far, the discoveries based on Lichens are very rare, out of which chief inventions can be described as under:

(i) "*The discovery of the spermagonia*", describes them to be the male reproductive organs and the apothecia with their spores to be the female reproductive organs, and soredia, a fertilizing element, or real pollen proposed by Hedwig in 1784 [1, 3].
(ii) In 1825, "*The discovery of the gonidial*" given by Wallroth, which received general recognition [3], concludes that the most important discovery thus far made. He called them gonidia because of their resemblance to spores. He believed that in lichens, gonidia, as well as spores are actually the only reproductive organs.

In 1856, Lindsay published a monograph, "*A Popular History of British Lichens*" [4]. He stated in the book, "They (the gonidia) may be considered as interlude in function between the vegetative and reproductive cell. Their great peculiarity is a desire of living together with their neighbor tissues to be in a union. They frequently break through the cortex region, appears on the upper surface like a tangle denominated with soredia." Further, Lindsay to propel the position of Lichens, says: "The thallus are grow by their reproductive systems and the gonidia are essentiality of them".

(iii) Although, lichens had not been widely recognized as an organism till 1867. In 1867, DeBary firstly pointed the true nature of Lichens as gelatinous lichens and published "*Morphology and Physiology of the Fungi, Lichens and Myxomycetes*", In his notes, he expresses the belief that "some species of lichens are similar with algal colonies such as Nostoc and Chroococcus in premature stage that are insinuate/ penetrate by hyphae of fungi to make complete thalli" [3]. Similarly, in 1867 (published 1868), Simon Schwendener (Swiss Botanist) introduced his theory of duality about lichens. According to his theory, the lichens are an inter-linkage between fungi and algae or cyanobacteria that form a real association", which suggests the same opinion as DeBary.

At that time, the theory of "*Lichen' duality*" had neglected by top Botanists and lichenologists (i.e., Nylander and Crombie) due to the fact that all living organisms were autonomous [3]. Over time, Schwendener's concept soon comes under considerations by other prominent biologists in another segment of their researches, for example, pathology, animal/plant/microbe [5]. Several botanists favor partially or whole Schwendener's concept such as Krempelhuber, Famintzin, Bornet, Korber, and Baranetzky [5]. Furthermore, in 1939, Eugen Thomas prosperously highlighted Schwendener's "*Lichen' duality Concept*" through his re-synthesis experiment [5]. However, Crombie [6], like Nylander, never believed if Lichens formed a distinct class of plants.

Albert Schneider, in 1994, published II edition of his famous book "*A Guide to the Study of Lichens*". Schneider says in the book that "Lichen form a group of plants that have less expectation and more misunderstood and abuse. Usually, we can compare lichens to *moss*. Generally, lichen term is uncommon, but the term moss is quite famous among both laymen and poet. Consequently, lichens, moss, and liverworts are all genetically different and generate their own categories on reproduction" [7].

(iv) In the early 20th century, Toby Spribille and co-workers (2010) discovered a latest flank of lichen's fungi-algae relation. They described a number of the symbiotic nature of fungi (ascomycete) with algal associate. Rarely, they found another *trios* form consisting of ascomycete-basidiomycete-algae [6, 8, 9].

4.2 Lichens and Symbiosis

Lichens are a slow growing organism and are the typical combination of two biological components, fungi and algae. However, the question arose as to the nature and relationship between these two components toward each other. Parasitism is called by Schwendener *et al.* unanimously integrated lichens as fungi. DeBary proposed the relationship as a mutual union for living each other in which neither of the components could exist alone [3, 5].

Latest views delivered as well as accepted extensively by various biologists toward symbiosis between two components can be discussed as below:

The lichen symbiosis (Figure 4.1) typically involves a close physiological integration between mycobiont (fungi) and photobiont or phycobiont (algae), in which the mycobiont is dominant and has heterotrophic nature, and obtains inorganic minerals and organic nutrients (carbohydrate as

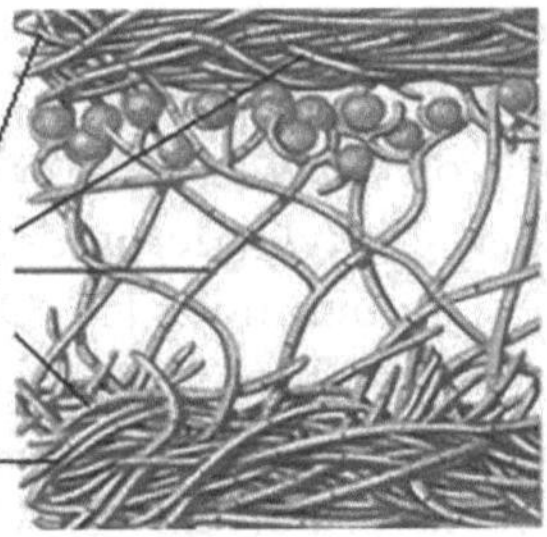

Figure 4.1 Schematic representation of lichen symbiosis.

polyols in case of green algae lichen and simple monosaccharide-glucose in case of cyanolichens) from phycobiont. At exchange cost, mycobiont gives protection to phycobiont [5, 10].

4.3 Economic Prospects

Limited numbers of lichen extracts and their derived products are being analyzed for their potent role in medicinal, therapeutic, and economic viabilities. This is possibly because of the stiffness faced in bulky sample collection, recognizance of particular species, and peculiar chemical structure elucidation of pure compound. The economic significance of lichens can be understood by the following facets:

4.3.1 Food Relation

4.3.1.1 As Food

Lichens contain certain metabolites/nutrients such as polysaccharides, enzymes, and some vitamins. Biological functionalities of secondary metabolites present in lichens are especially due to their ecological and climatic conditions of propagation. They serve as an important source of healthy food for vertebrates (human, few reptiles) as well as invertebrates, for example, mites, snails, slugs, caterpillars, termites, etc. *Cetraria islandica*, in common called as Iceland moss, has been consumed as food in many countries such as Scandinavian countries, Sweden, Norway, Iceland, Bosnia, etc. Incidentally, lichens are considered very slow growing organisms, but this lichen propagates as tiny nodules and quickly grows in proper and congruent circumstances [11]. Another species, *Lecanora esculenta*, is known locally as "manna" of the Israelites and used as food in Israel and

Umbilicaria esculenta is utilized in Japan. Manna lichen is termed because in rainshower season, the sudden appearance of large quantities of this lichen blown universally on rock's upper surface.

In India, *Peltigera canina* and *Parmelia sp.* (known as "rock flower" in India) are used as food since long times ago. Table 4.1 depicts lichen species used in several countries as food for vertebrates as well as invertebrates. *Ramalina sinensis* is used as food in India and Nepal, whereas the people of France have used lichens in confectionary for making chocolates and pastries [3, 5, 9–12].

4.3.1.2 As Fodder

Excavations of prehistoric cave dwellings have shown the existence of lichens (i.e., *Cladonia rangiferind*) inside the digestive part of various animals. It indicated that people of early epoch had already utilized lichens

Table 4.1 Some lichen species used in several countries as food for vertebrates as well as invertebrates [3, 5, 9–12].

Lichen species	Associated countries
Cetraria islandica	Scandinavian countries, Sweden, Norway, Bosnia, Iceland, North Africa
Bryoria fuscescens	South Eastern Europe
Bryoria fremontii	North American countries and Columbia
Umbilicaria mühlenbergii	North Africa and Deserts of West Central Asia
Umbilicaria esculenta	Japan, North American countries
Peltigera canina	India
Parmelia sp.	India
Ramalina sinensis	India and Nepal
Ramalina farinacea	South Eastern Europe
Pseudevernia furfuracea	South Eastern Europe
Evernia prunastri	South Eastern Europe
Lobaria pulmonaria	South Eastern Europe
Usnea barbata	South Eastern Europe
Lecanora esculenta	Israel

as fodder for their pets [4, 9]. In polar countries, *Cladonia rangiferina*, the local name "Reindeer moss", is the main food for reindeers while *Cetraria islandica* used as fodder for horses [5]. Species of *Stereocaulon*, *Evernia*, *Parmelia*, and *Lecanora* are also reported to be used as fodder [10]. Some well-recognized fodder species of lichen include *Cladonia* genus such as *C. rangiferina*, *C. alperstris*, and *C. sylvatica* and the species of *Cetraria*, *Evernia*, *Parmelia*, *Lecanora*, *Stereocaulon*, and *Alectoria* genera [10–12].

4.3.1.3 As Natural Food Flavoring Agent

Natural food flavoring substances are defined as "Those chemical components isolated from a natural aromatic raw material chiefly in the form of essential oils or extracts, are often known as natural food flavoring agents/substances". However, there is lack of documentation onto the lichens as food flavoring agents possibly due to cost, production, and availability [5, 12]. In India, the mixture of some lichens such as *Pyxine*, *Parmelia*, *Physcia*, and *Heterodermia* have been broadly employed to add the special flavor to the Indian spicy "*Biryani—a dish of rice*". Consequently, in Madhya Pradesh (a state in middle India), the thallus of *Heterodermia tremulans* are utilized as spice and flavoring agent in various dishes of vegetables and meat. Further, *H. diademata* is together with other lichen species chiefly *Parmotrema* species. In Karnataka, *Parmotrema* is traditionally utilized as a famous flavoring agent in meat and other food commodities [13, 14].

4.3.2 Paper and Textiles Relation

4.3.2.1 Litmus Dye Paper

Litmus is a water-soluble mixture containing approximate 12–15 natural dyes, which is extracted from lichen species. The mixture is commonly adsorbed onto filter paper to make one of the oldest acid-base pH indicating paper. Neutral litmus paper is purple in color/appearance. It turns red when immersed in an acidic solution while it turns blue when dipped into a basic solution (Figure 4.2) [15].

The litmus mixture with CAS No. 1393-92-6 (another related mixture known as orcein) contains 10 to 15 different dyes having an average molecular mass of 3,300 Daltons, and the mixture owes their properties to a 7-hydroxyphenoxazone (Figure 4.2) chromophore [16]. Litmus mixture is found in several lichen's species. Primarily, these were obtained from *Roccella tinctoria* homotypic synonym of *Lecanora tinctoria.* Table 4.2 shows major sources of litmus mixture around the world.

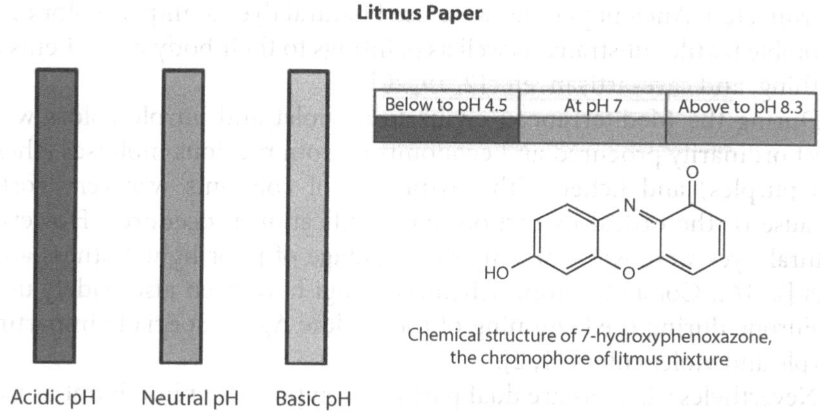

Figure 4.2 Schematic litmus paper color change with respect to pH of solution.

4.3.2.2 As Textile Dyes

From ancient times, dyes have been derived from minerals (soil generated), animals, and plants and have been widely utilized as primary sources for textile coloration as well as other substrates. For example, Alizarin, purpurin, and munjistin (hydroxyl anthraquinone) derived from *Rubia tinctorum* and *R. Cordifolia* were used to get red color [17], whereas Lawsone (2-hydroxy-1,4-naphthoquinone) [18] used to obtain yellow to brown color using an appropriate fixing agent (mordant, i.e., salts of Fe, Cu, Al, Cr,

Table 4.2 Major sources of litmus mixture around the world.

Lichen species	Distribution
Roccella tinctoria syn. *Lecanora tinctoria*	South America
Roccella montagnei	Mozambique
Roccella fuciformis	Angola and Madagascar
Roccella pygmaea	Algeria
Roccella phycopsis	Norway, Sweden
Dendrographa leucophoea	California
Lecanora tartarea	Norway, Sweden
Variolaria dealbata, Ochrolechia parella, Parmotrema tinctorum and *Parmelia spp.*	North America

Sn, Mn, etc.). Ancient peoples were used natural dyes to impart color onto colorable textile substrates as well as paintings to their body parts, items of clothing, and cave-artisan, etc. [2, 19, 20].

During the Mediterranean civilization, violet and purple colors were most ordinarily procured and economized from precious molluscs (shellfish purples) and lichens. The extraction of colorants was very costly because of the ordeal extraction and purification procedures. However, natural dyes were with the only disadvantage of poor light-fastness abilities [2, 16]. Colorants from lichenized fungi have been also widely used in Europe during the beginning of the Middle Ages, especially imparting purple and violet colors [1, 2].

Nevertheless, lichens are dual partners composed of photobiont as well as mycobiont. So, on account of the symbiotic relationship, the lichens have found enriched with almost familiar metabolites or dyes and pigments [1, 20–22]. Table 4.3 indicates the chemical structures of several coloring components present in many lichen species [23, 24].

4.3.3 In Perfumery

Some lichen species have specific and subtle herbal fragrance. Some species of *Heterodermia* have certain economic value and used in the form of medicine and perfumes. In south-central Europe, a lichen, *Evernia prunastri* (oak moss), has been collected and utilized for making perfumes as well as high-quality cosmetic substances [5, 14].

Oak moss also exploited in the formation of perfumes and a number of scents in France. Other main lichen species are *Pseudevernia furfuracea* (tree moss) and *Ramalina spp.*, which were also used as an ingredient in cosmetics/body lotion/perfume/soaps to add mossy fragrances [13]. *Evernia prunastri* var. *prunastri* (L.) and *Pseudevernia furfuracea* (L.) are commonly known as tree mosses and especially propagate on the woods of pine and cedar trees. The extracts of both the lichen species are currently used in the perfumery industry in France. It is estimated that 700 tons (approx.) of these species processed each year. The Chemical Abstracts Service (CAS) Registry has assigned the three following numbers to these oakmosses derived extracts: CAS-68917-10-2 and CAS-90028-68-5, both referred to "extractives and their physically modified derivatives without resin acids and their esters". Additionally, CAS-9000-50-4 is referred to the same directory and may contain resinous acids and esters thereof [25]. *Evernia mesomorpha* Nyl. and *Cetrariastrum nepalensie* were both known as "Chinese oakmoss" in China to obtained essential oil with a peculiar odor similar to perfumes [26].

Table 4.3 Chemical structures of several coloring components present in lichen species.

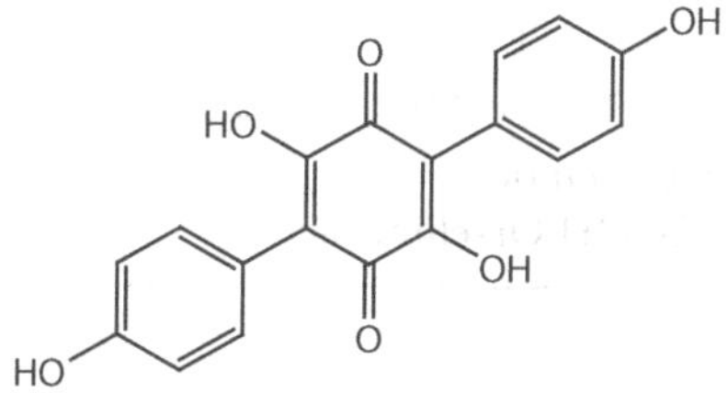 Atromentin	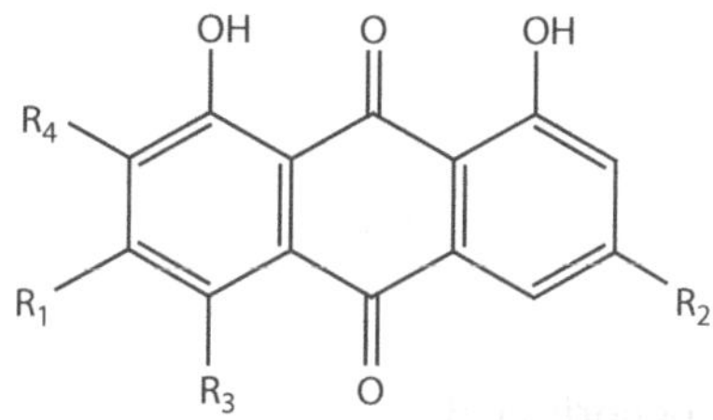 R_1=CH_3, R_2=OCH_3, R_3=R_4 Physcion R_1=OH, R_2=CH_3, R_3=R_4 Emodin R_1=OCH_3, R_2=CH_3, R_3=R_4=OH Dermocybin R_1=OH R_2=CH_3, R_3=H, R_4=Cl Chloro-emodin
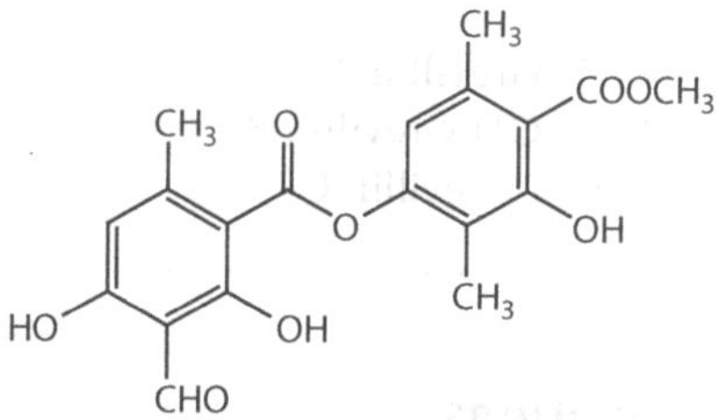 Atranorin	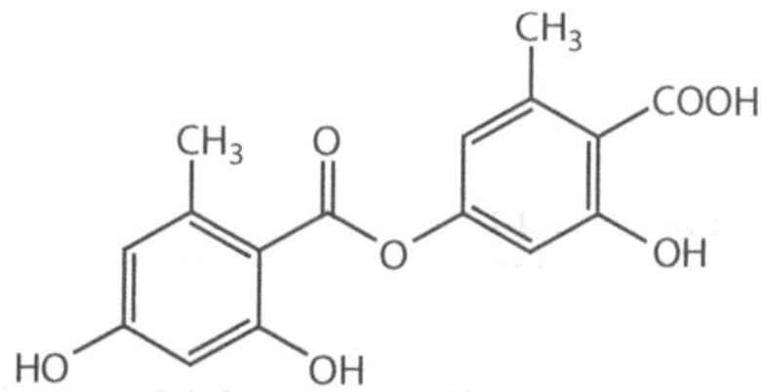 Lecanoric acid
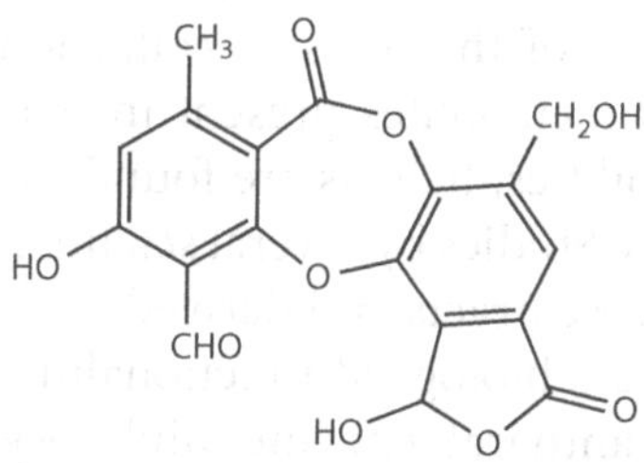 Salazinic acid	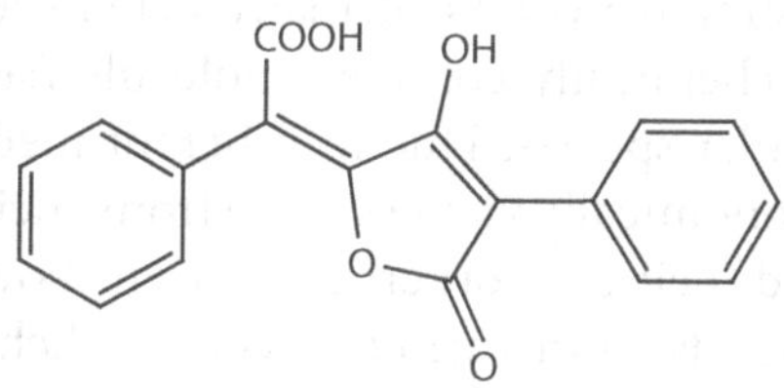 Pulvinic acid
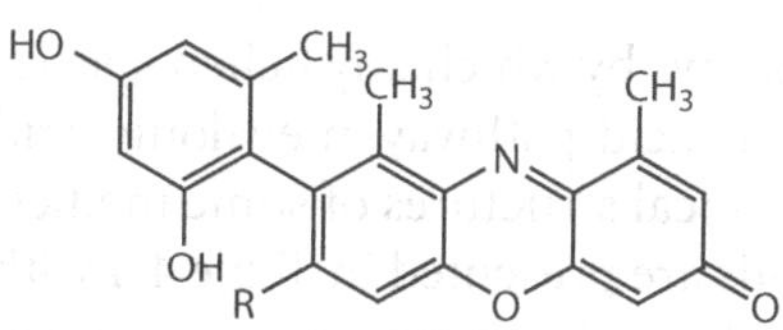 R=NH_2 α-Amino-orecin R=OH α-Hydroxy-orecin	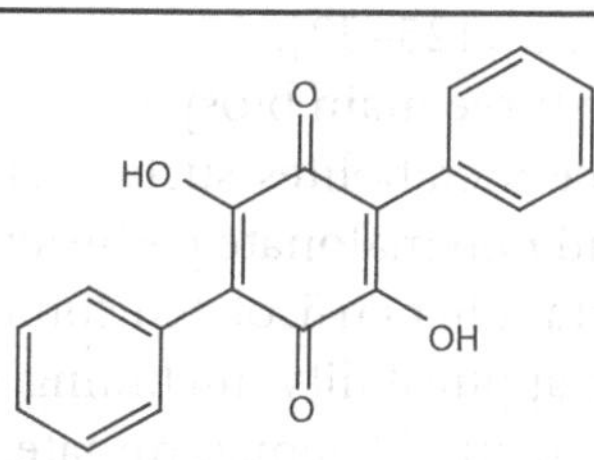 Polyphoric acid

(*Continued*)

Table 4.3 Chemical structures of several coloring components present in lichen species. (*Continued*)

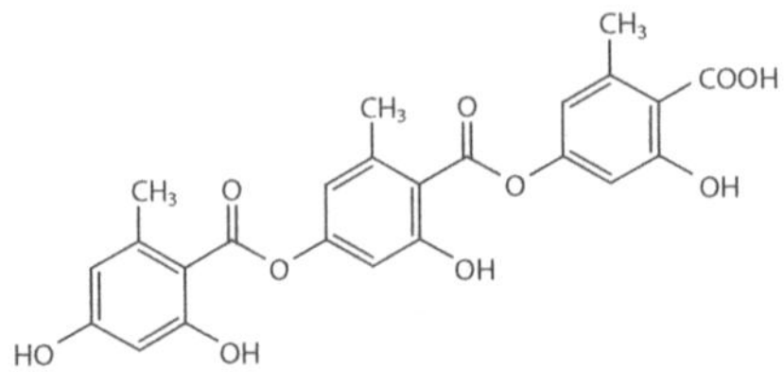 Gyrophoric acid	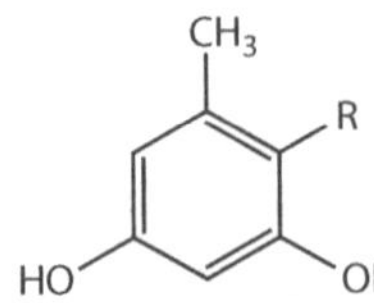 R=H Orcinol R=COOH Orsellinic acid
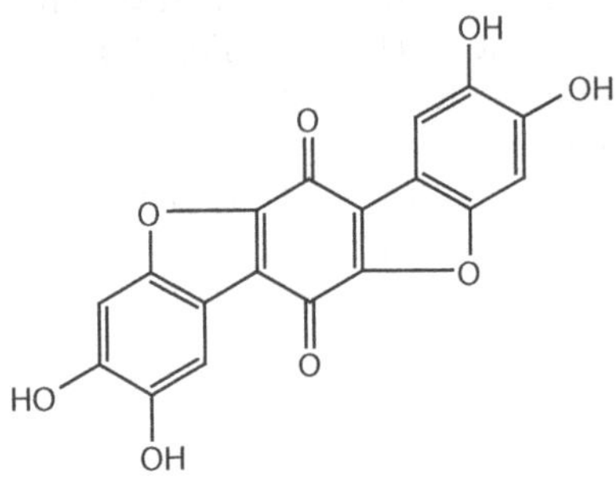 Telephoric acid	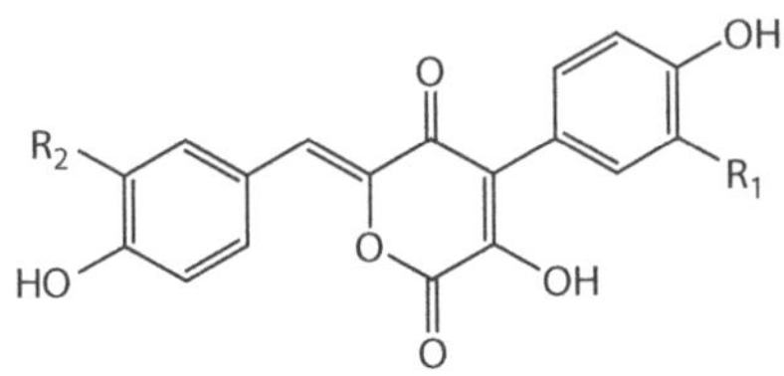 R_1=R_2= H Grevillin A R_1=H, R_2=OH Grevillin B R_1=R_2=OH Grevillin C

4.3.4 In Medicine and Medical Formulations

In nature, with abundant lichen species, we can see a colorful habitat. The distinctive colors of many lichens are because of the massive collection of chemically complex molecules/secondary metabolites present in particular species. Due to inherent medicinal abilities, lichens are found the economically viable source from ancient times. Studies by several scientists and their co-workers have been shown that the extracts and derived compounds from various species of lichens possess biological functionalities such as antimicrobial, analgesic, antipyretic, antiviral, enzyme inhibitory, insecticidal, neuroactive, anthelmintic, anti-inflammatory, and hepatoprotective activities [25–32].

There are three main biosynthetic pathways by which any lichen species may produce metabolites such as shikimic acid pathway, mevalonic acid pathway, and polymalonate pathway. Chemical structures of some medicinally important lichen-isolated compounds are presented in Table 4.4 with their major applicability, including Lobaric acid, benzofuran (sakisacaulon A) [33], methyl haematommate, methyl-β-orcinolcarboxylate, orcinol and their derivatives [14], quinones (parietin, emodin) [1, 2], pulvinic acid derivatives (pulvinic acid [34], dibenzofuran derivative (usnic acid) [35],

Table 4.4 Chemical structures of some medicinally important lichen-derived compounds with their major medicinal applicability.

Secondary metabolites	Major application in medicine	Ref.
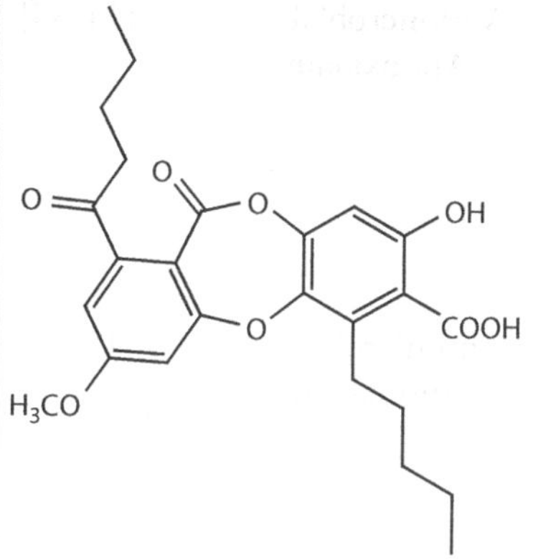 Lobaric acid	Antimicrobial, Antioxidant, Immunomodulator	[29]
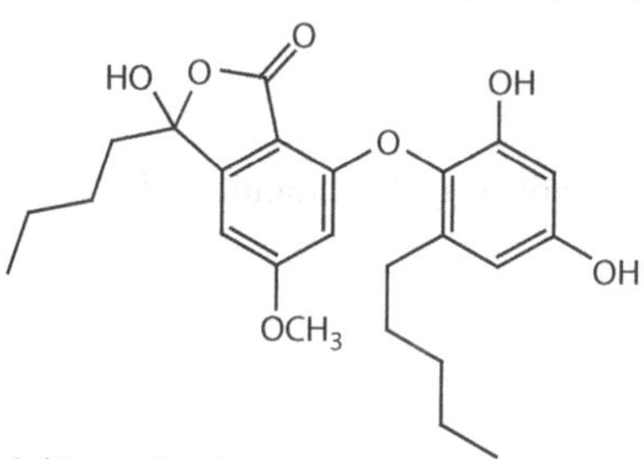 Sakisacaulon A	Antimitotic	[33]
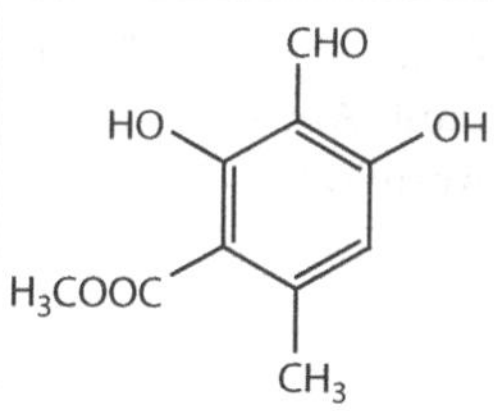 Methyl haematommate	Immunomodulator	[29]
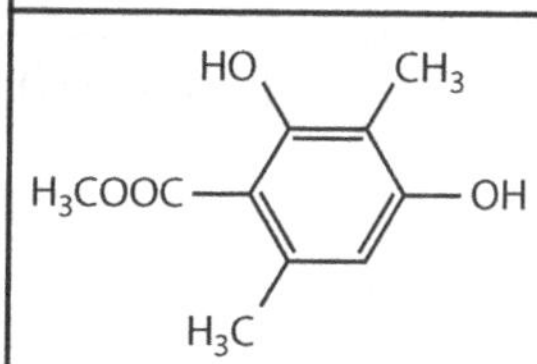 Methyl-β-orcinolcarboxylate	Immunomodulator	[29]

(*Continued*)

Table 4.4 Chemical structures of some medicinally important lichen-derived compounds with their major medicinal applicability. (*Continued*)

Secondary metabolites	Major application in medicine	Ref.
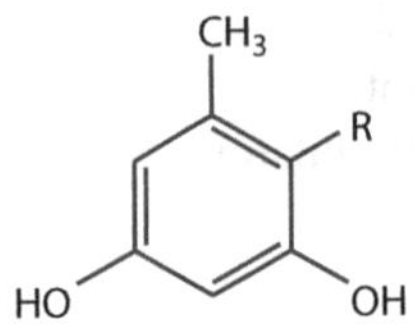 R=H Orcinol; R=COOH Orsellinic acid	Antimicrobial, Antioxidant	[14, 35]
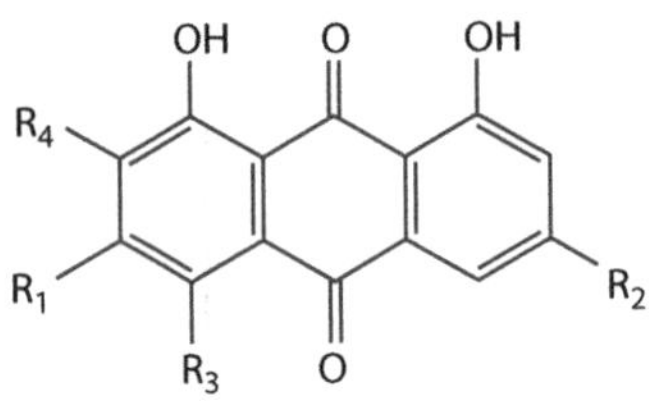 R_1=CH_3, R_2=OCH_3, R_3=R_4 Physcion R_1=OH, R_2=CH_3, R_3=R_4 Emodin R_1=OCH_3, R_2=CH_3, R_3=R_4=OH Dermocybin	Antimicrobial, Antioxidant	[1, 2]
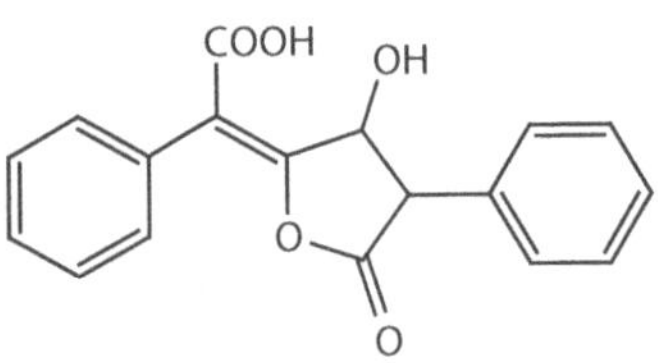 Pulvinic acid	Antifungal, Pesticidal	[34]
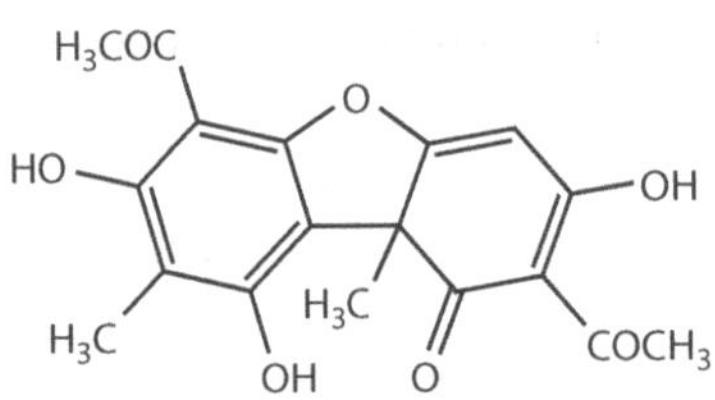 Usnic acid	Antimicrobial, Antioxidant, Anti-tumor, Antiviral	[35, 36]

(*Continued*)

Table 4.4 Chemical structures of some medicinally important lichen-derived compounds with their major medicinal applicability. (*Continued*)

Secondary metabolites	Major application in medicine	Ref.
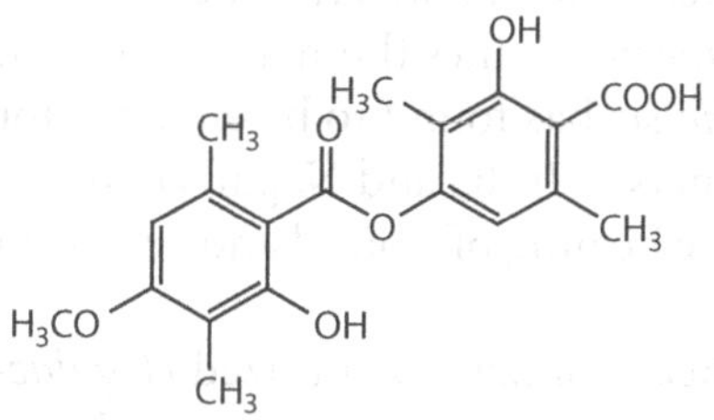 Barbatic acid	Antimicrobial, Antioxidant, Fungicidal	[36, 37]
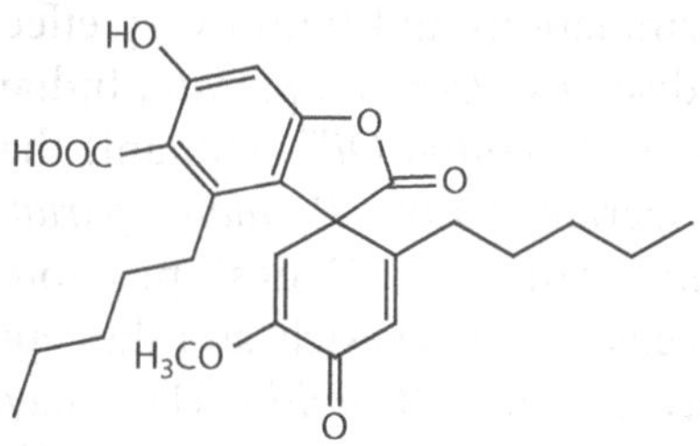 Picrolichenic acid	Antimicrobial, Antibiotic, Antiseptic	[38]
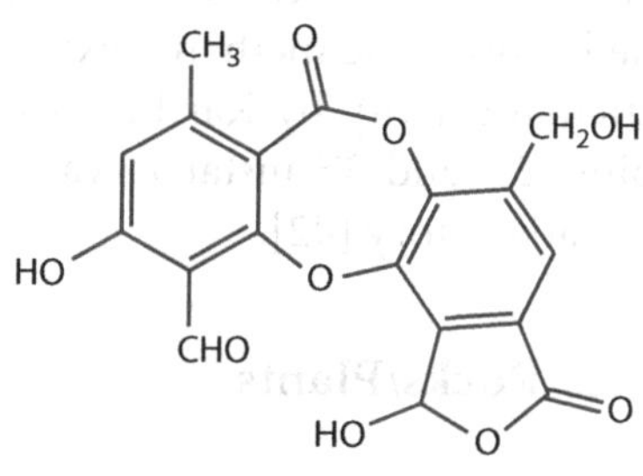 Salazinic acid	Antimicrobial, Antibiotic, Antiseptic	[14, 35]
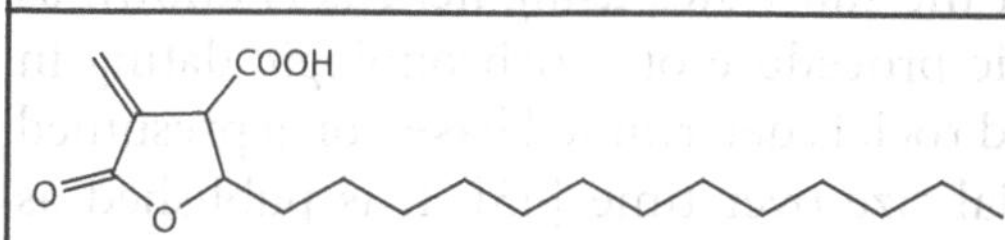 Protolichesterinic acid	Antimicrobial, Antioxidant	[39]

depsides (barbatic acid) [36, 37], depsones (picrolichenic acid) [38], depsidones (salazinic acid) [14, 35], and lactones (protolichesterinic acid) [39].

Whole thalli extracts are also found to have excellent medicinal abilities. *Usnea spp.* extract possesses strong anti-cancer and antioxidant properties, while *Umbilicaria esculenta* is enriched with anti-HIV metabolites. Wound healing ability of certain lichen species makes them an important economic resource. For example, *Parmelia sp.* has found to be effective for healing the wounds in parts of Eastern Ghats of India and Nepal. *H. diademata* is used by ethnic communities for the healing of wound and cessation of bleeding by external injury [24, 40].

In addition, Pankararu people of Northeast-Brazil have utilized *H. galactophylla* to strengthen digestive system and in the treatment of epilepsy, vomiting, and diarrhea [41]. Furthermore, some lichen species are employed in the Ayurvedic/Sidha/Homeopathic formulations and found very effective remedy toward various ailment and diseases. *Permalia parlata*, Indian name-Sailaja, is an important ingredient of "*Navratna Oil*" (Himani) that relieves headache with cooling and refrigerant action. *Permalia parlata* have antipotent characteristics. "*Speman*" (Himalaya Drugs) promotes spermatogenesis by raising the testicular cells, seminal vesicle and skin epidermal cells. Another non-hormonal medication is "*Confido*" (Himalaya Drugs) that acts through the neuro-endocrine pathway and regulates the process of ejaculation. "*Laboob-e-Kabir*" (Hamdard) also contains *P. parlata* which is used for sexual stimulation medication. The methanol extract of *H. leucomela* was assessed for its insecticidal activity by Karthik *et al.* (2011) in terms of its larvicidal effect against 2nd and 3rd instar larvae of *Aedes aegypti* and the result was found to be satisfactory [42].

4.3.5 Age Determination of Substrates/Rocks/Plants or Lichenometric Dating

The determination of the age of the substrates using lichens is known as lichenometry. It is a geomorphic procedure of geochronologic dating in which the growth rate of exposed rock is determined based on a presumed specific rate of increase in radial size over time [43]. It is published as lichenometric paper by Austrian Roland Beschel focused on ALPS (1928–1971) and firstly invented by Knut Faegri in 1933 [44]. Lichenometry has superior value than radiocarbon dating method that cannot detect the data accurately with respect to the surfaces less than 500 years old. Two genus of lichens, for example, *Rhizocarpon spp.* (i.e., *Rhizocarpon geographicum*) and *Xanthoria* are the most favorable and consecutively applicable in lichenometry.

4.3.6 Environmental Benefits/Biomonitoring

Lichens are believed very sensible organisms which noted the minute environmental modification like changes in climate, forest shape, air quality, and concentrations of pollutants. So, they could be used as a reliable biological indicator of pollution. Possibly, lichens absorb/utilize environmental elements along with heavy metals, causative to serious pollution problems in considerable amount and they can be predicted through spectroscopic techniques. Analysis of lichen species provides an important data based on a qualitative scale that has been devised for the estimation of average aerial pollutants such as sulphur oxides (SOx), heavy metal concentrations [45, 46]. Samples of air filter dust and lichen thalli can be used to assess the organic as well as radionuclide pollutants and air quality. Thus, lichens possess the ability to be a marked pollution indicator/monitor for the ecosystem.

4.4 Future Outlook and Conclusion

Lichens are being used for centuries on account of their ergastic medicinal values as well as other value-added opportunities. Also, in the present scenario, lichens have been utilized in a large number of applicable sectors, for example, pharmaceuticals, perfumery, medicinal, paper, textile, lichenometry, and many more. The extracts and derived substances from lichens possess large number of biologically explored properties. Moreover, lichens are recognized as potent biomonitoring agent toward environmental adulteration. Apart, they are often used in many industrial sectors, yet some of the species have toxic secondary metabolites [47]. Therefore, in depth research needed to explore new future dimensions and R&D opportunities.

References

1. Rather, L.J., Jameel, S., Ganie, S.A., Bhat, K.A., Lichen Derived Natural Colorants: History, Extraction, and Applications, in: *Handbook of Renewable Materials for Coloration and Finishing*, M. Yusuf (Ed.), pp. 103–114, John Wiley & Sons and Scrivener Publishing LLC Beverly, MA, USA, 2018.
2. Yusuf, M., Shabbir, M., Mohammad, F., Natural colorants: Historical, processing and sustainable prospects. *Nat. Prod. Bioprospect.*, 7, 1, 123–145, 2017.
3. Plitt, C.C., A Short History of Lichenology. *The Bryolog.*, 22, 6, 77–85, 1919.

4. Lindsay, W.L., *A Popular History of British Lichens: Comprising an Account of their structure, reproduction, uses, distribution, and classification*, Lovell Reeve, London, 1856.
5. Honegger, R., Simon Schwender (1829–1919) and the dual hypothesis in lichens. *The Bryolog.*, 103, 2, 307–313, 2000.
6. Crombie, R.J.M., On the Algo-Iichen Hypothesis, Linn. *J. Bot.*, 21, 259–282, 1885.
7. Schneider, A., *A Guide to the Study of Lichens*, 2nd Ed., pp. 25–50, Knight and Millet Press, Boston, 1994.
8. Spribille, T., Tuovinen, V., Resl, P., Vanderpool, D., Wolinski, H., Aime, M.C., Schneider, K., Stabentheiner, E., Toome-Heller, M., Thor, G., Mayrhofer, H., Basidiomycete yeasts in the cortex of ascomycete macrolichens. *Sci.*, 353, 6298, 488–492, 2016.
9. Thorne, S., *Learning Lichens*, vol. 84, 9, pp. 45–49, The Sci. Teach, Washington, 2017.
10. Lichen Biology. http://www.britishlichensociety.org.uk/about-lichens/lichen-biology (Retrieved on 23.06.2019).
11. Lawrey, J.D., Lichen Allelopathy: A Review, in: *Allelopathy: Organisms, Processes, and Applications, Inderjit*, vol. 582, K.M.M. Dakshini and F.A. Einhellig (Eds.), pp. 26–38, ACS Symposium Series, American Chemical Society, Salem MA, US, 1994.
12. Redzic, S., Barudanovic, S., Pilipovic, S., Wild mushrooms and lichens used as human food for survival in war conditions; Podrinje-Zepa Region (Bosnia and Herzegovina, W. Balkan). *Human Ecol. Rev.*, 17, 2, 175–187, 2010.
13. Upreti, D.K., Divakar, P.K., Nayaka, S., Commercial and ethnic use of lichens in India. *Econ. Bot.*, 59, 3, 269–273, 2005.
14. Kekuda, P.T.R., Vinayaka, K.S., Sachin, M.B., Chemistry, ethnobotanical uses and biological activities of the lichen genus *Heterodermia Trevis.*(*Physciaceae*; Lecanorales; Ascomycota): A comprehensive review. *J. Appl. Pharm. Sci.*, 8, 5, 148–155, 2018.
15. Crampton, L., Lichens and People: Uses, Benefits, and Potential Dangers, 2018. https://owlcation.com/stem (Retrieved on 08.07.2019).
16. Beecken, H., Gottschalk, E.M., v Gizycki, U., Krämer, H., Maassen, D., Matthies, H.G., Musso, H., Rathjen, C., Zdhorszky, U.I., Orcein and litmus. *Biotech. Histochem.*, 78, 6, 289–302, 2003.
17. Yusuf, M., Khan, M.A., Mohammad, F., Investigations of the colourimetric and fastness properties of wool dyed with colorants extracted from Indian madder using reflectance spectroscopy. Optik. *Int. J. Light Elect. Optics*, 127, 15, 6087–6093, 2016.
18. Yusuf, M., Ahmad, A., Shahid, M., Khan, M.I., Khan, S.A., Manzoor, N., Mohammad, F., Assessment of colorimetric, antibacterial and antifungal properties of woollen yarn dyed with the extract of the leaves of henna (*Lawsonia inermis*). *J. Clean Prod.*, 27, 42–50, 2012.

19. Yusuf, M., Mohammad, F., Shabbir, M., Eco-friendly and effective dyeing of wool with anthraquinone colorants extracted from *Rubia cordifolia* roots: Optimization, colorimetric and fastness assay. *J. King Saud Univ. Sci.*, 29, 2, 137–144, 2017.
20. Yusuf, M., Shahid, M., Khan, S.A., Khan, M.I., Islam, S.U., Mohammad, F., Khan, M.A., Eco-dyeing of wool using aqueous extract of the roots of Indian madder (*Rubia cordifolia*) as natural dye. *J. Nat. Fibers*, 10, 1, 14–28, 2013.
21. Cardon, D., *Natural Dyes: Sources, Tradition, Technology and Science*, Archetype Publications, London, 2007.
22. Schweppe, H., *Handbuch der Naturfarbstoffe*, Ecomed, Landsberg/Lech, 1993.
23. Shukla, P., Upreti, D.K., Nayaka, S., Tiwari, P., Natural dyes from Himalayan lichens. *Ind. J. Trad. Knowl.*, 13, 195–201, 2014.
24. Upreti., D.K., Tiwari, P., Shukla, P., Dwivedi, A., Natural thalli and cultured mycobiont of U. ghattensisG. Awasthi: A potential source of purple dye yielding lichen from India. *Ind. J. Nat. Prod. Resour.*, 3, 489–492, 2012.
25. Joulain, D. and Tabacchi, R., Lichen extracts as raw materials in perfumery. Part 1: oakmoss. *Flav. Frag. J.*, 24, 2, 49–61, 2009.
26. Kumar, S.P., Kekuda, T.P., Vinayaka, K.S., Sudharshan, S.J., Anthelmintic and antioxidant efficacy of two macrolichens of Ramalinaceae. *Pharmacog. J.*, 1, 4, 238–242, 2009.
27. Brisdelli, F., Perilli, M., Sellitri, D., Piovano, M., Garbarino, J.A., Nicoletti, M., Bozzi, A., Amicosante, G., Celenza, G., Cytotoxic activity and antioxidant capacity of purified lichen metabolites: An *in vitro* study. *Phytother. Res.*, 27, 3, 431–437, 2013.
28. Córdova, M.M., Martins, D.F., Silva, M.D., Baggio, C.H., Carbonero, E.R., Ruthes, A.C., Iacomini, M., Santos, A.R., Polysaccharide glucomannan isolated from *Heterodermia obscurata* attenuates acute and chronic pain in mice. *Carbohydr. Polym.*, 92, 2, 2058–2064, 2013.
29. Thadhani, V.M., Mesaik, A.M., Asif, M., Karunaratne, V., Choudhary, I.M., Immunomodulatory activities of some common lichen metabolites. *Int. J. Pharm. Pharm. Sci.*, 7, 11, 144–147, 2015.
30. White, P., Oliveira, R., Oliveira, A., Serafini, M., Araújo, A., Gelain, D., Moreira, J., Almeida, J., Quintans, J., Quintans-Junior, L., Santos, M., Antioxidant activity and mechanisms of action of natural compounds isolated from lichens: a systematic review. *Molecul.*, 19, 9, 14496–14527, 2014.
31. Prabhu, S.S. and Sudha, S.S., In vitro study on anthelmintic activity of *Heterodermia boryi* macrolichen collected from the Nilgiris, Tamilnadu, India. *Int. J. Curr. Microbiol. Appl. Sci.*, 5, 9, 764–768, 2016.
32. Rawat, S., North West region of Nanda Devi biosphere reserves, Garhwal Himalaya with unexplored lichen wealth and its medicinal lichen distribution in different aspect. *New York Sci. J.*, 9, 9, 28–40, 2016.
33. Morita, H., Tsuchiya, T., Kishibe, K., Noya, S., Shiro, M., Hirasawa, Y., Antimitotic activity of lobaric acid and a new benzofuran, sakisacaulon A from *Stereocaulon sasakii*. *Bioorg. Med. Chem. Lett.*, 19, 13, 3679–3681, 2009.

34. Dayan, F.E. and Romagni, J.G., Lichens as a potential source of pesticides. *Pesticide Outlook*, 12, 6, 229–232, 2001.
35. Cocchietto, M., Skert, N., Nimis, P., Sava, G., A review on usnic acid, an interesting natural compound. *Naturwissenschaften*, 89, 4, 137–146, 2002.
36. Boustie, J. and Grube, M., Lichens—A promising source of bioactive secondary metabolites. *Plant Genetic Resour.*, 3, 2, 273–287, 2005.
37. Shukla, V., Joshi, G.P., Rawat, M.S.M., Lichens as a potential natural source of bioactive compounds: A review. *Phytochem. Rev.*, 9, 2, 303–314, 2010.
38. Xu, M., Heidmarsson, S., Olafsdottir, E.S., Buonfiglio, R., Kogej, T., Omarsdottir, S., Secondary metabolites from cetrarioid lichens: Chemotaxonomy, biological activities and pharmaceutical potential. *Phytomed.*, 23, 5, 441–459, 2016.
39. Sasidharan, N.K., Sreekala, S.R., Lakshmanan, R., Jacob, J., Amma, S., Nair, D.K.B., Nambisan, B., Protolichesterinic acid: a prominent broad spectrum antimicrobial compound from the lichen *Usnea albopunctata. Int. J. Antibiotics*, 302182, 1–8, 2014. pageshttp://dx.doi.org/10.1155/2014/302182.
40. Devkota, S., Chaudhary, R.P., Werth, S., Scheidegger, C., Indigenous knowledge and use of lichens by the lichenophilic communities of the Nepal Himalaya. *J. Ethnobiol. Ethnomed.*, 13, 1, 1–10, 2017.
41. Londoño-Castañeda, P.A., Buril, M.L.L., Rego-Cunha, I.P., Silva, N.H., Honda, N.K., Pereira, E.C., Andrade, L.H.C., Lichens used in the traditional medicine by the Pankararu Indigenous community, Pernambuco-Brazil. *Global J. Sci. Front. Res.: C Biol. Sci.*, 17, 4, 15–22, 2017.
42. Karthik, S., Nandini, K.C., Kekuda, P.T.R., Vinayaka, K.S., Mukunda, S., Total phenol content, insecticidal and amylase inhibitory efficacy of *Heterodermia leucomela* (L). *Annals Biol. Res.*, 2, 4, 38–43, 2011.
43. Benedict, J.B., A Review of Lichenometric Dating and Its Applications to Archaeology. *Am. Antiquity*, 74, 1, 143–172, 2017.
44. doi:10.1017/S0002731600047545
44. Bradwell, T., Lichenometric dating: A commentary, in the light of some recent statistical studies. *Geografiska Annaler: Series A, Phys. Geography*, 91, 2, 61–69, 2009.
45. Nimis, P.L., Scheidegger, C., Wolseley, P.A., Monitoring with lichens—monitoring lichens, in: *Monitoring with Lichens—Monitoring Lichens*, pp. 1–4, Springer, Dordrecht, 2002.
46. Jeran, Z., Mrak, T., Jaćimović, R., Batič, F., Kastelec, D., Mavsar, R., Simončič, P., Epiphytic lichens as biomonitors of atmospheric pollution in Slovenian forests. *Environ. Poll.*, 146, 2, 324–331, 2007.
47. Emmerich, R., Giez, I., Lange, O.L., Proksch, P., Toxicity and antifeedant activity of lichen compounds against the polyphagous herbivorous insect *Spodoptera littoralis. Phytochem.*, 33, 6, 1389–1394, 1993.

PART 2

TRENDS AND APPLICATIONS

5

Biomedical Aspects of Lichen-Derived Products: Roadmap to Alternative Sources for Synthetic Drugs

Anjali Gupta*, Pooja Agarwal, Kavita Khatana and Shafat Ahmad Khan

Department of Chemistry, School of Basic and Applied Sciences, Galgotias University, Greater Noida, Uttar Pradesh, India

Abstract

Lichen, a distinct organism in plant kingdom, arises as a result of a symbiotic relationship between cyanobacteria and fungus. The medicinal use of lichen has been known in traditional systems including herbal, Chinese, and homeopathic medicines. The wide range of pharmacological activities and imaginable applications in medicinal and pharmaceutical chemistry in lichen are attributed to the presence of secondary metabolites like atranorin and usnic acid. They have been used for treating varied kidney diseases, arthritis, infections, leprosy, digestive and respiratory problems, and infestation and gynaecological issues. Several investigations on animals also proved lichens to possess antitumor, antibacterial, antiviral, analgesic, antipyretic, antiproliferative, antioxidant, and immunodilator activities. This review emphasizes the progress and development of lichens derived products in various biological activities with an update of current research findings.

Keywords: Lichen, usnic acid, antibacterial, antiviral, antioxidant

5.1 Introduction

In today's era, green chemistry has become a necessity due to fast depleting resources and the growing demand for raw materials. The use of plants and plant products have been used since ancient times as therapeutic agents for treating even crucial diseases. The discovery of new plants with potential

**Corresponding author*: anjaligupta@galgotiasuniversity.edu.in

Mohd Yusuf (ed.) Lichen-Derived Products: Extraction and Applications, (121–140) © 2020 Scrivener Publishing LLC

biological activities has now become a challenging task for most of the researchers. India is rich in plant resources possessing a medicinal value which are either used as herbal traditional medicines or also can be used in modern pharmaceutics. Lichen, a distinct organism in plant kingdom, arises as a result of the symbiotic relationship between cyanobacteria and fungus [1]. They have a unique characteristic to stand the severe change in weather conditions. Most of the lichen compounds possess phenolic structure like orcinol derivatives, depsides, dibenzofurans, depsidones, lactones, quinones, depsones, and derivatives of pulvinic acid [2].

Lichens species have been used since ancient times in dyes, folk medicines, perfumes, and spices [3]. Since ages, lichens are used in Ayurveda and Unani medications for the treatment of heart diseases, asthma, leprosy, piles, liver problems, and blood diseases. They are also commonly used in the treatment of wounds, digestive problems, skin disorders, and respiratory and gynecological issues. The different medicinal properties of lichen and its constituents have been reviewed by varied researchers [4–8]. This review emphasizes the progress and development of lichens in various biological activities with an update of current research findings.

5.2 Biological Activities

A wide range of lichen species and their secondary metabolites have been known for their various biological activities, *viz.*, antioxidant, anticancer, insecticidal, antidiabetic, antiviral, and antimicrobial. The secondary metabolites (Figure 5.1), extracts, and their activities have been enlisted in Table 5.1 and Table 5.2, respectively.

5.2.1 Anticancer Activity

Various studies have been carried out to evaluate the potential of lichens as anticancer agents. Shanmugam *et al.* studied the efficiency of varied lichen species on animal model *Albino wistar* rats induced with different cancer cell lines under *in vivo* conditions. *Parmotrema reticulatum* was found to exhibit the highest activity against cervical cancer cell line. The anti-cancer lichen treated rats showed 50% reduction in tumor proliferation with IC_{50} of 50 μg/ml [9]. Colorectal cancer is associated with abnormalities in Wnt signaling, and inhibition of Wnt signal is a potential therapeutic strategy in colon carcinogenesis. Jarosław *et al.* studied the effects of depsides, depsidones, and caperatic acid derived from lichens on Wnt signaling in DLD-1 and HCT116 colorectal cancer cell (CRC) lines. Among the tested compounds, physodic acid exhibited strongest inhibitory activity against both DLD-1 and HCT116 cells at MIC 50 μM while caperatic acid was found possess potential cytotoxicity at 100 μM similar to the reference compound atranorin [10]. Furthermore, methanolic, ethyl acetate, and acetone extracts of lichens *Platismatia glauca* and *Pseudevernia furfuracea* were explored for their effects on SW-480 and HCT 116 cell lines. The extracts showed significant cytotoxic effects on cancerous cells without any visible cytotoxic effect on healthy cells. The highest activity after 72 h was observed in case of acetone and ethyl acetate extracts of *P. furfuracea*

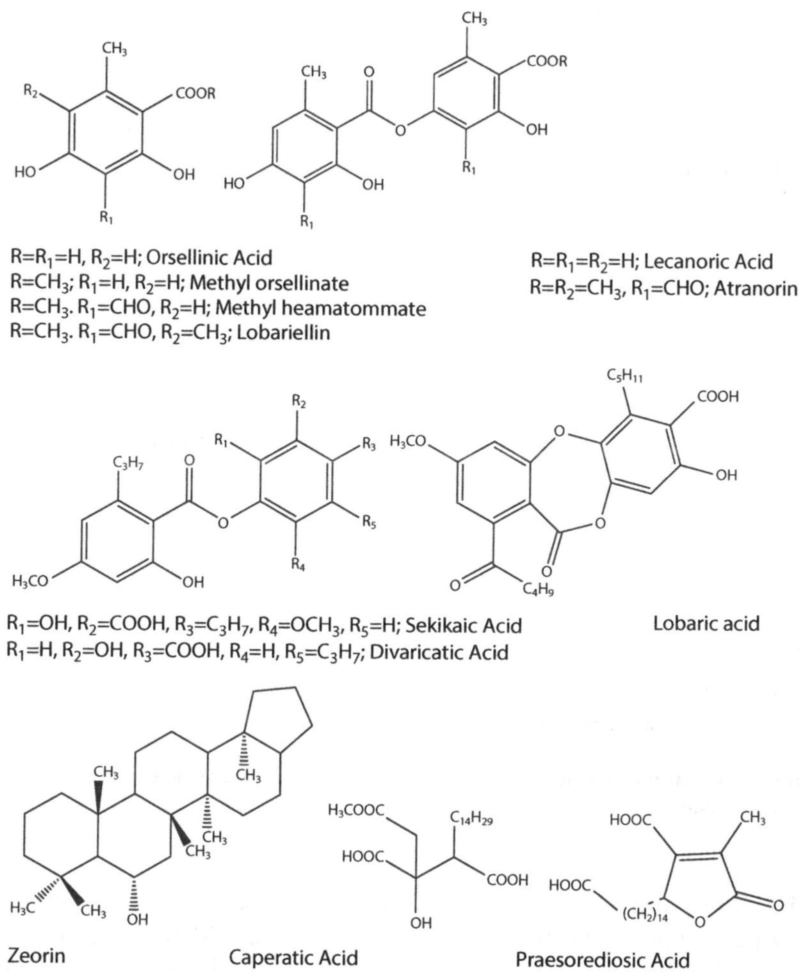

Figure 5.1 Chemical structures of biologically active lichen derived secondary metabolites.
(*Continued*)

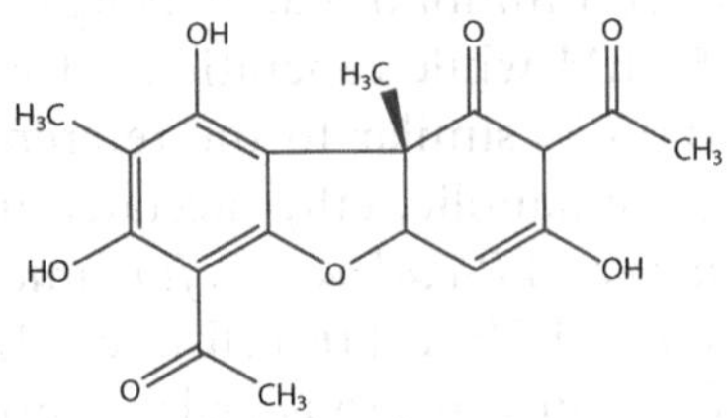

Usnic Acid

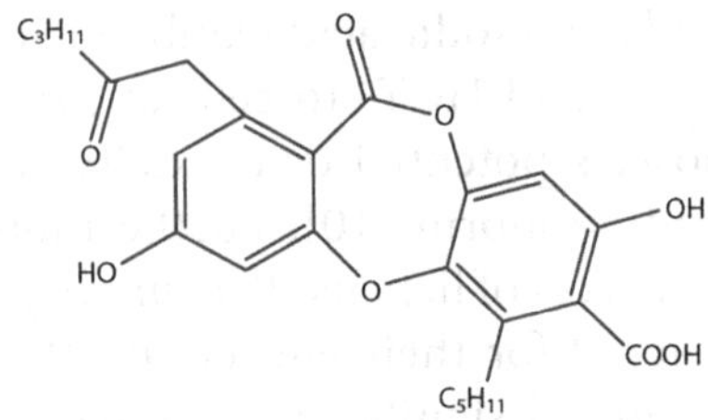

Physodic Acid

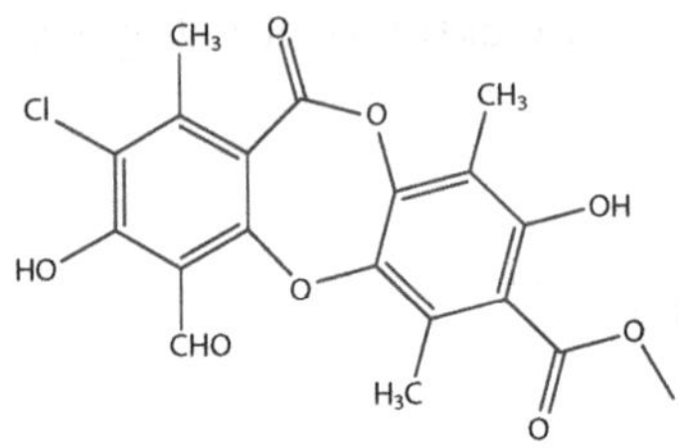

Physciosporin

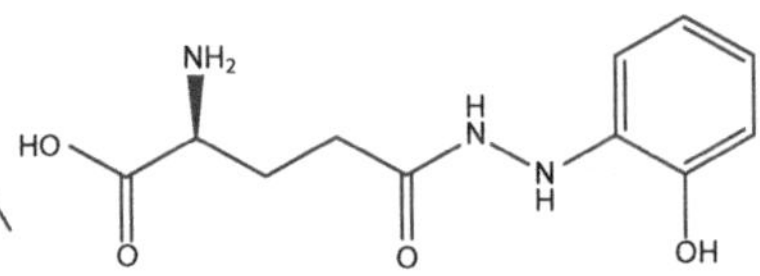

Ramalin

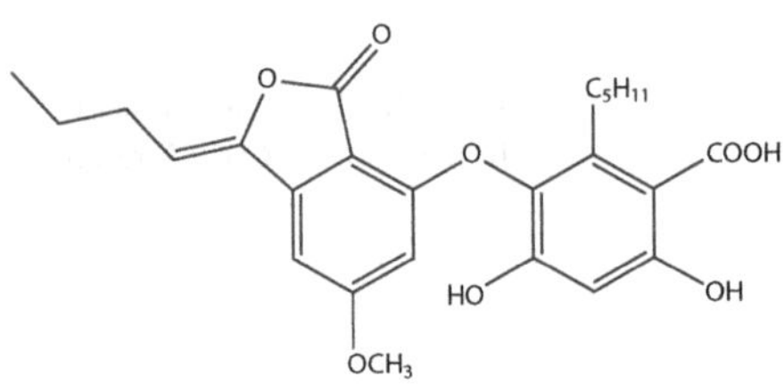

Lobarstin

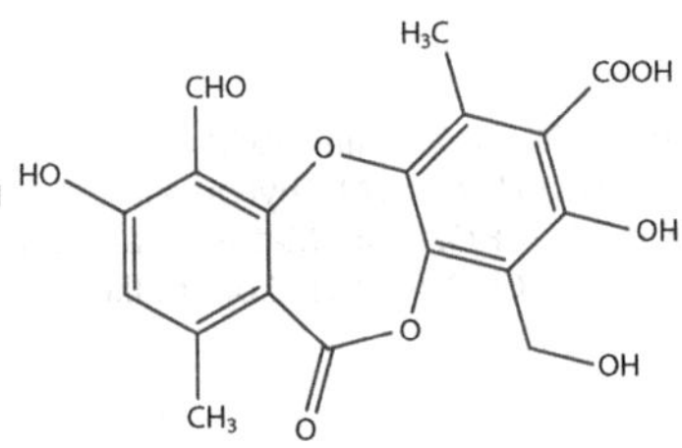

Protocetraric Acid

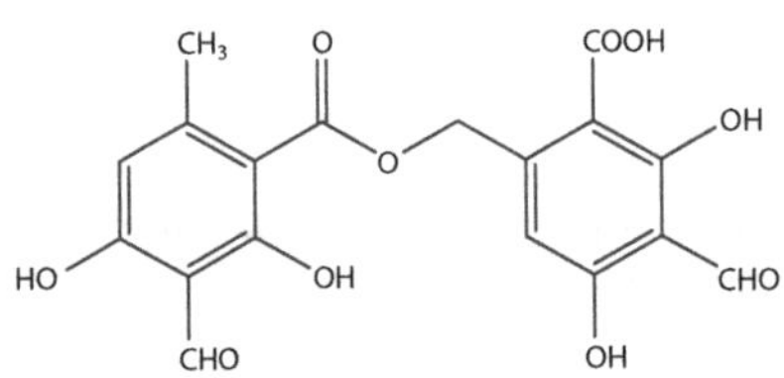

Barbatolic acid

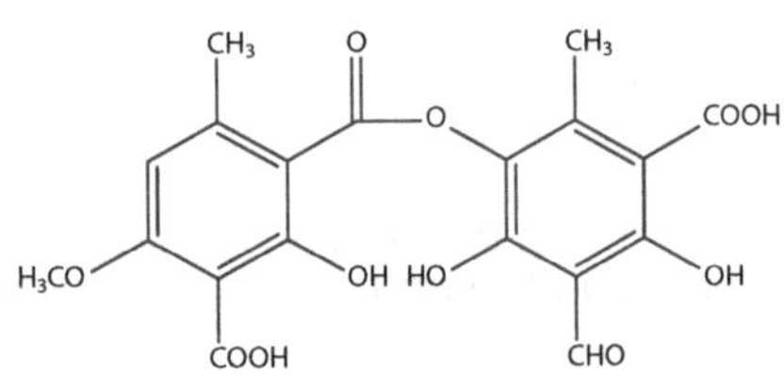

Thamnolic acid

Figure 5.1 (Continued) Chemical structures of biologically active lichen derived secondary metabolites.

on HCT 116 and SW-480 cells with IC_{50} value of 21.2 ± 1.3 and 51.3 ± 0.8 μg/ml, respectively with potential apoptotic effects. Methanolic extracts of *P. glauca* also showed cytotoxic effects on HCT-116 cells with $IC_{50} < 40$ μg/ml

Table 5.1 Biological activities of secondary metabolites isolated from lichen species.

Alectoria samentosa, Alectoria ochroleuca, Flavocetraria nivalis, Usnea florida	Usnic acid	Anticancer
Bryoria capillaris	Barbatolic acid	Antimicrobial
Bryoria fuscescens, Anaptychia ciliaris Xanthoria candelaria, Parmotrema	–	Anticancer
Cladonia substellata, Cladonia verticillaris, and Cladia aggregata	Usnic acid, fumarprotocetraric acid and barbatic acid	Insecticidal activity
Evernia mesomorpha	Divaricatic acid	Antimicrobial
Gynostemma pentaphyllum	Gynostemma pentaphyllum Makino polysacchrides	Immunodialator
Hypogymnia physodes Platismatia glauca	Physodic acid caperatic acid	Anticancer
Letharia vulpine, Lecanora muralis and Peltigera rufescen	–	Insecticidal
Parmotrema rampoddense, P. praesorediosum and P. tinctorum	Usnic acid, Protocetraric acid and praesorediosic acid	Antibacterial, antioxidant
Parmotrema reticulatum P. hababianum Rocella montagnei	Benzoic acid, 2, 4 dihydroxy, 6 methyl-methyl ester	Anticancer
Pseudocyphellaria coriacea	Physciosporin	Anticancer
Ramalina terebrata	Ramalin	Anticancer

(Continued)

Table 5.1 Biological activities of secondary metabolites isolated from lichen species. (*Continued*)

Ramalina usnea	(+)-Usnic acid and 2-hydroxy-4-methoxy-6-propyl-methyl benzoate	Insecticidal
Stereocaulon alpnum	Lobarstin and lobaric acid	Anticancer
Stereocaulon evolutum	Atranorin derivatives	Antiviral
Stereocaulon strictum var. compressum and Lobariella pallida	Lobariellin, methyl orsellinate, stereocaullin, methyl haematommate, methyl β-orcinol carboxylate, porphyrilic acid and atranorin	Antioxidant
Tremella	Tremella polysaccharides	Immunodialator activity
Umbilicaria antarctica, Flavoparmelia caperata, Ramalina sinensis, Usnea sinensis, Physcia aipolia, Heterodermia leucomelos	–	Antidiabetic, antioxidant
Umbilicaria esculenta	*Umbilicaria esculenta* polysaccharide	Immunodialator activity
Usnea florida	Thamnolic acid	Antibacterial, antifungal, and antituberculosis

after 72 h, and its methanolic and acetone extracts showed cytotoxic effects after 24 h on SW-480 cells with proapoptotic activity [11].

In another research article, the anticancer activity of ramalin was evaluated for HCT116 cells, CRC line in humans. Ramalin is known to be a secondary metabolite obtained from *Ramalina terebrata*, an Antarctic lichen. It significantly suppressed the proliferation and induced apoptosis in a concentration dependent manner. Moreover, ramalin also induced the modulation of genes associated with both transcriptional and translational level causing cell cycle arrest in G2/M phase. Its capacity to inhibit migration of CRCs in this manner suggested ramalin as a therapeutic drug

Table 5.2 Biological activities of lichen extracts.

Arthothelium awasthii, Heterodermia podocarpa, Usnea ghattensis, and Parmotrema tinctorum	Methanolic	Antibacterial, antioxidant
Caloplaca pusilla Protoparmeliopsis muralis Xanthoria parietina	Acetone	Anticancer
Herpothallon sp., H. leucomelos, P. reticulatum, P. tinctorum, P. crinitum, R. celastri and Leptogium sp	-Methanolic	Antibacterial
Hypogymnia tubulosa	Chloroform-methanol-acetone	Antibacterial
Parmelia vagans	Ethyl acetate	Antibacterial
Parmotrema cristiferum and Dirinaria applanata, Parmotrema reticulatum, P. kamatti, P. tinctorum, Parmelia erumpens, Leptogium papilosum, and R. montagnei	Methanolic	Insecticidal
Parmotrema tinctorum	2-Propanol	Antibacterial, antifungal
Parmotrema hababianum	Ethanolic	Antioxidant, antidiabetic
Platismatia glauca Pseudevernia furfuracea	Methanolic and acetone Ethyl acetate	Anticancer
Ramalina subpusilla, R. conduplicans, Parmotrema reticulatum, Usnea cineraria and Parmotrema cristeferum	Methanolic and acetone	Antioxidant
Ramalinafastigiata	Acetone	Antimicrobial, antioxidant
Usnea and Parmotrema	Acetone and methanolic	Antibacterial, antifungal

for the treatment of colorectal cancer [12]. In another study, Anar *et al.* investigated the antimutagenic and antigenotoxic activities in *Bryoria fuscescens, Anaptychia ciliaris, Xanthoria candelaria,* and *Parmotrema,* four lichen species, and found inhibitory mutagenic activities against toxic aflatoxin B1 at effective concentration of 20 μg/ml. These lichen species were also found to increase the level of antioxidant enzyme and decrease the malondialdehyde level thereby witness strong antioxidative activity as well [13].

Recently, the anticancer activities of lobarstin and lobaric acid were determined which are the metabolites obtained from *Stereocaulon alpnum,* an Antarctic lichen. It was shown that on treatment of HeLa and HCT116 cells with lobarstin and lobaric acid, significant decrease in proliferation was observed in a time- and dose-dependent manner. The inhibition was suggested to achieve through apoptosis induction and termination of cell cycle in cervix and colon cancerous cells making lobarstin and lobaric acid, novel agents for the treatment of cervix and colon cancer [14].

Several lichen samples were tested against human lung cancer cells, and among them, physciosporin obtained from lichen *Pseudocyphellaria coriacea* was found to show potential inhibition. It was achieved through suppression of gene, Kitenin that enhances metastasis and simultaneously increasing the activity of the gene, KAI1 that suppresses metastasis [15]. Similar inhibitory activities were shown by other lichen species *Alectoria samentosa, Alectoria ochroleuca, Flavocetraria nivalis,* and *Usnea florida* collected from the Romanian Carpathian mountains against human lung cancer cells. Usnic acid was found to be the chief component in these lichens. The results suggested that usnic acid might emerge as a potential candidate for anti-cancer therapy [16].

Felczykowska *et al.* also revealed antiproliferative and antibacterial activities of acetone extracts of *Caloplaca pusilla, Protoparmeliopsis muralis,* and *Xanthoria parietina.* The anticancer activity was tested against human breast cancer cells (MCF-7), human prostate cancer cells (PC-3), and HeLa cancer cells. Potential anticancer activity was observed in case of *X. parietina* that decreased MCF-7 and HeLa cancer cells sustainability having IC_{50} value 8 μg/ml. Highest anticancer activity was shown in the case of *C. pusilla* that decreased cancer cells activity with IC_{50} values of about PC-3 (7.96 μg/ml), MCF-7 (7.29 μg/ml), and HeLa (6.57 μg/ml) cancer cells viability. *P. muralis* was also found to possess strong antibacterial activity with MIC value ranging from 6.67 to 100.00 μg/ml [17].

5.2.2 Antimicrobial Activity

With growing resistance to antibiotics, the discovery of new antimicrobial agents has become really crucial. Thus, fabrications of anti-microbial agents with less toxicity, natural, and sustainable resource with multiple modes of action are very much required. In a report, lichen *Bryoria capillaris* and its constituent barbatolic acid were investigated for antibacterial, antifungal, and antitubercular activities. The extracts showed potential antimicrobial activity against the tested microorganisms, *Pseudomonas aeruginosa, E. coli,* and *Acinetobacter* sp. Moreover, barbatolic acid showed a considerable effect against *Mycobacterium tuberculosis* H37Rv having MIC value 31.25 μg/ml [18].

In another research article, 32 lichen extracts obtained from the species; *Herpothallon* sp., *H. leucomelos, P. reticulatum, P. tinctorum, P. crinitum, R. celastri,* and *Leptogium* sp. were assessed for their bactericidal potential for gram positive and gram negative pathogens. Methanolic extracts of lichens were found to possess maximum bactericidal potential. Among all, the most effective antibacterial species was found to be *P. reticulatum* which led to the inference that lichen species could be explored further for their potency as novel antibiotics [19].

Another article revealed the antibacterial activity of divaricatic acid isolated from lichen species: *Evernia mesomorpha* against *Enterococcus faecium* and *Staphylococcus aureus.* Divaricatic acid was found to possess potential antimicrobial activities against gram positive bacteria with MIC in the range of 7.0 to 64.0 μg/ml. The results proved lichen isolate as potential antibiotic for MRSA (methicillin-resistant *Staphylococcus aureus*) infections treatment [20]. Another study also showed the potential antimicrobial activity of 2-propanol extract of *Parmotrema tinctorum* (Despr. ex Nyl.) Hale against 8 fungal and 10 bacterial pathogens. The maximum zones of inhibition were noted for *Bacillus subtilis* (13.0 ± 2.99), *E. coli* (14.66 ± 0.57), *Corynebacterium rubrum* (11.33 ± 0.57), and *Salmonella abony* (12.33 ± .51), while lowest inhibition zones were found in case of *Bacillus cereus* (8.66 ± 1.15) and *Streptococcus pyogenes* (9.66 ± 0.57). The highest zones of inhibition were observed in case of fungal pathogens against *Aspergillus flavus* (10.0 ± 1.0) followed by *Colletotrichum falcatum, Fusarium oxysporum* and *Penicillium chrysogenum* (7.33 ± 0.57 each), *Fusarium moniliforme* (6.0 ± 5.2), and *Trichoderma lignorum* (7.0 ± 1.53) [21].

In another report, antifungal and antimicrobial activities of *Usnea* and *Parmotrema* lichen species were explored. Better inhibitory activity was found in case of acetone and methanolic extracts of *Parmotrema* lichen species as compared to that of *Usnea* species. Inhibition was found by the methanolic extract of *Parmotrema* species against Staphylococcus species having a zone of 0.3 cm (0.2 mg/ml conc.). Furthermore, its acetone extract also showed the greatest activity against *S. aureus*at a concentration of 0.2 mg/ml with a zone of inhibition of 1.4 cm. Moreover, *Parmotrema* methanolic extract showed better antifungal activity against *Aspergillus flavus*, *Aspergillus niger*, *Candida*, and *Trichophyton* species with maximum activity for *Candida* species (0.9 cm) and minimum activity for *Trichophyton* (0.6 cm) [22].

In another study, the methanol, acetone, and chloroform extracts of the lichen *Usnea florida* (L.) were tested for antibacterial, antifungal, and antituberculosis potential. The disk diffusion method was used for the investigation of antimicrobial activity and the minimal inhibitory concentration was identified against thirteen species of bacteria, four species of yeasts, and eight species of fungi. Strong activities toward bacteria and yeasts were determined in all extracts. However, there was no strong activity toward filamentous fungi. Thamnolic acid was found to be an active substance with MIC 200–400 μg/ml. Hence, the lichen may be used as a possible antimicrobial agent in various applications in the food industry and for the purpose of controlling different diseases [23].

Bondarenko *et al.* determined the antibacterial properties of extracts of lichen *Parmelia vagans*. It was observed that its ethyl acetate extract exhibited maximum antimicrobial activity against *S. aureus*, *Bacillus subtilis*, *Streptococcus pneumoniaei* while no inhibition was shown in case of *Klebsiella pneumoniae*, and *E. coli* [24]. In another report, chloroform-methanol-acetone extracts of lichen *Hypogymnia tubulosa* was tested against *Enterococcus faecalis*-ATCC 29212 and *S. aureus*-ATCC 25923. Strong inhibition as that of antibiotics was observed for *S. aureus* and *E. faecalis* and evidently revealed *H. tubulosa* having great antibacterial effect and strong antibiofilm potential [25]. In another research article, acetone, methanolic chloroform, and petroleum ether extracts of lichen species *Heterodermia boryi*, *Parmotrema melanothrix*, *Parmotrema stuppeum*, and *Usnea nilgirica* were found to possess antibacterial potential against *E. coli*, *Klebsiella*, *Pseudomonas* species, *Acinetobacter* species, *S. aureus*, and *Streptococcus viridians*. These observations indicated lichens have shown their prospective antimicrobial activities and so can be used under clinical trials after more research [26].

5.2.3 Antiviral Activity

An influenza virus is a serious threat which causes mortality due to epidemics worldwide. Antiviral drugs are divided into three groups: NA inhibitors, M2 inhibitors, and polymerase inhibitors which are currently proven to be effective against the influenza virus. Even after the success of these drugs, the main concern for drugs being their efficacy, cost, and resistance. Secondary metabolites produced by lichen species with varied biological activities may be used as potential antiviral drugs but these compounds are yet to be explored.

Shtro *et al.* studied the inhibitory effect of usnic acid derivatives against influenza viruses and save the mice from deadly influenza infection [27, 28]. Usnic acid was also found to be active against different types of viruses, *viz.*, Epstein-Barr virus, respiratory syncytial virus, human papilloma virus, and polio virus. According to the study of Lai *et al.*, potential antiviral activities of sekikaic acid obtained from Nigerian lichen farinacea were observed for the respiratory syncytial virus; rg (IC_{50} = 5.69 μg/ml) and A2 strain (IC_{50} = 7.73 μg/ml) [29]. In another report, Vu *et al.* also showed the inhibitory activity of atranorin and derivatives (lichen metabolites) for Hepatitis C virus (IC_{50} range from 12.8 to 100 μM) [30]. It was shown that atranorin derivatives isolated from *Stereocaulon evolutum* were found to interfere with the hepatitis C virus lifecycle; however, the effective inhibitors were added at different steps of the lifecycle. Most of the tested compounds inhibited the activity of the hepatitis C virus, with a half-MIC of 10–70 μM. The antiviral activity of lichen constituents against influenza virus has been extensively reviewed by Cuong *et al.* in 2018 [5].

5.2.4 Antioxidant Activity

Many biological properties of lichens and their secondary metabolites have been related to their antioxidant potential. In a research article, the antioxidant, anticancer, and antimicrobial activities of lichen species *Melanelia fuliginosa and Melanelia subaurifera* were investigated. These lichens were found to contain depsides (lecanoric acid, gyrophoric acid, atranorin, anziaic acid, and 2′-*O*-methyl anziaic acid), and dibenzofurane (usnic acid). Their antioxidant activity (IC_{50}) ranged from 121.52 to 424.51 μg/ml. The highest antimicrobial activity was exhibited by 2′-*O*-Methyl anziaic acid with MIC ranging from 0.0625 to 1 mg/ml. Moreover, *M. subaurifera* extract was found to exhibit highest anticancer activity with IC_{50} = 9.88–31.64 μg/ml) [31].

In another study, the antioxidant and antibacterial activity of methanolic and acetone lichen extracts (*Parmotrema rampoddense*, *P. praesorediosum*, *P. reticulatum*, and *P. tinctorum*) were evaluated. The compounds isolated from these lichens areprotocetraric acid, praesorediosic acid, α–collatolic acid, usnic acid, β-alectoronic acid, atranorin, and chloroatranorin.

Among the tested extracts, acetone extracts of *P. rampoddense*, *P. praesorediosum*, and *P. tinctorum* showed good inhibition against *B. subtilis* and *S. aureus* having MIC values 500–125 μg/ml, however, methanolic extracts did not show any activity. Methanolic extract of *P. praesorediosum* was found to exhibit more than 50% DPPH free radical scavenging activity. Among the isolated chemical constituents, usnic acid strongly inhibited the activity of *B. subtilis* and *S. aureus* having MIC value of 7.81 μg/ml. The isolates, protocetraric acid and praesorediosic acid, also exhibited more than 50% radical scavenging activity (63.97% and 57.57%, respectively) [32]. Another study revealed the antioxidant and antimicrobial potential of acetone extracts of *Ramalina fastigiata*. The extract containing glycosides, flavonoids, and phenols exhibited potential antimicrobial activity against *Candida krusei* and *Klebsiella pneumonia* with the inhibition rate of 1.3 and 2.1 mm, respectively. The extract also confirmed significant antioxidant activity with 61.53% inhibition [33].

Ramya *et al.* also studied the antioxidant potential of ethanolic, methanolic, and acetone extracts of lichen species *Ramalina subpusilla*, *R. conduplicans*, *Parmotrema reticulatum*, *Usnea cineraria*, and *Parmotrema cristeferum* through superoxide anion radical scavenging and DPPH radical scavenging. Potential antioxidant activity was shown by all tested lichens and highest DPPH radical scavenging activity was observed by methanol extract of *R. conduplicans*, whereas acetone extract of *U. Cineraria* exhibited highest superoxide dismutase radical scavenging activity [34]. In another article, moderate free radical scavenging activity was observed in case of acetone extract of lichen *Pleurosticta acetabulum* (IC_{50} of 151.7301 μg/ml). Secondary metabolites present in the lichen species were identified as norstictic acid, salazinic acid, evernic acid, protocetraric acid, and atranorin. MIC values were found to be in the range 1.25 to 20 mg/ml and cytotoxic activity ranged from 24.09 to 45.94 μg/ml. Hence, it was concluded that the lichen possessing anticancer, antioxidant, and antimicrobial potential could be developed as raw materials in food supplements and pharmaceutical industries [35]. Furthermore, the antioxidant ability of the lichen extracts *Stereocaulon strictum* var. compressum and *Lobariella pallida* and was evaluated. The evaluation was also carried out for the chemical constituents isolated: lobariellin, methyl orsellinate, stereocaullin, methyl haematommate, methyl β-orcinol carboxylate, porphyrilic acid, and atranorin. The results clearly indicated that the isolated compounds lobariellin, methyl haematommate, and porphyrilic acid inhibited lipid peroxidation, scavenged free radicals, and reduced ferric ions to its maximum as compared with other chemical constituents and extracts. Moreover, lobariellin preserved DNA and compounds, lobariellin, and methyl haematommate secured cells from oxidative destruction and hence thought to prevent neurodegenerative damage [36]. Leela *et al.* isolated the secondary constituents from lichen *Parmelia perlata*, purified them, and proved the antioxidant, antimicrobial, and antidiabetic potential of the alkaloid and glycoside isolates [37].

Recently, it has been reviewed by Thadhani *et al.* that most of the lichen components which are already established as antioxidants also exhibits antidiabetic activity [38]. Lichen species, *viz.*, *Umbilicaria antarctica*, *Flavoparmelia caperata*, *Ramalina sinensis*, *Usnea sinensis*, *Physciaaipolia*, and *Heterodermia leucomelos* were found to be α-glucosidase and α-amylase (digestive enzymes) inhibitors, protein tyrosinephosphatase1B activity inhibitor, and glycation end products inhibitor [39–42]. It was also inferred that bioactive compounds derived from lichen species could be promising antidiabetic agents. Ethanolic extracts of *Parmotrema hababianum*, which was already proved to be an antioxidant, had also shown antihyperglycemic potential *in vivo* [43]. Interestingly, it was also noted that most of the lichen species contain zeorin, a triterpenoid that possessed significant antidiabetic activity [44, 45].

5.2.5 Immunomodulator Activity

Immunomodulators interact with immune system and alter host response. The popularity of immunodialators in the medicinal field is increasing due to their applications in various treatments like rejection to transplant, patient recovery, stabilizing the immune system in HIV positive people. Plant derived products have been drawing great attention as an alternative to various treatments. Thadhani *et al.* evaluated the immunomodulatory potential of some of the lichen metabolites through luminol or lucigenin-based chemiluminescence probes. The chemical constituents were explored for their effects on isolated human polymorphonuclear leukocytes, respiratory burst of human blood phagocytes, and murine macrophages using chemiluminescence-based cellular assays. The results clearly indicated inhibitory effect on phagocytosis response of some of the tested lichen substances, *viz.*, methyl orsellinate, orsellinic acid, lecanoric acid, methyl haematomate, and lobaric acid as compared to the standards. The compound lobaric acid was found to suppress both the myloperoxidase independent as well as myloperoxidase dependent ROS production in the oxidative burst of isolated polymorphonuclear neutrophils with MIC value 3.1 μg/ml. While lecanoric acid inhibited only myloperoxidase dependent production of ROS with $IC_{50} < 3.1$ μg/ml. Myloperoxidase independent pathway was observed in case of orsellinic acid, methyl haematomate, and methyl orsellinate with IC_{50} values: <3.1 μg/ml, 3.3 ± 0.1 μg/ml, 6.1 ± 1.0 μg/ml, respectively [46].

In another report, the immune effects of Umbilicaria esculenta polysaccharide (UEP) were studied on murine macrophages RAW264.7. It was observed that UEP promoted their phagocytosis and proliferation to 2.5 and 1.4 times respectively in comparison with negative group at 600 μg/ml concentration. Moreover, the induction of NO synthase (2.7 times enhancement), nitric oxide (25.2 times enhancement), tumor necrosis factor-α, interleukin factors, and interferon-γ was carried out in a concentration dependent manner [47]. Furthermore, the effect of tremella polysaccharides was evaluated in immuno-deficient mice induced by cyclophosphamide. Significant results were noticed in this case and tremella polysaccharides at dose; 80 mg/kg·BW showed best results in mice treatment like, increase the spleen and thymus index andrelieve pathological characteristics of immunosuppression. Moreover, it was found to stimulate mRNA expression of IL-12, IL-4, and IL-1βin spleen and liver and inhibits mRNA expression in TGF-β [48]. In another research article, the immunomodulatory activities of *Gynostemma pentaphyllum* Makino polysacchrides (GPMPP) were determined. It was observed in *in vivo* studies that GPMPP enhanced the thymus and spleen indices significantly andstimulated the macrophage phagocytosis. Furthermore, GPMPP dose dependently elevated $CD4^+$ T lymphocyte counts and $CD4^+/CD8^+$ ratio as well, and it enhanced IL-2 level in the spleen and sera of Cy-immunosuppressed mice. Moreover, GPMPP notably increased the SOD, T-AOC, GSH-Px, CAT, and GSH level and inhibited the MDA level. The results indicated that GPMPP seems to be an active natural immunomodulatory agent [49]. Similar activity was shown by licorice polysaccharide in tumor bearing mice. Licorice polysaccharide was found to suppress the growth of tumor and augment the levels of IL 2, IL 6, and IL 7 (serum antitumor cytokines) while inhibiting pro-tumor cytokine TNFα [50].

In Umbilicaria esculenta, polysaccharide UP2 was analyzed for immunostimulatory activity and it was shown that UP2 could enhance RAW264.7 cells proliferation in a dose-dependent manner. It was found that entire samples (20–500 μg/ml) stimulated NO production and the maximum phagocytic activity was detected at 200 μg/ml [51].

5.2.6 Insecticidal Activity

In most of the developing countries, insect invasion results in severe destruction to crops and huge economic losses to farmers. Moreover, insects are also considered to be vectors of plant diseases transmission responsible for a human illness like dengue, malaria, chikungunya, and yellow fever [52, 53]. Hence, insect control has become an important key factor in crop management and in the prevention of human diseases [54–56]. Insect control is often achieved through harmful synthetic chemicals which are now replaced with natural products. Lichens and their metabolites especially coleopterans (e.g., *Sitophilus* and *Leptinotarsa*) appear to be promising biopesticides and their potential as insecticides has been extensively reviewed by Sachin *et al.* [57].

The insecticidal activity of extracts of three lichen species, *Letharia vulpine*, *Lecanora muralis*, and *Peltigera rufescens* was evaluated by Emsen *et al.* against *Sitophilus granaries* adults (wheat weevil) [58]. The extracts showed potential activity with LC_{50} values of *L. muralis* (0.666 mg/ml), *L. vulpine* (0.505 mg/l), and *P. rufescens* (0.328 mg/ml). In another report, the insecticidal activity of lichen metabolites; atranorin, (+)-usnic acid, gyrophoric acid, and 3-hydroxyphysodic acid were evaluated against *Culiseta longiareolata's* second and third instar larvae. The maximum activity was observed in the case of gyrophoric acid followed by (+)-usnic acid, atranorin and 3-hydroxyphysodic acid [59]. In another study, compounds (+)-usnic acid and 2-hydroxy-4-methoxy-6-propyl-methyl benzoate isolated from *Ramalina usnea* methanolic extract were found to possess larvicidal activity against *Aedes aegypti* III in star larvae with LC_{50} value 4.48 and 4.85 μg/ml, respectively [60]. Recently, it was shown by Martins *et al.* that compounds usnic acid, fumarprotocetraric acid, and barbatic acid extracted from lichen species *Cladonia substellata*, *Cladonia verticillaris*, and *Cladia aggregata*, respectively, exhibited dose-dependent insecticidal activity against the termite *Nasutitermes corniger* [61]. Another study also proved synergistic activity of usnic acid derivative and fungus *Beauveria bassiana* against larvae of Colorado potato beetle [62].

Kekuda *et al.* also revealed that methanolic extracts of lichen species *Parmotrema cristiferum* and *Dirinaria applanata* possessed insecticidal activity against *A. aegypti* II in star larvae [63]. In another study, methanolic extracts of *Parmotrema reticulatum*, *P. kamatti*, *P. tinctorum*, *Parmelia erumpens*, *Leptogium papilosum*, and *R. montagnei* were found to show insecticidal activity against *A. aegypti*, *Anopheles stephensi*, and *Culex quinquefasciatus* III in star larvae [64].

5.3 Conclusion

Though the therapeutic potential of lichens is insurmountable, it seems that lichen extracts and their secondary metabolites have great potential to find importance in pharmaceutics. These natural products represent a promising lead molecule to be developed as anticancer agents and useful in the inhibition of cell proliferation. It can be deployed as antiviral agents and effective immunomodulators, insecticidal, and antimicrobial agents.

References

1. Dayan, F.E. and Romagni, J.G., Lichens as a potential source of pesticides. *Pest. Outlook*, 12, 6, 229–232, 2001.
2. Huneck, S., The significance of lichens and their metabolites. *Naturwissenschaften*, 86, 12, 559–570, 1999.
3. Ingolfsdottir, K., Molecules of interest: Usnic Acid. *Phytochem.*, 61, 7, 729–736, 2002.
4. Bhattacharyya, S., Deep, P.R., Singh, S., Nayak, B., Lichen secondary metabolites and its biological activity. *Am. J. PharmTech Res.*, 6, 6, 1–17, 2016.
5. Van Cuong, T. and Thoa, N.T., Bioactive Compounds from Lichens as Promising Biomaterial for the Treatment of Influenza Virus: A review. *J. Scientific Res. Rep.*, 18, 4, 1–15, 2018.
6. Shukla, V., Joshi, G.P., Rawat, M.S.M., Lichens as a potential natural source of bioactive compounds: A review. *Phytochem. Rev.*, 9, 2, 303–314, 2010.
7. Studzinska-Sroka, E., Galanty, A., Bylka, W., Atranorin-an interesting lichen secondary metabolite. *Mini Rev. Med. Chem.*, 17, 17, 1633–1645, 2017.
8. Kekuda, T.R.P., Vinayaka, K.S., Sachin, M.B., Chemistry, ethnobotanical uses and biological activities of the lichen genus *Heterodermia Trevis* (Physciacea; Lecanorales; Ascomycota): A comprehensive review. *J. App. Pharm. Sci.*, 8, 148, 2018.
9. Poornima, S., Ponmurugan, P., Ahmed, K.S.Z., Ayyappadasan, G., Aldhafiri, F.K., Santhanaraj, B., Evaluation of Anti-Cancer Properties of Lichens Using Albino wistar Rats as an Animal Model. *Cancer Res. J.*, 4, 6, 84–89, 2016.
10. Paluszczak, J., Kleszcz, R., Studzińska-Sroka, E., Krajka-Kuźniak, V., Lichen-derived caperatic acid and physodic acid inhibit Wnt signaling in colorectal cancer cells. *Mol. Cellular Biochem.*, 441, 1–2, 109–124, 2018.
11. Šeklić, D.S., Obradović, A.D., Stanković, M.S., Živanović, M.N., Mitrović, T.L., Stamenković, S.M., Marković, S.D., Proapoptotic and Antimigratory Effects of Pseudevernia furfuracea and Platismatia glauca on Colon Cancer Cell Lines. *Food Technol. Biotechnol.*, 56, 3, 421–430, 2018.
12. Suh, S.S., Kim, T.K., Kim, J.E., Hong, J.M., Nguyen, T.T.T., Han, S.J., Youn, U.J., Yim, J.H., Kim, I.C., Anticancer activity of ramalin, a secondary metabolite from the antarctic lichen ramalina terebrata, against colorectal cancer cells. *Molecules*, 22, 8, 1–12, 2017.
13. Anar, M., Aslan, A., Agar, G., Ozgencli, I., Antigenotoxic and antioxidant activity of lichens Anaptychia ciliaris, Bryoria fuscescens, Parmotrema

chinensa and Xanthoria candelaria: An *in vitro* study. *Med. Aromat. Plants*, 5, 233, 2167–0412, 2016.

14. Hong, J.M., Suh, S.S., Kim, T.K., Kim, J.E., Han, S.J., Youn, U.J., Yim, J.H., Kim, I.C., Anti-cancer activity of lobaric acid and lobarstin extracted from the Antarctic lichen Stereocaulon alpnum. *Molecules*, 23, 3, 658, 2018.
15. Yang, Y., Park, S.Y., Nguyen, T.T., Yu, Y.H., Van Nguyen, T., Sun, E.G., Udeni, J., Jeong, M.H., Pereira, I., Moon, C., Ha, H.H., Kim, K.K., Hur, J.S., Kim, H., Lichen secondary metabolite, physciosporin, inhibits lung cancer cell motility. *PloS one*, 10, 9, e0137889, 2015.
16. Yang, Y., Nguyen, T.T., Jeong, M.H., Crişan, F., Yu, Y.H., Ha, H.H., Choi, K.H., Jeong, H.G., Jeong, T.C., Lee, K.Y., Kim, K.K., Hur, J.S., Kim, K., Inhibitory activity of (+)-usnic acid against non-small cell lung cancer cell motility. *PloS one*, 11, 1, e0146575, 2016.
17. Felczykowska, A., Pastuszak-Skrzypczak, A., Pawlik, A., Bogucka, K., Herman-Antosiewicz, A., Guzow-Krzemińska, B., Antibacterial and anticancer activities of acetone extracts from *in vitro* cultured lichen-forming fungi. *BMC Compl. Alternat. Med.*, 17, 1, 300, 2017.
18. Sariözlü, N.Y., Cankiliç, M.Y., Mehmet, C., Tay, T., Antimicrobial activity of lichen Bryoria capillaris and its compound barbatolic acid. *Biomed. Res.*, 419–423, 2016.
19. Hengameh, P. and Rajkumar, G.H., Assessment of bactericidal activity of some lichen extracts by disc diffusion assay. *Int. J. Drug Develop. Res.*, 9, 1, 9–19, 2017.
20. Oh, J.M., Kim, Y.J., Gang, H.S., Han, J., Ha, H.H., Kim, H., Antimicrobial Activity of Divaricatic Acid Isolated from the Lichen *Evernia mesomorpha* against Methicillin-Resistant *Staphylococcus aureus*. *Molecules*, 23, 12, 3068, 2018.
21. Anjali, D.B., Mohabe, S., Reddy, A.M., Nayaka, S., Antimicrobial activities of 2-Propanol crude extract from lichen Parmotrema tinctorum (Despr. ex. Nyl.) Hale, collected from Eastern Ghats, India. *Curr. Res. Environ. Appl. Mycol.*, 5, 3, 160–168, 2015.
22. Dixit, A., Mishra, M., Singh, A.N., Singh, M., Antimicrobial and antifungal activity of Indian lichen (Usnea sps and Parmotrema sps.) against human pathogenic bacterial and fungal sps. *World J. Pharm. Pharmaceut. Sci.*, 7, 3, 711–720, 2017.
23. Cankiliį, M.Y., Sariözlü, N.Y., Mehmet, C., Tay, F., Screening of antibacterial, antituberculosis and antifungal effects of lichen Usnea florida and its thamnolic acid constituent. *Biomed. Res.*, 28, 7, 3108–3113, 2017.
24. Bondarenko, V., Korczynski, M., Techathaveewat, W., The Antimicrobial Properties of Extracts Isolated from Lichen *Parmelia vagans*. *The FASEB Journal*, 31, 1, 939–13, 2017.
25. Ozyigitoglu, G., Acikgoz, B., Tahiroglu, G., Sesal, N.C., Comparison of antibacterial and antibiofilm activity properties of Hypogymnia tubulosa

(Schaer.) Hav. lichen extracts from different locations in Turkey. *Mycosphere*, 8, 8, 994–1002, 2017.
26. Prabhu, S.S. and Sudha, S.S., Evaluation of the antibacterial properties of some lichen species against human pathogens. *Int. J. Adv. Res. Biol. Sci.*, 2, 81–177, 20152017.
27. Shtro, A.A., Zarubaev, V.V., Luzina, O.A., Sokolov, D.N., Kiselev, O.I., Salakhutdinov, N.F., Novel derivatives of usnic acid effectively inhibiting reproduction of influenza A virus. *Bioorg. Med. Chem.*, 22, 24, 6826–6836, 2014.
28. Shtro, A.A., Zarubaev, V.V., Luzina, O.A., Sokolov, D.N., Salakhutdinov, N.F., Derivatives of usnic acid inhibit broad range of influenza viruses and protect mice from lethal influenza infection. *Antiviral Chem. Chemother.*, 24, 3–4, 92–98, 2015.
29. Lai, D., Odimegwu, D.C., Esimone, C., Grunwald, T., Proksch, P., Phenolic compounds with *in vitro* activity against respiratory syncytial virus from the Nigerian lichen Ramalina farinacea. *Planta Med.*, 79, 15, 1440–1446, 2013.
30. Vu, T.H., Le Lamer, A.C., Lalli, C., Boustie, J., Samson, M., Lohézic-Le Dévéhat, F., Le Seyec, J., Depsides: Lichen metabolites active against hepatitis C virus. *PloS one*, 10, 3, e0120405, 2015.
31. Ristić, S., Ranković, B., Kosanić, M., Stanojković, T., Stamenković, S., Vasiljević, P., Manojlović, I., Manojlović, N., Phytochemical study and anti-oxidant, antimicrobial and anticancer activities of Melanelia subaurifera and Melanelia fuliginosa lichens. *J. Food Sci. Technol.*, 53, 6, 2804–2816, 2016.
32. Rajan, V.P., Gunasekaran, S., Ramanathan, S., Murugaiyah, V., Samsudin, M.W., Din, L.B., Biological activities of four Parmotrema species of Malaysian origin and their chemical constituents. *J. Appl. Pharmaceut. Sci.*, 6, 8, 36–43, 2016.
33. Soundararajan, S., Shanmugam, P., Nagarajan, N., Palanisamy, D., Ponnusamy, P., *In vitro* study on screening antimicrobial and anti-oxidant potential of Ramalina fastigiata. *J. Drug Delivery Therapeut.*, 9, 1, 216–219, 2019.
34. Ramya, K. and Thirunalasundari, T., Evaluation of antioxidant activity among the epiphytic lichens of Kodaikanal. *World J. Pharm. Pharmaceut. Sci.*, 6, 828–841, 2017.
35. Tomović, J., Kosanić, M., Ristić, S., Ranković, B., Stanojković, T., Manojlović, N., Chemical composition and bioactive properties of the lichen, Pleurosticta acetabulum. *Trop. J. Pharmaceut. Res.*, 16, 12, 2977–2984, 2017.
36. Perico-Franco, L.S., Rojas, J.L., Cerbón, M.A., González-Sánchez, I., Valencia-Islas, N.A., Antioxidant activity and protective effect on cell and DNA oxidative damage of substances isolated from lichens of Colombian Páramo. *UK J. Pharm. Biosci.*, 3, 4, 9–17, 2015.
37. Leela, K. and Devi, A., Isolation, Purification and Application of Secondary Metabolites from Lichen Parmelia Perlata. *Biosci. Biotechnol. Res. Asia*, 14, 4, 1413–1428, 2017.

38. Thadhani, V.M. and Karunaratne, V., Potential of lichen compounds as anti-diabetic agents with antioxidative properties: A review. *Oxidat. Med. Cellular Long.*, 2079697, 1–10, 2017.
39. Hengameh, P., Rashmi, S., Rajkumar, H.G., *In vitro* inhibitory activity of some lichen extracts against α-amylase enzyme. *Eur. J. Biomed. Pharmaceut. Sci.*, 3, 5, 315–318, 2016.
40. Valadbeigi, T. and Shaddel, M., Amylase inhibitory activity of some macrolichens in Mazandaran province, Iran. *Physiol. Pharmacol.*, 20, 4, 215–219, 2016.
41. Shivanna, R., Parizadeh, H., Garampalli, R.H., Screening of lichen extracts for *in-vitro* antidiabetic activity using alpha amylase inhibitory assay. *Int. J. Biolog. Pharmacet. Res.*, 6, 5, 364–367, 2015.
42. Paliya, B.S., Bajpai, R., Jadaun, V., Kumar, J., Kumar, S., Upreti, D.K., Singh, B.R., Nayaka, S., Joshi, Y., Singh, B.N., The genus Usnea: A potent phytomedicine with multifarious ethnobotany, phytochemistry and pharmacology. *RSC Adv.*, 6, 26, 21672–21696, 2016.
43. Ganesan, A., Mahesh, A., Sundararaj, J.P., Mani, K., Ponnusamy, P., Antihyperglycemic and anti-oxidant activity of various fraction of Parmotrema hababianum in streptozotocin-induced diabetic rat. *Bangladesh J. Pharmacol.*, 11, 4, 935–939, 2016.
44. Karunaratne, V., Thadhani, V.M., Khan, S.N., Choudhary, M.I., Potent α-glucosidase inhibitors from the lichen Cladonia species from Sri Lanka. *J. National Sci. Foundation Sri Lanka*, 42, 1, 95–98, 2014.
45. Jiang, C.S., Liang, L.F., Guo, Y.W., Natural products possessing protein tyrosine phosphatase 1B (PTP1B) inhibitory activity found in the last decades. *Acta Pharmacol. Sinica*, 33, 10, 1217–1247, 2012.
46. Thadhani, V.M., Mesaik, M.A., Asif, M., Karunaratne, V., Choudhary, I.M., Immunomodulatory activities of some common lichen metabolites. *Int. J. Pharm. Pharmaceut. Sci.*, 7, 11, 144–147, 2015.
47. Wang, Y., Zhang, Y., Shao, J., Ren, X., Jia, J., Li, B., Study on the immunomodulatory activity of a novel polysaccharide from the lichen Umbilicaria Esculenta. *Int. J. Biological Macromol.*, 121, 846–851, 2019.
48. Zhou, Y., Chen, X., Yi, R., Li, G., Sun, P., Qian, Y., Zhao, X., Immunomodulatory effect of tremella polysaccharides against cyclophosphamide-induced immunosuppression in mice. *Molecules*, 23, 2, 239, 2018.
49. Shang, X., Chao, Y., Zhang, Y., Lu, C., Xu, C., Niu, W., Immunomodulatory and antioxidant effects of polysaccharides from Gynostemma pentaphyllum Makino in immunosuppressed mice. *Molecules*, 21, 8, 10852016.
50. Ayeka, P.A., Bian, Y., Githaiga, P.M., Zhao, Y., The immunomodulatory activities of licorice polysaccharides (Glycyrrhiza uralensis Fisch.) in CT 26 tumor-bearing mice. *BMC Compl. Alternat. Med.*, 17, 1, 536, 2017.
51. Zhang, B.W., Xu, J.L., Zhang, H., Zhang, Q., Lu, J., Wang, J.H., Structure elucidation of a polysaccharide from Umbilicaria esculenta and its immunostimulatory activity. *PloS one*, 11, 12, e0168472, 2016.

52. Patole, S.S., Review on Beetles (Coleopteran): An Agricultural Major Crop Pests of the World. *Int. J. Life. Sci. Scientific Res.*, 3, 6, 1424–1432, 2017.
53. Popp, J., Pető, K., Nagy, J., Pesticide productivity and food security. A review. *Agr. Sus. Develop.*, 33, 1, 243–255, 2013.
54. Oberemok, V.V., Laikova, K.V., Gninenko, Y.I., Zaitsev, A.S., Nyadar, P.M., Adeyemi, T.A., A short history of insecticides. *J. Plant Protect. Res.*, 55, 221–230, 2015.
55. Ricci, I., Valzano, M., Ulissi, U., Epis, S., Cappelli, A., Favia, G., Symbiotic control of mosquito borne disease. *Pathogens Global Health*, 106, 7, 380–385, 2012.
56. Singh, N., Shukla, S., Gupta, V., Tandia, N., Singh, P., Mosquito borne zoonotic diseases: A review. *J. Livestock Sci.*, 6, 65–72, 2015.
57. Sachin, M.B., Mahalakshmi, S.N., Kekuda, T.P., Insecticidal efficacy of lichens and their metabolites—A mini review. *J. Appl. Pharmaceut. Sci.*, 8, 10, 159–164, 2018.
58. Emsen, B., Yildirim, E., Aslan, A., Insecticidal Activities of Extracts of Three Lichen Species on *Sitophilus granarius (L.) (Coleoptera: Curculionidae). Plant Protect. Sci.*, 51, 155–161, 2015.
59. Cetin, H., Tufan-Cetin, O., Turk, A.O., Tay, T., Candan, M., Yanikoglu, A., Sumbul, H., Larvicidal activity of some secondary lichen metabolites against the mosquito Culiseta longiareolata Macquart (Diptera: Culicidae). *Nat. Prod. Res.*, 26, 4, 350–355, 2012.
60. Moreira, A.S., Fernandes, R.O., Lemos, F.J., Braz-Filho, R., Vieira, I.J., Larvicidal activity of Ramalina usnea lichen against Aedes aegypti. *Revista Brasileira de Farmacog.*, 26, 4, 530–532, 2016.
61. Martins, M.C., Lopes, R.S., Barbosa, P.S., Santiago, R., Rodrigues, B.R., de Albuquerque, A.C., Flacão, E.P.S., Lima, V.L.D.M., da Silva, N.H., Pereira, E.C., Efects of Usnic, Barbatic and Fumarprotocetraric acids on Survival of Nasutitermes corniger (Isoptera: Termitidae: Nasutitermitinae). *Sociobiol.*, 65, 1, 79–87, 2018.
62. Kryukov, V.Y., Tomilova, O.G., Luzina, O.A., Yaroslavtseva, O.N., Akhanaev, Y.B., Tyurin, M.V., Duisembekov, B.A., Salakhutdinov, N.F., Glupov, V.V., Effects of fluorine containing usnic acid and fungus Beauveria bassiana on the survival and immune-physiological reactions of Colorado potato beetle larvae. *Pest Managem. Sci.*, 74, 3, 598–606, 2018.
63. Kekuda, T.P., Dhanya, R., Dhatri, R., Sunita, C.M., Onkarappa, R., Vinayaka, K.S., Radical scavenging, Antimicrobial and Insecticidal Efficacy of Parmotrema cristiferum and Dirinaria applanata. *Sci. Technol. Arts Res. J.*, 4, 1, 95–102, 2015.
64. Khader, S.Z.A., Ahmed, S.S.Z., Venkatesh, K.P., Chinnaperumal, K., Nayaka, S., Larvicidal potential of selected indigenous lichens against three mosquito species-Culex quinquefasciatus, Aedes aegypti and Anopheles stephensi. *Chinese Herbal Med.*, 10, 2, 152–156, 2018.

6

Lichens as Sources of Antibacterial Compounds

Es-sadeqy Youness[1], Taha Chouati[2], Noura Aoussar[1], Imane Zalegh[1], Rajaa Ait Mhand[1], Naima Rhallabi[1] and Fouad Mellouki[1*]

[1]Research Unit Microbiology, Hygiene & Biomolecules, Laboratory of Virology, Microbiology, Quality & Biotechnology/Ecotoxicology and Biodiversity, University Hassan II of Casablanca, FSTM, Casablanca, Morocco
[2]Laboratory of Microbiology and Molecular Biology, UATRS-PGF, National Center for Scientific and Technical Research, Rabat, Morocco

Abstract
Lichens are defined as a symbiotic relationship between filamentous fungi, photosynthesis capable organism (algae and/or cyanobacteria), and bacterial community. Such a complex symbiotic system with a wide diversity suggests the existence of molecules that are used as a way to induce such symbiosis or as a protection against hostile elements. This has led to the search for potentially useful molecules in numerous fields, among which antibacterial properties. While still under-explored, lichens have demonstrated an interesting antibacterial activity. Therefore, the aim of this study was to assess the advancement of this field through the most prominent finding concerning the use of lichens, whether as crude extracts or their purified substances, against various pathogenic bacteria. As a result, these molecules, showed a significant inhibitory effect against pathogenic bacteria. Subsequently, this review highlighted the studies that searched for a potential synergy between lichenic acids and standard antibiotics. Consequently, the present work showed that lichens could be considered as a promising source of antibiotic molecules. Furthermore, this study emphasized the need for more in-depth research of lichen's molecules, by both screening for potentially useful molecules and understanding the mode of action of these bioactive compounds.

Keywords: Lichens, antibacterial activity, antibiotic resistance, secondary metabolites, synergy, lichenic acids

**Corresponding author*: fouad.mellouki@univh2c.ma

Mohd Yusuf (ed.) Lichen-Derived Products: Extraction and Applications, (141–178) © 2020 Scrivener Publishing LLC

6.1 Introduction

Classically, lichens are presented as an ecologically obligate symbiotic mode of life in which a filamentous fungus (mycobiont) associate with a photosynthetic partner (photobiont) which is either an alga and/or a cyanobacterium [1]. In this symbiosis system, the fungal partner provides shelter, water, and mineral nutrients for the photobiont, while the later undergoes, through photosynthesis, and provides the necessary organic compounds for the growth of the mycobiont [2]. However, it is important to note that from 2006, it has been proposed that this definition should also include a devoted bacterial community (bacteriobiont), and the impact of which is still to be less assessed [3]. Indeed, about 18,500 different species of lichens have been described around the world, growing under diverse (and sometimes extreme) ecological conditions such as polar latitudes or high altitudes (up to 7,400 m) [4].

Generally, lichens produce two types of bioactive compounds classified as primary and secondary metabolites, combinable in quantities in certain cases [5]. These chemically diverse substances (aliphatic and aromatic), after been produced by the mycobiont, are accumulated in the cortex (such as atranorin, usnic acid, and parietin) or in the medullary layer (such as physodic acid, protocetraric acid, and physodalic acid) as extracellular tiny crystals on the upper surface of the hyphae [6]. Primary metabolites mainly include amino acids, polyols, vitamins, and polysaccharides. Some of them have taxonomic significance such as cell wall polysaccharide compounds (lichenane and isolichenane) [5]. On the other hand, lichen's secondary metabolites, often called lichenic acids, are produced mainly by the mycobiont, and are secreted on the surface of lichen hyphae, either in amorphous forms or as crystals [7]. Very little is known of the role of these lichenic acids. It has, however, been speculated that they help the lichenic thalli to protect itself from its often-hostile environment, against pathogens, herbivores, and abiotic stress. In concordance to this, most isolated lichen compounds have been found to exhibit multiple biological activities, such as the atranorin, which showed antioxidant, antimicrobial, and anti-protective activities, or even the dibenzofuran usnic acid, which exerted antimicrobial and larvicidal effects, anticancer activities, and UV absorption (regarding the photo-protection activity) [8].

It is fully accepted that a large number of microorganism species (mainly bacteria) show a strong affinity for humans as hosts and while this cohabitation can prove neutral or even mutualistic, there is always a chance it can be harmful to the host, depending on the strain considered.

However, excessive use of antibiotics in hospitals, communities, animal husbandry, and agricultural systems has led to the apparition and widespread of antibiotic-resistant strains at an alarming fashion [9]. Recent studies have shown that resistant bacteria can be selected at concentrations several hundred-fold below the lethal concentrations for susceptible cells [10]. According to the World Health Organization (WHO), we are at the "brink of living" in post antibiotic era, in which a small injury or common infections could be lethal.

The antibiotic resistance already causes 700,000 deaths per year and the number of victims could reach 10 million in 2050 if left unchecked [11]. It is then clear that an urgent alternative must be found in order to avoid a total collapse of antibiotics as a treatment. Multiple alternatives have been, and are, explored [12–14], among which the development of new antibacterial molecules from natural sources, such as lichens.

To fight against pathogenic microbes that cause several threats to human health and develop constantly resistance against available therapeutically drugs. It is imperative to find out new alternatives capable to maintain pace evolution of the super-pathogens [15]. Among these alternatives, lichenic acids from lichens are of a special interest to researchers thanks to their antimicrobial properties [16]. Furthermore, several studies show that lichen's secondary metabolites have a broad variety of biological activities including antimicrobial activity (antibacterial, antifungal, and antiviral), antioxidant and UV protection, cytotoxic and antitumoral activity, anti-herbivore and anti-insecticidal activity, enzyme inhibitory activity, and some effects against the growth of parasite plants [2, 5].

The aim of this review is then to firstly summarize and report the main studies concerning the antibacterial activity of either the crude extracts of lichens or their isolated secondary metabolites, or both, against pathogenic bacteria. Secondly, this work aims to review the studies that tested the possible synergic, or antagonistic, interactions between lichen's secondary metabolites and the therapeutically available standard antibiotics.

6.2 Antibacterial Activity of Crude Extracts of Lichens

At first, the study of the bioactivity of lichen has been done by preparing the crude extracts of lichen thallus, then testing them directly against pathogenic bacteria. While, the evolution of the analytical tools for the isolation of different chemical compounds present in each crude extract allowed more detailed screening. Consequently, secondary metabolites have been the object of many studies as a new point of interest to the scientific community. In this respect, they are many studies whose succeeded to prove the efficacy of the crude extracts of lichens [17, 18] and of the lichen's secondary metabolites against a significant number of pathogenic bacteria including Gram-positive or Gram-negative bacteria and Mycobacteria [19–21].

To our knowledge, two recent studies have been interested in reviewing the antibacterial effect of lichens whether for their brute extracts or for pure molecules. Firstly, Zambrae and Christopher (2012) have reported in their review fourteen publications showing an antibiotic effect of lichens against pathogenic bacteria [5]. Secondly, Shrestha and Clair, a year later, discussed 18 publications presenting lichens substances as a promising source for antibacterial drugs [1]. The limited number of works in these two reviews can however be explained by the fact that both studies focused on a number of other biological activities. While, the present study was focused specifically on one biological activity of lichens and consists separately classifying the antibacterial activity of lichens in three levels: (1) antibacterial activity of crude extracts, (2) antibacterial effect of lichenic acids, and (3) antibacterial activity of both in the same studies.

Burkholder *et al.* (1944), the pioneers of the study of lichens as antibiotic agents, tested aqueous extracts of 42 lichen species against *Bacillus subtilis*, *Staphylococcus aureus*, *Alcaligenes faecalis*, *Proteus vulgaris*, and *Escherichia coli*. Results of this first study showed that 27 species had an antagonistic activity against either *S. aureus* or *B. subtilis*, while four species inhibited the growth of *P. vulgaris* or *A. faecalis*. On the other hand, none of the tested species showed antibacterial activity against *E. coli* [18]. It should be noted, however, that this study did not estimate the Minimal Inhibitory Concentrations (MICs) of the crude extracts (the antibacterial activity has being demonstrated by disk diffusion). In this respect, and since Burkholder's first work, many studies have screened and showed the antibacterial activity of crude extracts of lichens against various pathogenic bacteria only by disk diffusion method (Table 6.1) [22–26].

The above table shows that lichen's crude extracts have been found to be effective against a variety of pathogenic bacteria, some of which have been showing antibacterial activity only against Gram-positive bacteria. Thusly, Zambare *et al.* (2010), found that both acetone and methanol extracts of

Table 6.1 Antibacterial activity of lichen's crude extracts.

Lichens	**Zone of collect**	**Solvents**	**Sensitive bacteria to the crude extracts**	**MIC values (µg/ml)**	**References**
34 species belonging to 22 genera: *Alectoria, Bryoria, Evernia, Everniastrum, Flavocetraria, Hypogymnia, Letharia, Masonhalea, Parmelia, Parmotrema, Peltigera, Platismatia, Ramalina, Rhizoplaca, Sphaerophorus, Thamnolia, Tuckermannopsis, Umbilicaria, Usnea, Vulpicida, Xanthoparmelia*	Various locations throughout the United States	Acetone Methanol	*MRSA* *S. aureus* *E. coli* *P. aeruginosa*	7.8–500 3.9–500 125–500 3.9–500	[30]
Usnea ghattensis	India	Acetone Ethanol Methanol	*B. cereus* *S. aureus* *S. faecalis* *P. aeruginosa*	3.125 25 6.25 200	[31]
Ramalina menziesii *Usnea lapponica*	California (USA)	Acetone Methanol	*MRSA* *S. aureus* *P. aeruginosa*	15.6–500 15.6–125 15.6–62.5	[32]

(*Continued*)

Table 6.1 Antibacterial activity of lichen's crude extracts. (*Continued*)

Lichens	Zone of collect	Solvents	Sensitive bacteria to the crude extracts	MIC values (µg/ml)	References
Umbilicaria cylindrica	Serbia	Ethylacetate Methanol	*B. subtilis* *S. aureus* *E. coli* *K. pneumoniae* *P. mirabilis* *P. vulgaris*	15.62–31.25 15.62–31.25 15.62–31.25 31.25 31.25–62.5 62.5	[33]
Parmelia perlata	India	Acetone Ethylacetate Methanol	*S. aureus*	64–128	[28]
Usnea barbata	Former Yugoslav Republic of Macedonia	Five different extracts with the Supercritical Fluid Extraction (SFE) as technique for isolation of extracts	*E. faecalis* *MRSA* *S. aureus* *S. epidermidis* *S. pseudintermedius* *S. agalactiae* *S. uberis*	20 40–640 5–320 5–320 1.25–320 20–320 5–320	[29]

(*Continued*)

Table 6.1 Antibacterial activity of lichen's crude extracts. (*Continued*)

Lichens	Zone of collect	Solvents	Sensitive bacteria to the crude extracts	MIC values (µg/ml)	References
Cladonia furcata *Lecanora atra* *Lecanora muralis*	Kopaonik, Serbia	Acetone	*B. mycoides* *B. subtilis* *S. aureus* *E. cloaceae* *E. coli* *K. pneumoniae*	780–3120 780–6250 780–3120 780–3120 1,560 780–3120	[35]
Everniastrum Cirrhatum	Karnataka, India	Methanol	*B. cereus* *S. aureus* *S. epidermidis* *E. coli* *P. aeruginosa*	600 600 400 700 800	[36]
Ramalina pacifica	India	Ethanol Petroleum ether	*S. aureus* *E. coli* *K. pneumoniae* *P. aeruginosa* *S. typhi* *S. paratyphi*	ND	[23]

(*Continued*)

Table 6.1 Antibacterial activity of lichen's crude extracts. (*Continued*)

Lichens	Zone of collect	Solvents	Sensitive bacteria to the crude extracts	MIC values (μg/ml)	References
Usnea baileyi *Ramalina dendriscoides* *Stereocaulon massartianum* *Cladonia gracilis*	Luzon Island Philippines	Acetone	*B. subtilis* *S. aureus* *E. coli* *P. aeruginosa*	ND 156 ND ND	[22]
Xanthoparmelia somloensis	South Dakota, USA	Acetone Methanol	*S. aureus* *S. pyogenes* *S. agalactiae*	800 700 900	[27]
Cladonia furcata *Parmelia caperata* *Parmelia pertusa* *Hypogymnia physodes* *Umbilicaria polyphylla*	Kopaonik, Serbia	Acetone Ethanol Water	*B. mycoides* *B. subtilis* *S. aureus* *E. cloaceae* *E. coli* *K. pneumoniae*	780–25,000 390–25,000 780–25,000 780–25,000 1,560–6,250 780–25,000	[37]

(*Continued*)

Table 6.1 Antibacterial activity of lichen's crude extracts. (*Continued*)

Lichens	Zone of collect	Solvents	Sensitive bacteria to the crude extracts	MIC values (µg/ml)	References
Anaptychia ciliaris *Cetrelia olivetorum* *Lecanora muralis* *Peltigera polydactyla* *Peltigera praetextata* *Ramalina farinacea* *Rhizoplaca melanophthalma* *Umbilicaria vellea* *Xanthoria elegans* *Xanthoria parietina* *Xanthoparmelia tinctina*	Artvin, Giresun and Trabzon provinces, Turkey	Ethanol Water	*B. subtilis* *S. aureus* *S. epidermidis* *E. coli*	ND	[24]
Everniastrum cirrhatum *Flavoparmelia caperata* *Heterodermia leucomela* *Lecanora flavidorufa* *Leptogium pedicellatum* *Lobaria isidiosa* *Rimelia reticulata* *Phaeophyscia hispidula* *Stereocaulon foliolosum*	Uttranchal, India	Ethanol	*Mycobacterium tuberculosis* (H37Rv and H37Ra)	250–500	[40]

(*Continued*)

Table 6.1 Antibacterial activity of lichen's crude extracts. (*Continued*)

Lichens	Zone of collect	Solvents	Sensitive bacteria to the crude extracts	MIC values (µg/ml)	References
Lasallia pustulata, *Parmelia sulcata* *Umbilicaria crustulosa* *Umbilicaria cylindrica*	Kopaonik, Serbia	Acetone Methanol Water	*B. mycoides* *B. subtilis* *S. aureus* *E. cloaceae* *E. coli* *K. pneumoniae*	780–6,250 1,560–6,250 6,250–50,000 1,560 1,560–6,250 1,560–12,500	[38]
Rhizoplaca chrysoleuca *Rhizoplaca melanophthalma* *Rhizoplaca peltata*	Middle Anatolia, Turkey	Acetone	*B. megaterium* *B. subtilis* *E. faecalis* *S. aureus* *E. coli* *P. mirabilis*	ND	[26]
Usnea florida *Usnea barbata* *Usnea longissimi* *Usnea rigida, Usnea hirta* *Usnea subflorida*	Anatolia, Turkey	Acetone	*B. megaterium* *B. subtilis* *E. coli* *P. mirabilis*	ND	[25]

(*Continued*)

Table 6.1 Antibacterial activity of lichen's crude extracts. (*Continued*)

Lichens	Zone of collect	Solvents	Sensitive bacteria to the crude extracts	MIC values (µg/ml)	References
Parmelia saxatilis *Platismatia glauca* *Ramalina pollinaria* *Ramalina polymorpha* *Umbilicaria nylanderiana*	Artvin province, Turkey	Methanol	*B. macerans* *B. subtilis* *E. faecalis* *S. pyogenes* *Acinetobacter baumanii* *E. coli* *P. vulgaris*	31.25–62.50 15.62–62.5 15.62 15.62 15.62–31.25 15.62–31.25 15.62	[34]
Usnea barbata	North of Alice, South Africa	Acetone Methanol Water	*B. subtilis* *E. faecalis* *S. aureus* *E. coli* *P. vulgaris* *P. aeruginosa*	100 100–500 100 500 5,000 5,000	[39]
Cetraria islandica	Iceland	Acetone Methanol petroleum ether	*Helicobacter pylori*	16–64	[17]
42 species of lichens belonging to different genera	USA	Water	*B. subtilis* *S. aureus* *Alcaligenes faecalis* *P. vulgaris*	ND	[18]

ND, not defined (activity proved by disk diffusion only).

Xanthoparmelia somloensis showed a marked inhibition against *S. aureus*, *Straptococcus pyogenes*, and *Streptococcus agalactiae* with MIC values of 0.7–0.9 mg/ml [27]. In the same way, Vidyalakshmi *et al.* (2012) demonstrated that the crude extracts of *Parmelia perlata* have an inhibitory effect on *S. aureus* with MIC values of 64–128 μg/ml [28]. In the same year, Zizovic *et al.* tested five different crude extracts of *Usnea barbata* obtained with Supercritical Fluid Extraction (SFE) as a superior method for the isolation of extracts with strong bacterial activities. Results from this new technique showed that SFE showed a higher antibacterial activity compared to the extracts obtained by conventional methods, and the supercritical extract obtained at 40°C showed a stronger antibacterial activity than usnic acid (standard sample) and the rest of the extracts against the majority of examined strains (nine strains). The distribution of MIC values of this extract was 5 μg/ml for six bacteria of the investigated strains, 20 μg/ml for four strains, 40 μg/ml for two bacteria, and 160 μg/ml for one strain (Table 6.1) [29].

Crude extracts of lichens are also active against a significant number of pathogenic bacteria regardless of their Gram. Thus, the bacterial inhibitory effect has been registered with either lower or higher MICs values (3.125–500 to 100–50,000 μg/ml) (Table 6.1). Firstly, the lower MICs were reported against various species (*Bacillus macerans*, *B. cereus*, *B. subtilis*, *E. faecalis*, Methicillin-Resistant *Staphylococcus aureus* [MRSA], *S. aureus*, *S. pyogenes*, *Acinetobacter baumanii*, *E. coli*, *Proteus mirabilis*, *P. vulgaris*, *P. aeruginosa*, and *K. pneumoniae*) [30–34]. Secondly, the elevated MICs values were recorded against either Gram-positive (*Bacillus cereus*, *B. mycoides*, *B. subtilis*, *Enterococcus faecalis*, MRSA, and *Staphylococcus epidermidis*) or Gram-negative pathogenic bacteria (*Enterobacter cloaceae*, *E. coli*, *P. vulgaris*, *Pseudomonas aeruginosa*, and *Klebsiella pneumonia*) [35–39].

Furthermore, two important studies proved an antibacterial effect of lichen's crude extracts against a specific species of pathogenic bacteria. Firstly, as shown by Ingólfsdóttir *et al.* (1997), the crude extracts of *Cetraria islandica* were active against *Helicobacter pylori* with a MIC ranging from 16 to 64 μg/ml [17]. Secondly, in 2007, Gupta *et al.* worked on the ethanol extract of nine lichen species (*Everniastrum cirrhatum*, *Flavoparmelia caperata*, *Heterodermia leucomela*, *Lecanora flavidorufa*, *Leptogium pedicellatum*, *Lobaria isidiosa*, *Rimelia reticulata*, *Phaeophyscia hispidula*, and *Stereocaulon foliolosum*) and reported their effect on two strains of *Mycobacterium tuberculosis* which proved to have a MIC ranging from 250 to 500 μg/ml (Table 6.1) [40].

The numerous studies concerning lichen's crude extracts prove their activity against a wide range of microbes (Gram-positive, Gram-negative, or Mycobacteria). Consequently, the assessment of crude extracts of lichens against pathogenic bacteria as the first screening of antibiotic activity can be very useful for discovering new extracts with interesting bioactive compounds.

6.3 Antibacterial Activity of Lichenic Acids

According to Vartia *et al.* (1973), more than 50% of the lichen species possess antibiotic properties through bioactive molecules [41]. Indeed, lichen organisms produce various secondary metabolites that serve as a protection against many pathogens and then help the lichen thalli to adapt to their varying environmental conditions. The current literature provides various examples of such antibacterial activity from a large number of identified secondary metabolites, as summarized in Table 6.2.

It should be noted that, in concordance to Cocchietto *et al.* (2002), usnic acid is considered the most effective antibacterial lichen substance [42]. Many studies corroborate this statement, such as the work of Lauterwein *et al.* (1995) in which the efficacy of usnic acid against five Gram-positive bacteria (*Clostridium perfringens*, *E. faecalis*, *E. faecium*, MRSA, and Methicillin-Sensitive *Staphylococcus aureus* [MSSA]) was proven [21]. On the other hand, Safak *et al.* (2009) focused on *Helicobacter pylori*, and tested the usnic acid from *Usnea dasypoga* against 38 strains of this bacteria, with a resulting MIC values ranging from 0.016 to 32 μg/ml (Table 6.2) [43]. Furthermore, 10 studies from the 13 listed on the above table have demonstrated the antibacterial activity of usnic acid against a large number of pathogenic bacteria in recent years [44–47].

While usnic acid is the strongest antibiotic compound extracted from lichens to date, there are numerous secondary metabolites also characterized by their antibacterial activity. In this regard, Ingólfsdóttir *et al.* (1998) tested four different compounds (usnic acid from *Cladonia arbuscula*, atranorin, and lobaric acid from *Stereocaulon alpinum*, salazinic acid from *Parmelia saxatilis*, and protolichesterinic acid from *Cetraria islandica*) against *Mycobacterium aurum* and found that the usnic acid is the most active compound with a MIC value of 32 μg/ml while the MICs of others compounds were significantly higher than the usnic acid MIC (≥125 μg/ml) [48]. Moreover, in 2010, Honda *et al.* assessed various depsides, depsidones, xanthones, usnic acid, orsellinic acid esters, salazinic acid derivatives,

Table 6.2 Antibiotic activity of lichen's secondary metabolites.

Lichen's secondary metabolites/Lichens	Sensitive bacteria to the lichenic acids	MIC values (μg/ml)	References
Acetylportentol (*Roccella fuciformis*) Conhypoprotocetraric acid (*Ramalina siliquosa* var. x) Demethylbarbatic acid (*R. siliquosa* var. *druidarum*) Erythrin (*Roccella fuciformis*) Hypoprotocetraric acid (*Ramalina siliquosa* var. *zopfii*) Lepraric acid (*Roccella fuciformis*) Methyl-beta-orcinolcarboxylate (various lichens) Psoromic acid (*Squamarina cartilaginea*) Roccellic acid (*Roccella phycopsis*) Variolaric acid (*Ochrolechia parella*) Vulpinic acid (*Letharia vulpina*)	*Streptococcus gordonii* *Porphyromonas gingivalis*	11.72–750 5.86–1,000	[19]
Lobaric acid and lobastin isolated from *Stereocaulon alpinum*	*B. subtilis* *S. aureus*	20.09–40.17 16.07–18.08	[52]
Usnic acid and atranorin from lichens collected in Chilean territories	MSSA MRSA	2–128 2–128	[44]

(*Continued*)

Table 6.2 Antibiotic activity of lichen's secondary metabolites. (*Continued*)

Lichen's secondary metabolites/Lichens	Sensitive bacteria to the lichenic acids	MIC values (µg/ml)	References
Usnic acid from *Parmelia caperata* (used as a standard) Protocetraric acid from *Parmelia caperata* Salazinic acid from *Parmelia saxatilis*	*B. mycoides* *B. subtilis* *S. aureus* *E. coli* *K. pneumoniae*	15 15.31.2 1 500 15–500	[45]
Various depsides, depsidones, xanthones, usnic acid, orsellinic acid esters, salazinic acid derivatives andlichexanthone derivatives extracted from: *Parmotrema dilatatum* *Parmotrema tinctorum* *Pseudoparmelia sphaerospora* *Usnea subcavata*	*Mycobacterium tuberculosis*	15.6–250	[49]
Usnic acid (*Usnea dasypoga*)	*Helicobacter pylori*	0.016–32	[43]
Physodic acid (*Hypogymnia physodes*) Usnic acid (*Parmelia caperata*) Atranorin (*Physcia aipolia*) Gyrophoric acid (*Umbilicaria polyphylla*)	*B. mycoides* *B. subtilis* *S. aureus* *E. cloacae* *E. coli* *K. pneumoniae*	7.5–1,000 15–1,000 15–1,000 15–1,000 31–1,000 3.7–1,000	[46]

(*Continued*)

Table 6.2 Antibiotic activity of lichen's secondary metabolites. (*Continued*)

Lichen's secondary metabolites/Lichens	**Sensitive bacteria to the lichenic acids**	**MIC values (μg/ml)**	**References**
Usnic acid and its derivate dibenzoylusnic acid from lichen *Alectoria ochroleuca*	*B. subtilis* *S. aureus* *P. aeruginosa* *Shigella flexneri*	ND	[53]
Fumarprotocetraric (*Cladonia furcata*) Lecanoric acid (*Ochrolechia androgyna*) Protocetraric acid (*Parmelia caperata*) stictic acid (*Parmelia conspresa*)	*B. mycoides* *B. subtilis* *S. aureus* *E. cloacae* *E. coli* *K. pneumoniae*	62–500 62–500 62–500 62–500 62–500 31–250	[20]
Atranorin (*Lepraria lobificans*) Evernic acid (*Evernia prunastri*) Hybocarpone (*Lecanora conizaeoides*) Lobaric acid (*Sterocaulon dactylophyllum*) Physodic acid (*H. physodes*) Rhizocarpic acid (*Psilolechia lucida*) Usnic acid (*Lecanora albescens*) 3-hydroxyphysodic acid (*H. physodes*)	Methicillin- and Multidrug-Resistant *S. aureus* Two hospital isolates of resistant *S. aureus* Norfloxacin-Susceptible strain of *S. aureus* Two epidemic MRSA strains	8–128 4–128 4–64 4–64	[50]
Usnic acid (synthesized by chemical pathway)	*S. aureus* *P. aeruginosa*	32 256	[47]

(*Continued*)

Table 6.2 Antibiotic activity of lichen's secondary metabolites. (*Continued*)

Lichen's secondary metabolites/Lichens	Sensitive bacteria to the lichenic acids	MIC values (μg/ml)	References
Atranorin (*Stereocaulon alpinum*) Lobaric acid (*S. alpinum*) Protolichesterinic acid (*Cetraria islandica*) Salazinic acid (*P. saxatilis*) Usnic acid (*Cladonia arbuscula*)	*Mycobacterium aurum*	250 125 125 125 32	[48]
Usnic acid (Commercial) Usnic acid (*Cladonia stellaris*) Vulpinic acid (*Letharia vulpina*)	*Clostridium perfringens* *E. faecalis* *E. faecium* MRSA MSSA	4–16 4–>16 4–32 4–>16 2–>16	[21]

ND, not defined (activity proved by disk diffusion only).

and lichexanthone derivatives extracted from *Parmotrema dilatatum*, *Parmotrema tinctorum*, *Pseudoparmelia sphaerospora*, and *Usnea subcavata* against *Mycobacterium tuberculosis* and concluded that several compounds were found to inhibit growth of *M. tuberculosis*, of which the most active was diffractaic acid (MIC = 15.6 μg/ml), isolated from the lichen *U. subcavata*, followed by norstictic acid (MIC = 62.5 μg/ml), usnic acid (MIC = 62.5 μg/ml), hypostictic acid (MIC = 94.0 μg/ml), and the protocetraric acid (MIC = 125 μg/ml) [49].

Regarding the multidrug-resistant bacteria (MDR), Kokubun *et al.* (2007) have found that eight different lichen's compounds were active against Methicillin- and Multidrug-Resistant *S. aureus*. Interestingly, most of the tested secondary metabolites showed consistent activity against various strains of *S. aureus*, and amidst the tested molecules, hybocarpone (isolated from *Lecanora conizaeoides*) was the most active compound against all strains with MIC values of 4–8 μg/ml (Table 6.2) [50].

Another area of concern is the capacity for bacteria to aggregate in microcolonies and form biofilm structures. In this respect, a recent study by Sweidan *et al.* (2017) tested eleven natural lichen compounds belonging to a wide variety of chemical families against *Streptococcus gordonii* (Gram-positive) and *Porphyromonas gingivalis* (Gram-negative); two bacteria identified to be as a cause of dental biofilm. Results of this study have shown that all the tested molecules were found active except the acetylportentol and lepraric acids (on *S. gordonii*). More importantly, three compounds were shown to have promising antibacterial activities, namely, roccellic acid, demethylbarbatic acid, and psoromic acid in order of increasing efficiency (based on both MIC value and Minimal Bactericidal Inhibition [MBC]). The latter exhibiting a MIC value of 11.72 and 5.86 μg/ml against *S. gordonii* and *P. gingivalis*, respectively (Table 6.2) [19]. Furthermore, Francolini *et al.* (2004) have assessed the capacity of usnic acid (synthesized by chemical pathway) to control biofilm formation by *S. aureus* or *P. aeruginosa* (two major bacteria that are capable to protect themselves under biofilm framework [51]). This test was performed by loading the acid into the modified polyurethane. Interestingly, usnic acid loaded polymers did not inhibit the initial attachment of *S. aureus*, however, inhibited the formation of a biofilm by killing the attached cells. Contrarily, concerning *P. aeruginosa*, there was no inhibition of biofilm formation but its morphology was altered [47].

The *in vitro* activity of lichenic acids against pathogenic bacteria has thusly been confirmed throughout many works. In addition, the emergence of drug-resistant bacteria, e.g., *M. tuberculosis* or *S. aureus*, has currently led an increase in the number of clinical complications. Consequently, new antibacterial molecules are urgently needed, and as the search is being extended to new sources, lichen's secondary metabolites do clearly contain a large number of bioactive compounds to draw from.

6.4 Antibacterial Activity of Crude Extracts of Lichens and of Their Bioactive Compounds

After reporting the activity of crude extracts and lichenic acids separately, this third part will focus on works considering both. For this purpose, recently Tozatti *et al.* (2016) have assessed the effect of the acetone extract (AE) of the lichen *Usnea steineri* and its major metabolites (difractaic acid and usnic acid) against four multidrug-resistant Gram-positive bacteria (*E. faecalis, S. aureus, S. epidermidis, and Staphylococcus haemolyticus*). As a result, they have found that the AE showed strong activity (less than 10 µg/ml) against resistant strains of *S. epidermidis* and *E. faecalis*. Moreover, as purified molecules, the usnic acid exerted strong activity against *S. epidermidis* (MIC 3.12 µg/ml), *S. aureus*, and *S. haemolyticus* (MIC 12.5 µg/ml), while difractaic acid was inactive (Table 6.3) [54]. In the same way, concerning the multi-drug resistant bacteria, Martins *et al.* (2010) tested crude extracts of *Cladia aggregata* and its barbatic acid against four multi-drug resistant *S. aureus* strains and found that, whether in crude organic or in purified forms, it inhibited the growth of *S. aureus*. The MIC values for purified barbatic acid against different strains of *S. aureus* was 50 µg/ml, except for one highly resistant strain, who's MIC was 100 µg/ml (Table 6.3) [55].

Furthermore, Basile *et al.* (2015) have evaluated the effect of AEs of lichen *Xanthoria parietina* and its secondary metabolite parietin against various pathogenic bacteria (nine ATCC [American Type Culture Collection] and nine matched, clinically isolated strains). Thus, they have found that the AE exerted a strong antibacterial activity against the nine bacterial strains. Surprisingly, both AE and parietin inhibited all the tested bacteria at concentrations of 7.8–62.5 µg/ml. The Gram-positive *S. aureus*, from both ATCC and clinical source, showed the highest sensitivity to both AE and parietin (MIC = 7.8 µg/ml for standard strains and MIC = 15.6 µg/ml for clinical isolates). Moreover, both treatments demonstrated also a significant activity against Gram-negative bacteria, particularly against *P. vulgaris* and *P. mirabilis* (MIC = 15.6 µg/ml for both). Whereas, amidst the

Table 6.3 Antibacterial activity of crude extracts of lichens and of their bioactive compounds.

Lichens and their bioactive compounds	Zone of collect of lichens	Sensitive bacteria	MIC values (μg/ml)		References
			Crude extracts	Lichenic acids	
Acetone extract of the lichen *Usnea steineri* and its major metabolites (difractaic acid and usnic acid)	São Paulo State, Brazil	Four multidrug-resistant bacteria:			[54]
		E. faecalis	6.25	50	
		S. aureus	50	12.5	
		S. epidermidis	6.25	3.12	
		S. haemolyticus	25	12.5	
Acetone extract of lichen *Xanthoria parietina* and its secondary metabolite parietin	Pian del Lago, Siena, Italy	Nine ATCC and nine matched clinically isolated strains (belonging to the same species):			[56]
		S. aureus	7.8–15.6	7.8–15.6	
		S. faecalis	15.6–31.3	7.8–15.6	
		E. aerogenes	31.3–62.5	15.6–31.3	
		E. cloacae	31.3–62.5	31.3–62.5	
		K. pneumoniae	62.5–>100	31.3–62.5	
		P. mirabilis	15.6	15.6–31.3	
		P. vulgaris	15.6–31.3	15.6–31.3	
		P. aeruginosa	31.3–62.5	31.3–62.5	
		S. typhi	15.6–31.3	31.3–62.5	

(*Continued*)

Table 6.3 Antibacterial activity of crude extracts of lichens and of their bioactive compounds. (*Continued*)

Lichens and their bioactive compounds	Zone of collect of lichens	Sensitive bacteria	MIC values (µg/ml)		References
			Crude extracts	Lichenic acids	
Acetone extracts of the lichen *Cladonia furcata*, *Cladonia pyxidata* and *Cladonia rangiferina* and their atranorin and fumarprotocetraric acid constituents	Kopaonik, Serbia	*B. mycoides* *B. subtilis* *S. aureus* *E. coli* *K. pneumoniae*	780–12,500 780–12,500 780–12,500 1560–25,000 780–6,250	15–62 31.2–62 125–250 125–1,000 30–500	[57]
Crude extracts of *Evernia prunastri* and *Pseudevernia furfuracea* and their evernic acid and physodic acid constituents	Kopaonik, Serbia	*B. mycoides* *B. subtilis* *S. aureus* *E. coli* *K. pneumoniae*	1,560–6,250 3,120–6,250 6,250–12,500 12,500–25,000 1,560–6,250	1.6–250 0.8–250 31.2–500 500 7.5–250	[58]
Crude extracts of *Cladonia furcata*, *Hypogymnia physodes*, and *Umbilicaria polyphylla* and their fumarprotocetraric acid, gyrophoric acid, and physodic acid	Kopaonik, Serbia	*B. mycoides* *B. subtilis* *S. aureus* *E. cloaceae* *E. coli* *K. pneumoniae*	780–3,120 780–6,250 780–6,250 780–3,120 1560–6,250 780–6,250	62–1,000 62–1,000 62–1,000 62–1,000 62–1,000 31–1,000	[59]

(*Continued*)

Table 6.3 Antibacterial activity of crude extracts of lichens and of their bioactive compounds. (*Continued*)

Lichens and their bioactive compounds	Zone of collect of lichens	Sensitive bacteria	MIC values (μg/ml)		References
			Crude extracts	Lichenic acids	
Crude extracts of *Cladia aggregata* and its barbatic acid	Northeastern of Brazil	Four multi drug resistant *S. aureus* strains	ND	50–100	[55]
Methanol extract and usnic acid, usimine A, usimine B, usimine C, and ramalin from *Ramalina terebrata*	Korean Antarctic Research Station site (King Sejong Station) on King George Island	*B. subtilis* *S. aureus*	33.8 ± 2.3 69.1 ± 6.7	1.2 ± 0.1–26.4 ± 2.7 5.6 ± 0.8	[67]
Methanol extract and 6 isolates (Atranorin, (+)-usnic acid, divaric acid, 2,4-di-O-methyldivaric acid, Divaricatinic acid, and 2-O-Methylnordivaricatic acid) from *Evernia divaricata*	Qinghai Province, People's Republic of China	*B. subtilis* *S. aureus* *E. coli* *P. aeruginosa*	ND	7.8–>2,000 15.6–>2,000 250–>2,000 125–>2,000	[66]

(*Continued*)

Table 6.3 Antibacterial activity of crude extracts of lichens and of their bioactive compounds. (*Continued*)

Lichens and their bioactive compounds	Zone of collect of lichens	Sensitive bacteria	MIC values (µg/ml)		References
			Crude extracts	**Lichenic acids**	
Crude extracts of *Lecanora frustulosa* and *Parmeliopsis hyperopta* and their divaricatic acid and zeorin constituents	Kopaonik, Serbia	*B. mycoides*	780–1,560	390–1,560	[60]
		B. subtilis	780–3,120	390–1,560	
		S. aureus	780–1,560	780–1,560	
		E. cloacae	780–1,560	390–780	
		K. pneumoniae	780–1,560	390–780	
Acetone, chloroform, diethyl ether, methanol, and petroleum ether extracts of the lichen *Parmelia sulcata* and its salazinic acid constituent	Bozdağ, Eskişehir Province in Turkey	*B. cereus*	15–62	1,510.8	[61]
		B. subtilis	15–53	2,997.6	
		S. aureus	26–80	2,997.6	
		S. faecalis	15–40	Inactive	
		Aeromonas hydrophila	40–210	2,997.6	
		L. monocytogenes	40–80	Inactive	
		P. vulgaris	120–640	Inactive	
		P. aeruginosa	Inactive	5,995.2	
		S. typhimurium	Inactive	5,995.2	
		Yersinia enterocolitica	60–1,280	Inactive	

(*Continued*)

Table 6.3 Antibacterial activity of crude extracts of lichens and of their bioactive compounds. (*Continued*)

Lichens and their bioactive compounds	Zone of collect of lichens	Sensitive bacteria	MIC values (μg/ml)		References
			Crude extracts	Lichenic acids	
Crude extracts of chemical races of the lichen *Pseudevernia furfuracea* and their physodic acid, chloroatranorin, atranorin, and olivetoric acid constituents	Bozdağ, Eskişehir Province in Turkey	*B. cereus* *B. subtilis* *S. aureus* *S. faecalis* *A. hydrophila* *E. coli* *L. monocytogenes* *P. vulgaris* *S. typhimurium* *Y. enterocolitica*	8.3–453 8.3–113 17–907 19–800 19–907 Inactive 25–6,400 19–800 Inactive 8.3–100	623.5–6,240 623.5–6,240 623.5–25,000 10,000–25,000 2494–3,120 19976 2494–3,120 2494–25,000 19,976 623.5–6,240	[62]
Crude extracts of the lichen *Cladonia foliacea* and its (+)-usnic acid, atranorin, and fumarprotocetraric acid constituents	Eskişehir Province in Turkey	*B. cereus* *B. subtilis* *S. aureus* *S. faecalis* *A. hydrophila* *L. monocytogenes* *P. vulgaris*	1.9–46.8 0.48–7.8 0.73–15.6 0.24–3.9 3.9–46.8 0.12–3.9 3.9–46.8	0.15–31.2 0.61–15.6 2.4–500 0.15–250 1.2–150 0.31–15.6 0.15–37.5	[63]

(*Continued*)

Table 6.3 Antibacterial activity of crude extracts of lichens and of their bioactive compounds. (*Continued*)

Lichens and their bioactive compounds	Zone of collect of lichens	Sensitive bacteria	MIC values (µg/ml)		References
			Crude extracts	Lichenic acids	
Acetone extract of the lichen *Ramalina farinacea* and its (+)-usnic acid, norstictic acid, and protocetraric acid constituents	Bozdağ, Eskişehir Province in Turkey	*B. subtilis* *S. aureus* *S. faecalis* *A. hydrophila* *L. monocytogenes* *P. vulgaris* *Y. enterocolitica*	2.64 132 132 Inactive 132 132 132	12.48–1,250 49.6–2,506.7 25.6–312 312 6.24–156 12.48–625.3 6.24	[64]
Crude extracts of the lichen *Cetraria aculeata* and its protolichesterinic acid constituent	Eskişehir Province in Turkey	*B. cereus* *B. subtilis* *S. aureus* *S. faecalis* *A. hydrophila* *E. coli* *L. monocytogenes* *P. vulgaris* *P. aeruginosa*	2,960–8,460 1,480–8,460 425–1,215 1,480–8,460 212–2,330 850–2,330 741.5–8,460 1,480–8,460 1,480–8,460	Inactive 7,341 Inactive Inactive Inactive 7,341 3,670 Inactive 3670	[65]

(*Continued*)

Table 6.3 Antibacterial activity of crude extracts of lichens and of their bioactive compounds. (*Continued*)

Lichens and their bioactive compounds	Zone of collect of lichens	Sensitive bacteria	MIC values (µg/ml)		References
			Crude extracts	Lichenic acids	
Crude extracts and (-)-usnic acid, physodic acid, 8'-O-ethyl-p-alectoronic acid and alectosarmentin from *Alectoria sarmentosa*	NM	*S. aureus* *Mycobacterium smegmatis*	100 100	25–50 6.25–100	[68]
Crude extracts of 17 lichen species and 5 fractions from 3 species of these lichens (*Stereocaulon alpinum*, *Peltigera aphthosa*, and *Thamnolia subuliformis*): Methyl β-orsellinate, methyl and ethyl orsellinates, chlorocresol, methyl p-hydroxybenzoate, and ethyl and propyl p-hydroxybenzoate	various locations in Iceland	*B. subtilis* *S. aureus* *E. coli* *P. aeruginosa*	ND	100–1,500 80–2,000 200–1,000 30–2,000	[69]

ND, not defined (activity proved by disk diffusion only); NM, not mentioned.

others, *S. typhi* and *E. aerogenes* were particularly sensitive to AE and parietin, respectively (Table 6.3) [56].

Both lichen's crude extracts and their purified components showed a relatively strong antibacterial activity against pathogenic bacteria, but comparatively, the lichenic acids, in most of cases, were more active than the crude extracts. This observation is perfectly illustrated by a large number of studies. For example, researchers from a Serbian team, through a series of studies assessed the effect of crude extracts of nine lichens from Serbia (*Cladonia furcata*, *Cladonia pyxidata*, *Cladonia rangiferina*, *Evernia prunastri*, *Pseudevernia furfuracea*, *Hypogymnia physodes*, *Umbilicaria polyphylla*, *Lecanora frustulosa*, and *Parmeliopsis hyperopta*), in addition to their secondary metabolites (atranorin, fumarprotocetraric acid, evernic acid, physodic acid, gyrophoric acid, divaricatic acid, and zeorin) against an interesting range of pathogenic bacteria (Table 6.3). Results of these studies demonstrated that the lichen's secondary metabolites were more active than crude extracts [57–60]. In contrast, regarding the works of a team testing various Turkish-grown lichens with their purified substances against multiple pathogenic bacteria (Table 6.3), the crude extracts have shown a surprisingly higher activity than the lichenic acids [61–65].

In addition, regarding the divaric acid, Yuan *et al.* (2010) first reported its antibacterial activity. In their study, they assessed the effect of methanol extract and six isolates (Atranorin, usnic acid, divaric acid, 2,4-di-O-methyldivaric acid, divaricatinic acid, and 2-O-Methylnordivaricatic acid) from *Evernia divaricata* against pathogenic bacteria. As a result, they have concluded that crude extracts showed a significantly high inhibition zone against three species (*B. subtilis*, *S. aureus*, and *P. aeruginosa*). Whereas, as a purified compounds, the usnic acid and divaric acid showed very potent inhibitory activity against the above-mentioned bacteria [66].

6.5 Interactions Between Lichen's Secondary Metabolites and Standard Antibiotics

As one of the predominant MDR, *S. aureus* strains are largely incriminated in clinical complications [70–72]. More specifically, amidst the resistant *S. aureus*, the MRSA strains are generally known to be associated with the hospital-acquired infections (HA-MRSA). Indeed, as of 2002, the first report of Vancomycin-Resistant strains of *S. aureus* (VRSA) (due to the acquisition of vanA gene from Vancomycin-Resistant Enterococci [VRE]) has been published in the United States [73]. Since this first discovery, VRSA had spread around the globe [74–76], sparking a worldwide, ever-growing concern considering its potential impact on Health Care System, it should be mentioned that vancomycin is considered the last resort against MRSA infections.

Lately, a considerable effort has been put in the research for a possible interaction between natural molecules and standard antibiotics, indeed such a synergy would boost the activity of the available treatments. In this respect, several studies have revealed a substantial number of natural products which, albeit being unsuccessful as antimicrobials, are able to dramatically enhance the effectiveness of chemotherapeutics against Gram-positive and Gram-negative bacteria [77–82].

Concerning MRSA strains specifically, three studies were recently undergone and can be considered as a very interesting initiative. Hence, in this work, we report these pioneer works that assessed the capacity of different natural purified compounds of lichens, whether alone or in combinations to five available antibiotics (clindamycin, erythromycin, gentamycin, levofloxacin, and oxacillin) against 20 different MRSA clinical isolates.

Firstly, Bellio *et al.* (2015) evaluated five compounds, alone, isolated from five different lichens (collatolic acid [COL] from *Lecanora atra*, epiphorellic acid [EPI] from *Cornicularia epiphorella*, lobaric acid [LOB] from *Stereocaulon alpinum*, perlatolic acid [PER] from *Stereocaulon sp.*, and protolichesterinic acid [PRO] from *Cornicularia aculeata*), and in combination with the above antibiotics. Results of this study shown that the MICs values of lichenic acids were ranging from 4 to 128 μg/ml (Table 6.4), and amidst the tested compounds, the most active were PER and PRO with the same range of MICs values (from 4 to 16 μg/ml). Whereas, regarding the combinations of lichen's secondary metabolites and antibiotics, a synergetic action was observed in combination with gentamicin. While, antagonism was observed for some lichen compounds in combination with levofloxacin. All combinations with erythromycin were indifferent, while variability was observed for clindamycin and oxacillin combinations (Table 6.4) [83].

Secondly, Celenza *et al.* (2012) tested the depsidones pannarin alone from four *Psoroma* species (*P. dimorphum*, *P. pallidum*, *P. pulchrum*, and *P. reticulatum*) and in combination with the five precited antibiotics. As a result, the growth of the 50% of the strains (MIC_{50}) was inhibited by pannarin at a concentration of 4 μg/ml (Table 6.4), while a concentration of 8 μg/ml of pannarin was enough to inhibit the growth of the 90% of the strains (MIC_{90}). Moreover, a moderate synergistic action was observed in combination with gentamicin, antagonism was observed in combination with levofloxacin. All combinations with erythromycin were indifferent,

Table 6.4 Different interactions between lichen's secondary metabolites and standard antibiotics.

Lichen's secondary metabolites/Lichens	Standard antibiotics	Used bacteria	MIC values (µg/mg) of lichenic acids alone	Interpretation of the different interactions	References
Collatolic acid (COL) from *Lecanora atra* Epiphorellic acid (EPI) from *Cornicularia epiphorella* Lobaric acid from (LOB) *Stereocaulon alpinum* Perlatolic acid (PER) from *Stereocaulon* sp. Protolichesterinic acid (PRO) from *Cornicularia aculeata*	Clindamycin Erythromycin Gentamycin Levofloxacin Oxacillin	Twenty Methicillin-Resistant *S. aureus* (MRSA) strains	4–128	SYN (EPI, LOB, PER, PRO) IND (COL) IND (All combinations) SYN (All combinations) ANT (COL-LOB-PER) IND (EPI-PRO) SYN (COL-EPI-LOB-PRO) IND (PER)	[83]

(*Continued*)

Table 6.4 Different interactions between lichen's secondary metabolites and standard antibiotics. (*Continued*)

Lichen's secondary metabolites/Lichens	Standard antibiotics	Used bacteria	MIC values (µg/mg) of lichenic acids alone	Interpretation of the different interactions	References
Usnic acid from *Cladonia lepidophora*	Clindamycin Erythromycin Gentamycin Levofloxacin Oxacillin	Twenty Methicillin-Resistant *S. aureus* (MRSA) strains	1–8	SYN IND SYN ANT SYN	[77]
Pannarin from four *Psoroma* species: *P. dimorphum* *P. pallidum* *P. pulchrum* *P. reticulatum*	Clindamycin Erythromycin Gentamycin Levofloxacin Oxacillin	Twenty Methicillin-Resistant *S. aureus* (MRSA) strains	4–8	IND IND SYN ANT SYN	[78]

ANT, antagonism; SYN, synergy; IND, indifference.

while variability was recorded for clindamycin and oxacillin combinations [78]. Thirdly, and following the same protocol, Segatore *et al.* (2012) assessed the effect of usnic acid from *Cladonia lepidophora*. Consequently, the benzofuran derivative usnic acid alone showed a strong antibacterial activity against all the tested strains with MICs values ranging from 1 to 8 μg/ml. Combination of usnic acid with gentamycin gave synergistic action, while antagonism was observed with levofloxacin, whereas there was no difference with erythromycin. However, variability was reported with clindamycin and oxacillin (Table 6.4) [77].

These works have shown that a combination between the available antibiotics and natural molecules (namely, lichen's secondary metabolites) is a possible approach that might help extend the usefulness of some said antibiotics.

6.6 Future Visions

Lichenic acids are synthetized by one of the oldest and more adaptable organisms and, thusly, could provide us with new templates for the development of more efficient and potent antibacterial agents. The numerous studies mentioned in this review determine that lichens substances can be classed as a promising source of bioactive compounds. However, the main issue related to the limited exploration of these compounds in the modern pharmaceutical industry is affiliated with their slow growth rate. These limits were interestingly reduced with the evolution of the technological tools aiming either to cultivate lichens and their mycobionts *in vitro* [7, 84–86] or to use molecular techniques as an alternative approach for the exploration of the diversity of polyketide biosynthetic pathways in lichens [87].

Indeed, there are many studies that showed promising results concerning the antibacterial properties of lichens, either in their brute forms or as purified molecules. However, the understanding of the modes of action of this biological inhibitory effect is very limited. Presently, there are only two studies that have specifically addressed the mechanism of action of lichen's secondary metabolites. Firstly, Shrestha *et al.* (2016) have reported that the vulpinic acid from lichen *Letharia vulpina* compromises membrane integrity of the MRSA isolates and disrupts cell division processes [88]. Secondly, Maciąg-Dorszyńska *et al.* (2014) have proposed that the usnic acid causes rapid and strong inhibition of RNA and DNA synthesis in Gram-positive bacteria, represented by *B. subtilis* and *S. aureus* [89].

6.7 Conclusion

In conclusion, according to Boustie and Grube 2005, all the reported secondary molecules as antibacterial agents still represent a very limited number among the totality of classified lichens compounds [4]. This, coupled to the promising antibiotic activity of lichen's molecules, led us suggest that there is an urgent need for: (1) continued screening of lichen secondary metabolites through the assessment of new species of lichens worldwide, (2) amplified endeavors for boosting standard antibiotics with various lichenic acids, (3) move to clinical trials for the select substances that have shown notable antibacterial activity for commercial production and implementation of effective new antibiotic agents, and finally (4) strengthen researches that aim to explain mechanisms of action of diverse secondary metabolites of lichens for more comprehension of the mode of action of those bioactive molecules against pathogenic bacteria.

References

1. Shrestha, G. and Clair, L.L., St., Lichens: A promising source of antibiotic and anticancer drugs. *Phytochem. Rev.*, 12, 229–244, 2013.
2. Nguyen, K.H., Chollet-Krugler, M., Gouault, N., Tomasi, S., UV-protectant metabolites from lichens and their symbiotic partners. *Nat. Prod. Rep.*, 30, 1490–508, 2013.
3. Cardinale, M., Puglia, A.M., Grube, M., Molecular analysis of lichen-associated bacterial communities. *FEMS Microbiol. Ecol.*, 57, 484–495, 2006.
4. Boustie, J. and Grube, M., Lichens - a promising source of bioactive secondary metabolites. *Plant Genet. Resour.*, 3, 273–287, 2005.
5. Zambare, V.P. and Christopher, L.P., Biopharmaceutical potential of lichens. *Pharm. Biol.*, 50, 778–798, 2012.
6. Huneck, S., The significance oflichens and their metabolites. *Naturwissenschaften*, 88, 559–570, 1999.
7. Stocker-Wörgötter, E., Metabolic diversity of lichen-forming ascomycetous fungi: Culturing, polyketide and shikimate metabolite production, and PKS genes. *Nat. Prod. Rep.*, 25, 188–200, 2008.
8. Molnár, K. and Farkas, E., Current results on biological activities of lichen secondary metabolites: A review. *Z. Naturforsch. C*, 65, 157–173, 2010.
9. Andersson, D.I. and Hughes, D., Evolution of antibiotic resistance at non-lethal drug concentrations. *Drug. Resist. Updat.*, 15, 162–172, 2012.
10. Sandegren, L., Selection of antibiotic resistance at very low antibiotic concentrations. *UPSALA J. Med. Sci.*, 119, 103–107, 2014.

11. Résistance aux antibiotiques, 2018. Inserm, https://www.inserm.fr/information-en-sante/dossiers-information/resistance-antibiotiques.
12. Gyawali, R. and Ibrahim, S.A., Natural products as antimicrobial agents. *Food Control*, 46, 412–429, 2014.
13. Taylor, P.W., Alternative natural sources for a new generation of antibacterial agents. *Int. J. Antimicrob. Agents*, 42, 195–201, 2013.
14. Iwu, M.M., Duncan, A.R., Okunji, C.O., New antimicrobials of plant origin, in: *Perspectives on new crops and new uses*, J. Janick (Ed.), pp. 457–462, ASHS Press, Alexandria, VA, 1999.
15. Hughes, J.M. and Tenover, F.C., Approaches to limiting emergence of antimicrobial resistance in bacteria in human populations. *Clin. Infect. Dis.*, 24, S131–S135, 1997.
16. Lawrey, J.D., Biological role of lichen substances. *Bryologist*, 89, 111–122, 1986.
17. Ingolfsdottir, K., Hjalmarsdottir, M.A., Sigurdsson, A., Gudjonsdottir, G.A., Brynjolfsdottir, A., Steingrimsson, O., *In vitro* susceptibility of *Helicobacter pylori to* protolichesterinic acid from the lichen *Cetraria islandica*. *Antimicrob. Agents Chemother.*, 41, 215–217, 1997.
18. Burkholder, P.R., Evans, A.W., McVeigh, I., Thornton, H.K., Antibiotic activity of lichens. *Proc. Natl. Acad. Sci. U S A*, 30, 250–255, 1944.
19. Sweidan, A., Chollet-Krugler, M., Sauvager, A., Van de Weghe, P., Chokr, A., Bonnaure-Mallet, M., Tomasi, S., Bousarghin, L., Antibacterial activities of natural lichen compounds against *Streptococcus gordonii* and *Porphyromonas gingivalis*. *Fitoterapia*, 121, 164–169, 2017.
20. Ranković, B. and Mišić, M., The antimicrobial activity of the lichen substances of the lichens *Cladonia Furcata, Ochrolechia Androgyna, Parmelia Caperata*, and *Parmelia Conspresa*. *Biotechnol. Biotec. EQ.*, 22, 1013–1016, 2008.
21. Lauterwein, M., Oethinger, M., Belsner, K., Peters, T., Marre, R., *In vitro* activities of the lichen secondary metabolites vulpinic acid, (+)-usnic acid, and (-)-usnic acid against aerobic and anaerobic microorganisms. *Antimicrob. Agents Chemother.*, 39, 2541–2543, 1995.
22. Santiago, K.A.A., Borricano, J.N.C., Canal, J.N., Marcelo, D.M.A., Perez, M.C.P., Antibacterial activities of fruticose lichens collected from selected sites in Luzon Island, Philippines. *Philipp. Sci. Lett.*, 3, 18–22, 2010.
23. Hoskeri, H.J., Krishna, V., Amruthavalli, C., Effects of extracts from lichen *Ramalina pacifica* against clinically infectious bacteria. *Nat. Sci. J.*, 3, 81–85, 2010.
24. Karagöz, A., Doğruöz, N., Zeybek, Z., Aslan, A., Antibacterial activity of some lichen extracts. *J. Med. Plants Res.*, 3, 1034–1039, 2009.
25. Cansaran, D., Kahya, D., Yurdakulola, E., Atakol, O., Identification and quantitation of usnic acid from the lichen *Usnea* species of Anatolia and antimicrobial activity. *Z. Naturforsch. C*, 61, 773–776, 2006.

26. Cansaran, D., Cetin, D., Halici, M.G., Atakol, O., Determination of usnic acid in some *Rhizoplaca* species from Middle Anatolia and their antimicrobial activities. *Z. Naturforsch. C*, 61, 47–51, 2006.
27. Zambare, V.P., Zambare, A.V., Christopher, L.P., Antioxidant and antibacterial activity of extracts from lichen *Xanthoparmelia somloensis*, native to the Black Hills, South Dakota, USA. *Int. J. Med. Sci. Tec.*, 3, 46–51, 2010.
28. Vidyalakshmi, A. and Kruthika, K., Antibacterial activity of *Parmelia perlata*. Asian. *Pac. J. Trop. Biomed.*, 2, S892–S894, 2012.
29. Zizovic, I., Ivanovic, J., Misic, D., Stamenic, M., Djordjevic, S., Kukic-Markovic, J., Petrovic., S.D., SFE as a superior technique for isolation of extracts with strong antibacterial activities from lichen *Usnea barbata* L. *J. Supercrit. Fluid.*, 72, 7–14, 2012.
30. Shrestha, G., Raphael, J., Leavitt, S.D., Clair, L.L., St., *In vitro* evaluation of the antibacterial activity of extracts from 34 species of North American lichens. *Pharm. Biol.*, 52, 1262–1266, 2014.
31. Srivastava, P., Upreti, D.K., Dhole, T.N., Srivastava, A.K., *In-vitro* evaluation of the antimicrobial activities of lichen. *Int. J. Curr. Microbiol. App. Sci.*, 2, 271–279, 2013.
32. Shrestha, G. and Clair, L., Antimicrobial activity of extracts from two lichens *ramalina menziesii* and *Usnea lapponica*. *Bull. California Lichen Soc.*, 20, 5–10, 2013.
33. Manojlovic, N.T., Vasiljevic, P.J., Maskovic, P.Z., Juskovic, M., Bogdanovic-Dusanovic, G., Chemical composition, antioxidant, and antimicrobial activities of lichen *Umbilicaria cylindrica* (L.) Delise (Umbilicariaceae). *J. Evid. Based Complement. Altern. Med.*, 2012, 1–8, 2012.
34. Gulluce, M., Aslan, A., Sokmen, M., Sahin, F., Adiguzel, A., Agar, G., Sokmen, A., Screening the antioxidant and antimicrobial properties of the lichens *Parmelia saxatilis, Platismatia glauca, Ramalina pollinaria, Ramalina polymorpha* and *Umbilicaria nylanderiana*. *Phytomed.*, 13, 515–521, 2006.
35. Ranković, B.R., Kosanić, M.M., Stanojković, T.P., Antioxidant, antimicrobial and anticancer activity of the lichens *Cladonia furcata, Lecanora atra* and *Lecanora muralis*. *BMC Complement. Altern. Med.*, 11, 1–8, 2011.
36. Swathi, D., Suchitha, Y., Prashith-Kekuda, T.R., Venugopal, T.M., Vinayaka, K.S., Mallikarjun, N., Raghavendra, H.L., Antimicrobial, anthelmintic and insecticidal activity of a macrolichen *Everniastrum cirrhatum* (Fr.) Hale. *Int. J. Drug Dev. Res.*, 2, 780–789, 2010.
37. Ranković, B., Mišić, M., Sukdolak, S., Antimicrobial activity of extracts of the lichens *Cladonia furcata, Parmelia caperata, Parmelia pertusa, Hypogymnia physodes* and *Umbilicaria polyphylla*. *Biologia*, 64, 53–58, 2009.
38. Ranković, B., Mišić, M., Sukdolak, S., Evaluation of antimicrobial activity of the lichens *Lasallia pustulata, Parmelia sulcata, Umbilicaria crustulosa*, and *Umbilicaria cylindrica*. *Microbiol.*, 76, 723–727, 2007.
39. Madamombe, I.T. and Afolayan, A.J., Evaluation of antimicrobial activity of extracts from South African *Usnea barbata*. *Pharm. Biol.*, 41, 199–202, 2003.

40. Gupta, V.K., Darokar, M.P., Saikia, D., Anirbal, P., Fatima, A., Khanuja, S.P.S., Antimycobacterial activity of lichens. *Pharm. Biol.*, 45, 200–204, 2007.
41. Vartia, K.O., Antibiotics in lichens, in: *The Lichens*, V. Ahmadjian and M. E. Hale (eds.), pp. 547–561, Academic Press, New York, 1973.
42. Cocchietto, M., Skert, N., Nimis, P., Sava, G., A review on usnic acid, an interesting natural compound. *Naturwissenschaften*, 89, 137–146, 2002.
43. Safak, B.1., Ciftci, I.H., Ozdemir, M., Kiyildi, N., Cetinkaya, Z., Aktepe., O.C., Altindis, M., Asik, G., *In vitro* anti-*Helicobacter pylori* activity of usnic acid. *Phytother. Res.*, 23, 955–957, 2009.
44. Pompilio, A., Pomponio, S., Di Vincenzo, V., Crocetta, V., Nicoletti, M., Piovano, M., Garbarino, J.A., Di Bonaventura, G., Antimicrobial and antibiofilm activity of secondary metabolites of lichens against Methicillin-Resistant *Staphylococcus aureus* strains from cystic fibrosis patients. *Fut. Microbiol.*, 8, 281–292, 2013.
45. Manojlović, N., Ranković, B., Kosanić, M., Vasiljević, P., Stanojković, T., Chemical composition of three *Parmelia* lichens and antioxidant, antimicrobial and cytotoxic activities of some their major metabolites. *Phytomed.*, 19, 1166–1172, 2012.
46. Ranković, B., Mišić, M., Sukdolak, S., The antimicrobial activity of substances derived from the lichens *Physcia aipolia, Umbilicaria polyphylla, Parmelia caperata* and *Hypogymnia physodes. World J. Microbiol. Biotechnol.*, 24, 1239–1242, 2008.
47. Francolini, I., Norris, P., Piozzi, A., Donelli, G., Stoodley, P., Usnic acid, a natural antimicrobial agent able to inhibit bacterial biofilm formation on polymer surfaces. Antimicrob. *Agents Chemother.*, 48, 4360–4365, 2004.
48. Ingólfsdóttir, K., Chung, G.A., Skúlason, V.G., Gissurarson, S.R., Vilhelmsdóttir, M., Antimycobacterial activity of lichen metabolites *in vitro*. *Eur. J. Pharm. Sci.*, 6, 141–144, 1998.
49. Honda, N.K., Pavan, F.R., Coelho, R.G., de Andrade Leite, S.R., Micheletti, A.C., Lopes, T.I.B., Misutsu, M.Y., Beatriz, A., Brum, R.L., Leite, C.Q.F., Antimycobacterial activity of lichen substances. *Phytomed.*, 17, 328–332, 2010.
50. Kokubun, T., Shiu, W., Gibbons, S., Inhibitory activities of lichen-derived compounds against Methicillin- and Multidrug-Resistant *Staphylococcus aureus. Planta Med.*, 73, 176–179, 2007.
51. Macia, M.D., Rojo-Molinero, E., Oliver, A., Antimicrobial susceptibility testing in biofilm-growing bacteria. *Clin. Microbiol. Infect.*, 20, 981–990, 2014.
52. Bhattarai, H.D., Kim, T., Oh, H., Yim, J.H., A new pseudodepsidone from the Antarctic lichen *Stereocaulon alpinum* and its antioxidant, antibacterial activity. *J. Antibiot.*, 66, 559–561, 2013.
53. Melgarejo, M., Sterner, O., Vila Castro, J., Mollinedo, P., More investigations in potent activity and relationship structure of the lichen antibiotic (+)-usnic acid and its derivate dibenzoylusnic acid. *Rev. Bol. Quim.*, 25, 24–29, 2008.
54. Tozatti, M.G., Ferreira, D.S., Flauzino, L.G., Moraes Tda, S., Martins, C.H., Groppo, M., Andrade e Silva, M.L., Januário, A.H., Pauletti, P.M., Cunhaa,

W.R., Activity of the lichen *Usnea steineri* and its major metabolites against Gram-positive, Multidrug-resistant Bacteria. *Nat. Prod. Commun.*, 11, 493–496, 2016.

55. Martins, M.C.B., de Lima, M.J.G., Silva, F.P., Azevedo-Ximenes, E., da Silva, N.H., Pereira, E.C., *Cladia aggregata* (lichen) from Brazilian northeast: Chemical characterization and antimicrobial activity. *Braz. Arch. Biol. Technol.*, 53, 115–122, 2010.
56. Basile, A., Rigano, D., Loppi, S., Di Santi, A., Nebbioso, A., Sorbo, S., Conte, B., Paoli, L., De Ruberto, F., Molinari, A.M., Altucci, L., Bontempo, P., Antiproliferative, antibacterial and antifungal activity of the lichen *Xanthoria parietina* and Its secondary metabolite parietin. *Int. J. Mol. Sci.*, 16, 7861–7875, 2015.
57. Kosanić, M., Ranković, B., Stanojković, T., Rančić, A., Manojlović, N., Cladonia lichens and their major metabolites as possible natural antioxidant, antimicrobial and anticancer agents. *LWT-Food Sci. Technol.*, 59, 518–525, 2014.
58. Kosanić, M., Manojlović, N., Janković, S., Stanojković, T., Ranković, B., *Evernia prunastri* and *Pseudevernia furfuracea* lichens and their major metabolites as antioxidant, antimicrobial and anticancer agents. *Food Chem. Toxicol.*, 53, 112–118, 2012.
59. Kosanić, M. and Ranković, B., Antioxidant and antimicrobial properties of some lichens and their constituents. *J. Med. Food*, 14, 1624–1630, 2011.
60. Ranković, B., Kosanić, M., Sukdolak, S., Antimicrobial activity of some lichens and their components. *Curr. Contents Clin. Med.*, 279–286, 2010.
61. Candan, M., Yılmaz, M., Tay, T., Erdem, M., Türk, A.Ö., Antimicrobial activity of extracts of the lichen *Parmelia sulcata* and its salazinic acid constituent. *Z. Naturforsch. C*, 62, 619–921, 2007.
62. Türk, H., Yilmaz, M., Tay, T., Türk, A.O., Kivanç, M., Antimicrobial activity of extracts of chemical races of the lichen *Pseudevernia furfuracea* and their physodic acid, chloroatranorin, atranorin, and olivetoric acid constituents. *Z. Naturforsch. C*, 61, 499–507, 2006.
63. Yilmaz, M., Türk, A.O., Tay, T., Kivanç, M., The antimicrobial activity of extracts of the lichen *Cladonia foliacea* and its (-)-usnic acid, atranorin, and fumarprotocetraric acid constituents. *Z. Naturforsch. C*, 59, 249–254, 2004.
64. Tay, T., Türk, A.O., Yilmaz, M., Türk, H., Kivanç, M., Evaluation of the antimicrobial activity of the acetone extract of the lichen *Ramalina farinacea* and its (+)-usnic acid, norstictic acid, and protocetraric acid constituents. *Z. Naturforsch. C*, 59, 384–388, 2004.
65. Türk, A.O., Yilmaz, M., Kivanç, M., Türk, H., The antimicrobial activity of extracts of the lichen *Cetraria aculeata* and its protolichesterinic acid constituent. *Z. Naturforsch. C*, 58, 850–854, 2003.
66. Yuan, C., Zhang, X.J., Du, Y.D., Guo, Y.H., Sun, L.Y., Ren, Q., Zhao, Z.T., Antibacterial compounds and other constituents of *Evernia divaricata* (L.). *Ach. J. Chem. Soc. Pakistan.*, 32, 189–193, 2010.

67. Paudel, B., Bhattarai, H.D., Lee, H.K., Oh, H., Shin, H.W., Yim, J.H., Antibacterial activities of Ramalin, usnic acid and its three derivatives isolated from the Antarctic lichen *Ramalina terebrata*. *Z. Naturforsch. C*, 65, 34–38, 2010.
68. Gollapudi, S.R., Gollapudi, S.R., Telikepalli, H., Jampani, H.B., Mirhom, Y.W., Drake, S.D., Bhattiprolu, K.R., Vander Velde, D., Mitscher, L.A., Alectosarmentin, a new antimicrobial dibenzofuranoid lactol from the lichen, *Alectoria sarmentosa*. *J. Nat. Prod.*, 57, 934–938, 1994.
69. Ingólfsdóttir, K., Bloomfield, S.F., Hylands, P.J., *In vitro* evaluation of the antimicrobial activity of lichen metabolites as potential preservatives. *Antimicrob. Agents Chemother.*, 28, 289–292, 1985.
70. Johnson, A.P., Methicillin-Resistant *Staphylococcus aureus*: the European landscape. *J. Antimicrob. Chemother.*, 66, iv43–iv48, 2011.
71. Ippolito, G., Leone, S., Lauria, F.N., Nicastri, E., Wenzel, R.P., Methicillin-Resistant *Staphylococcus aureus*: the superbug. *Int. J. Infect. Dis.*, 14, S7–S11, 2010.
72. Navarro, M.B., Huttner, B., Harbarth, S., Methicillin-Resistant *Staphylococcus aureus* control in the 21st century: Beyond the acute care hospital. *Curr. Opin. Infect. Dis.*, 21, 372–379, 2008.
73. Limbago, B.M., Kallen, A.J., Zhu, W., Eggers, P., McDougal, L.K., Albrecht, V.S., Report of the 13th Vancomycin-Resistant *Staphylococcus aureus* isolate from the United States. *J. Clin. Microbiol.*, 52, 998–1002, 2014.
74. Hasan, R., Acharjee, M., Noor, R., Prevalence of Vancomycin Resistant *Staphylococcus aureus* (VRSA) in Methicillin Resistant *S. aureus* (MRSA) strains isolated from burn wound infections. *Tzu Chi Med. J.*, 28, 49–53, 2016.
75. Breves, A., Miranda, C.A.C., Flores, C., Filippis, I.D., Clementino, M.M., Methicillin- and Vancomycin-Resistant *Staphylococcus aureus in* health care workers and medical devices. *J. Bras. Patol. Med. Lab.*, 51, 143–152, 2015.
76. Sievert, D.M., Rudrik, J.T., Patel, J.B., McDonald, L.C., Wilkins, M.J., Hageman, J.C., Vancomycin-Resistant *Staphylococcus aureus* in the United States, 2002–2006. *Clin. Infect. Dis.*, 46, 668–674, 2008.
77. Segatore, B., Bellio, P., Setacci, D., Brisdelli, F., Piovano, M., Garbarino, J.A., Nicoletti, M., Amicosante, G., Perilli, M., Celenza, G., *In vitro* interaction of usnic acid in combination with antimicrobial agents against Methicillin-Resistant *Staphylococcus aureus* clinical isolates determined by FICI and ΔE model methods. *Phytomed.*, 19, 341–347, 2012.
78. Celenza, G., Segatore, B., Setacci, D., Bellio, P., Brisdelli, F., Piovano, M., Garbarino, J.A., Nicoletti, M., Perilli, M., Amicosante, G., *In vitro* antimicrobial activity of pannarin alone and in combination with antibiotics against Methicillin-Resistant *Staphylococcus aureus* clinical isolates. *Phytomed.*, 19, 596–602, 2012.
79. Hemaiswarya, S., Kruthiventi, A.K., Doble, M., Synergism between natural products and antibiotics against infectious diseases. *Phytomed.*, 15, 639–652, 2008.

80. Stavri, M., Piddock, L.J.V., Gibbons, S., Bacterial efflux pump inhibitors from natural sources. *J. Antimicrob. Chemother.*, 59, 1247–1260, 2007.
81. Tegos, G., Stermitz, F.R., Lomovskaya, O., Lewis, K., Multidrug pump inhibitors uncover remarkable activity of plant antimicrobials. *Antimicrob. Agents Chemother.*, 46, 3133–3141, 2002.
82. Gibbons, S. and Udo, E.E., The effect of reserpine, a modulator of multidrug efflux pumps, on the *in vitro* activity of tetracycline against clinical isolates of Methicillin Resistant *Staphylococcus aureus* (MRSA) possessing the tet (K) determinant. *Phytother. Res.*, 14, 139–140, 2000.
83. Bellio, P., Segatore, B., Mancini, A., Di Pietro, L., Bottoni, C., Sabatini, A., Brisdelli, F., Piovano, M., Nicoletti, M., Amicosante, G., Perilli, M., Celenza, G., Interaction between lichen secondary metabolites and antibiotics against clinical isolates Methicillin-Resistant *Staphylococcus aureus* strains. *Phytomed.*, 22, 223–230, 2015.
84. Behera, B.C., Adawadkar, B., Makhija, U., Tissue culture of selected species of the Graphis lichen and their biological activities. *Fitoterapia*, 77, 208–215, 2006.
85. Stocker-Wörgötter, E. and Elix, J.A., Secondary chemistry of cultured mycobionts: Formation of a complete chemosyndrome by the lichen fungus of *Lobaria spathulata. Lichenologist*, 34, 351–359, 2002.
86. Stocker-Wörgötter, E., Experimental lichenology and microbiology of lichens: Culture experiments, secondary chemistry of cultured mycobionts, resynthesis, and thallus morphogenesis. *Bryologist*, 104, 576–581, 2001.
87. Miao, V., Coëffet-LeGal, M.F., Brown, D., Sinnemann, S., Donaldson, G., Davies, J., Genetic approaches to harvesting lichen products. *Trends Biotechnol.*, 19, 349–355, 2001.
88. Shrestha, G., Thompson, A., Robison, R., Clair, L.L., St, *Letharia vulpina*, a vulpinic acid containing lichen, targets cell membrane and cell division processes in Methicillin-Resistant *Staphylococcus aureus. Pharm. Biol.*, 54, 413–418, 2016.
89. Maciąg-Dorszyńska, M., Węgrzyn, G., Guzow-Krzemińska, B., Antibacterial activity of lichen secondary metabolite usnic acid is primarily caused by inhibition of RNA and DNA synthesis. *FEMS Microbiol. Lett.*, 353, 57–62, 2014.

7

Lichens: Natural Alternatives in Textiles Finishing

Amrita Kaushik* **and Pavan Prakash Giri**

Department of Chemistry, Noida Institute of Engineering and Technology (NIET), Greater Noida, India

Abstract

Use of natural compounds in textiles industry for coloring and finishing is not an avant-grade practice but it has been a part of ancient practice in textile manufacturing. However, currently used practices in textile words majorly use synthetic products and compounds and everyday piling the environment with tons of hazardous chemicals. The readvent of natural materials for textile functionalization to replace synthetics and harmful products from textile industry is the need of hour. Natural dyes from plants (madder) and animals (lac) have been used since ancient times for textile dyeing. Many other natural dyes have been explored with various functional properties such as antimicrobial, antioxidant, and coloration simultaneously on textile materials. The search of new and eco-friendly products for fabrics finds lichens as a promising yet unconventional option to produce colorants. This chapter gives an insight of lichens producing compounds for textile industry and suggesting the new approaches for futuristic developments in lichen driven products.

Keywords: Lichens, textiles, coloration, functionalization

7.1 Introduction

Tourneforte in 1694 first found and segregated the lichens, and for a long time, early lichenologists assumed that these plants were normal and treated as single living flora entity of earth, not as an assembly of algae, fungi, mosses, and liverworts phases. Later, Schwendener based on his anatomical

**Corresponding author*: amrita27.kaushik@gmail.com

Mohd Yusuf (ed.) Lichen-Derived Products: Extraction and Applications, (179–188) © 2020 Scrivener Publishing LLC

studies declared the symbiotic existence of lichens and explained their colonial partners as two distinct units: fungi and algae [1–3]. Lichen has a mutualistic survival relationship between fungi and photosynthetic organisms: either an algae or a cyanobacteria. The photosynthesis partner is called as the photobiont or photosynthesizer, since it contains chlorophyll and the fungal partner referred to as the mycobiont. Generally, the photobiont surrounded by the growth of mycobiont and fulfill the need of growth for eachother and photobiont reproduce more rapidly than their destruction by fungi; thus, it is evident that, in this symbiotic relationship, fungus is actually parasitically growing on photosynthesis organisms. However, it is also reported that fungi share the metabolism of raw material uptake for algae and harvest the growth of lichens. Lichens are spread all around the world in approximately 20,000 species and more and every type of climatic conditions including the terrestrial world, from the ice-free polar areas to the tropics, from tropical rainforests to those desert areas free of mobile sand dunes. Unlikely, some aquatic lichens are also identified. It has remarkable ability of survival so that they can grow on any surfaces (or substrates) like the natural (such as soil, rock, wood, bone) or the man-made one (bitumen, concrete, glass, canvas, and metal, to give just a few examples). It also identified as exceptionally useful because its symbiotic culture can synthesize over 800 substances, many of them not found elsewhere in nature [3–8].

Since ancient times, lichens have been used as fodder for some animals and human also utilized them not only for food and medicine but also for the production of alcohol. Some species of lichens were also used for tanning and dyeing of various types of textiles [1, 3].

History of Indian subcontinent claims their key contribution of lichens usage and application in various forms. The first record was found as *Shipal* in *Atharveda* (1500 B.C.) where lichen were used as medicine in India. Moreover, the vernacular name *Charila* is widely used in Ayurveda, an ancient system of medicine in India. The Sanskrit synonyms for lichens are *Shailaya* and *ShilaPushp* (*Shila*, rock; *Pushp*, flower). Different species of lichens are used for perfumes, food ingredients, and Ayurvedic or Unani medicines [4–9].

7.2 Lichens as Dyes

Lichens produced some most incredible colors like neon yellow, electric magenta, blue-violet, yellow orange, brown, and a pink that changes to blue in the presence of sunlight. Many of them may not be appear in the fresh lichen, but it converted through the process of its extraction [10].

History of lichen's dyes was started for the color purple which was earlier extracted from particular species of molluscs. A verse from the Bible (Book of Ezekiel, chap. 27, v. 7) concerning the riches of the city of Tyre is usually quoted as the earliest reference to the usage of lichen purple made from *Roccella tinctoria*, and other lichens of the same genus. In past, there is a lot of history for dye extraction from lichens and dyeing of clothes for purple color [11]. The use of lichens as a source of purple dye was well-established in parts of the Mediterranean by Greek and Roman times. The *Roccellaeeae* lichen dyes among all attained great historical and commercial importance around the world. These dyes are known as *Orchella* Moss or Weed in English, as Orseille and as Persis in French and German, respectively. The art of making dye from Roccellae is discontinued during the Middle Ages and the markets of the world until the 17th and 18th centuries lost the dye, and again became an article of international trade. About 1300, Federigo (a Florentine merchant) found good results with the application of urine with some plants. Later, he experimented and gave to his native city a monopoly in the Orcellarii, Ruccellarii, or Rucellae dye industry that existed until the discovery of the Cape Verde Islands. This, or possibly the Spanish term for the plant, "*Orcigilia*", is thought to be the origin of the botanical name *Roccellaceae* [1, 12–15].

7.3 Coloring Components From Lichens

Presence of lichen acids, primary chromogens, provides dyeing properties to lichens. It is reported that the outermost part of the lichen thallus is the primary site for its development along with the upper side of the apothecia, or in general, those parts of the plant best exposed to the air throughout the year. Erythrin and lecanoric acid are coloring substances in the *Roccellaceae*. Acids apparently split to orcin and carbonic acid when treated with ammonia. Orcin yields orcein under the influence of oxygen, that is the coloring principle of orchil [3–8, 12, 14].

All the compounds extracted from lichens (Table 7.1) also have their biological importance in many ways such as antimicrobial, antioxidant, pro-oxidant, anticarcinogenic, cytoprotective, and anti-inflammatory (Figure 7.1), etc. [13–23].

Table 7.1 Chemical structures of some coloring compounds from lichens [14–16, 24, 25].

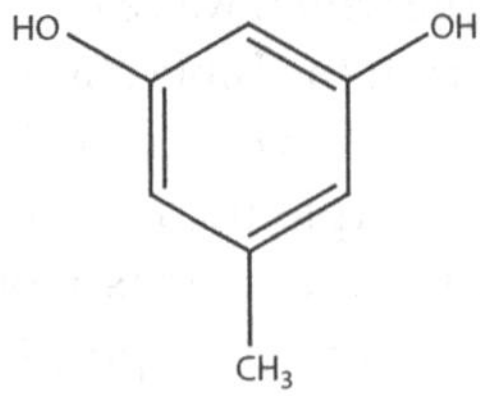 Orcinol Orcinol	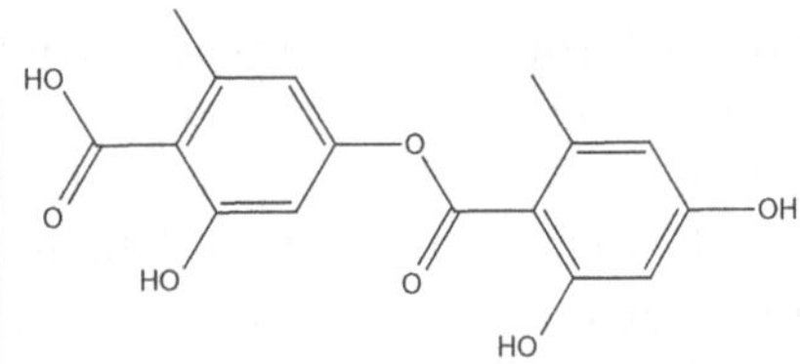 Lecanoric Acid Lecanoric Acid
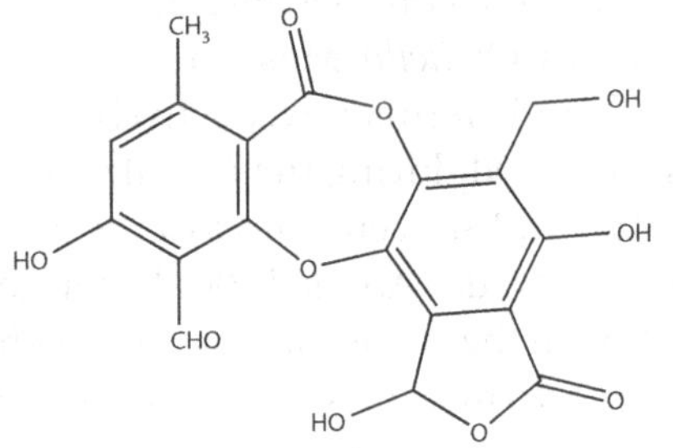 Salazinic acid Salazinic acid	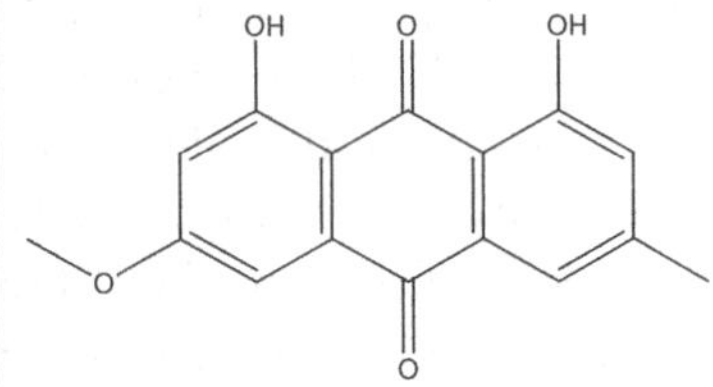 Parietin Parietin
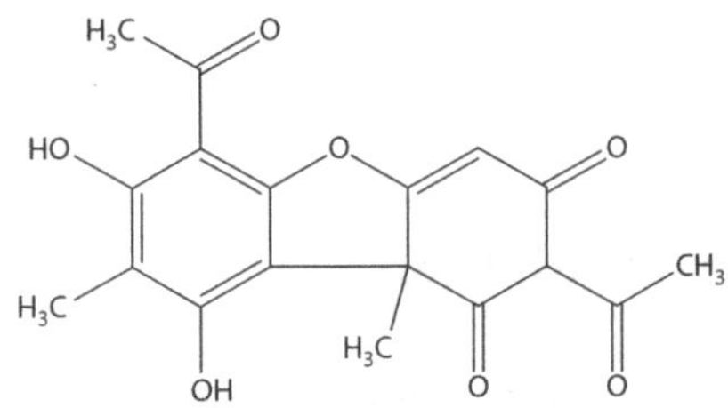 Usnic acid Usnic acid	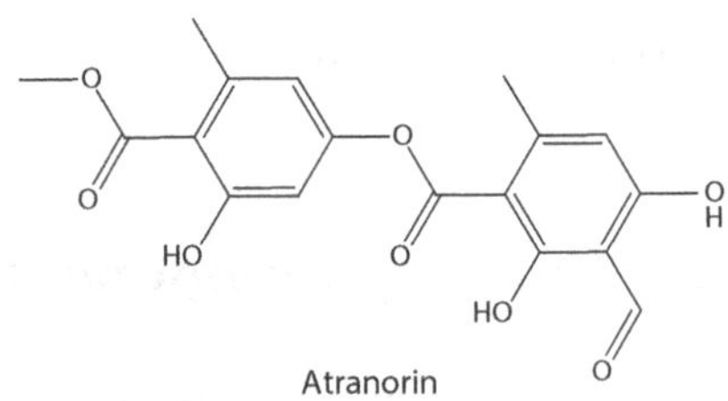 Atranorin Atranorin
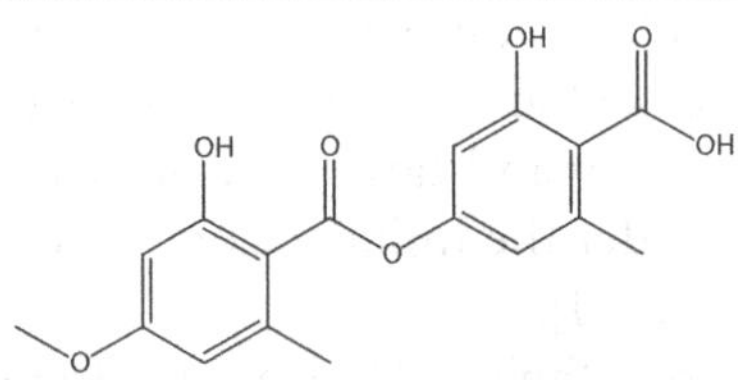 Evernic acid Evernic acid	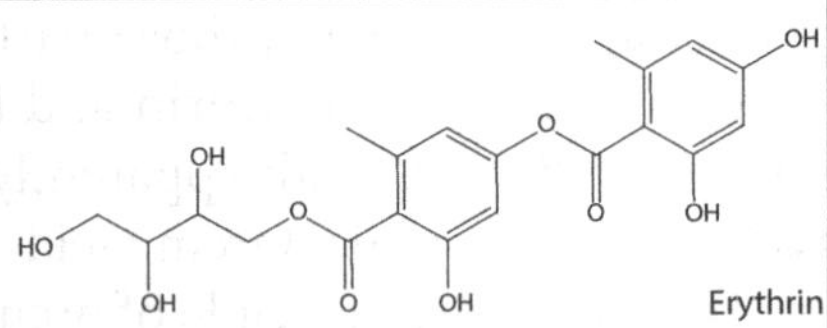 Erythrin Erythrin

(*Continued*)

Table 7.1 Chemical structures of some coloring compounds from lichens [14–16, 24, 25]. (*Continued*)

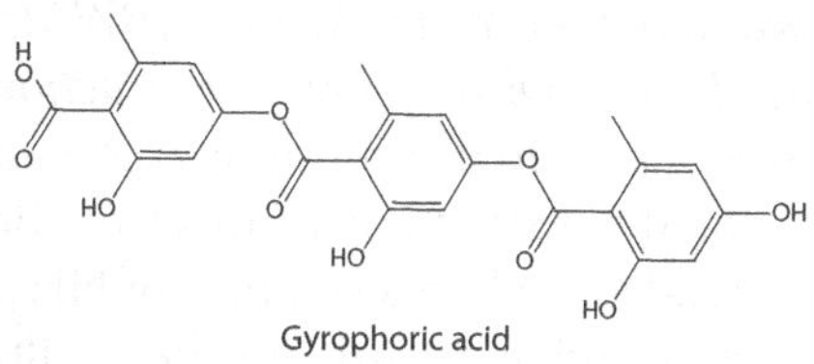 Gyrophoric acid	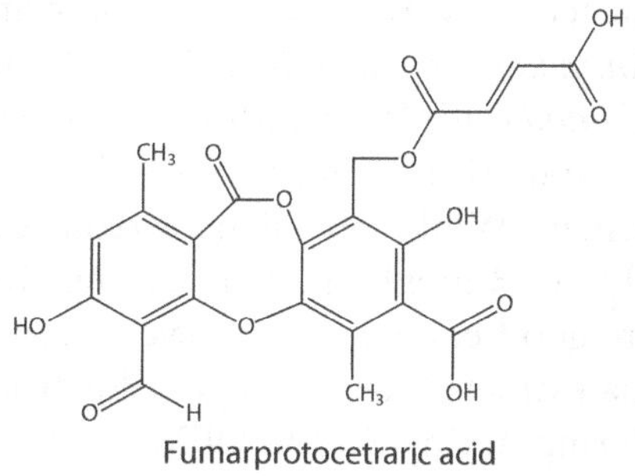 Fumarprotocetraric acid

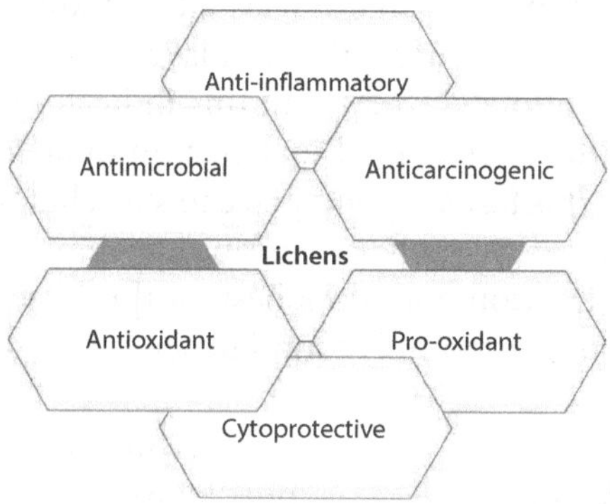

Figure 7.1 Biological activities by lichen compounds.

The most important orchil-yielding substances [3–6] are:

Erythrin: found in most species of *Roccella*, including *R. fuciformis*, *R. montagnei*, and *R. peruensis*.

Lecanoric acid: found in *Roccella tinctoria*, *R. portentosa*, *R. sinuensis*; all *Diploschistes* species; *Parmeliaglabratula*, and several other *Parmelia* species, etc.

Evernic acid: *Evernia prunastri*; *Ramalinapollinaria*.

Gyrophoricacid: found in *Umbilicaria* species; *Ochrolechiatartarea*, *O. frigida*, and *O. androgyna*; *Lobaria pulmonaria*, etc.

7.4 Dye Extraction From Lichens

Lichen dyes are extracted in following ways: *Boiling Water Method (BWM), Ammonia Fermentation Method (AFM), Di-methyl Sulphoxide Extraction Method (DEM)* depending on the presence of certain dye substances. About an hour simmering in water is needed for dye extraction in most of the lichens. While some lichens require an extended soak in a solution of NH_3-H_2O ($\geq$ 3 months) before dyeing. This may sound awkward, but the results are good enough, worth every minute of the wait. A subset of the ammonia extracted dyes change color from pink to blue on exposing to sunlight during drying. In the DEM, it requires 1 month of digestion in di-methyl sulphoxide solution. This method also produced good quality light stable colors. The method of extraction is determined by a simple test called as the C test (C for chlorine). Drop a tiny bit of bleach onto the medulla (a layer of white inner flesh) exposed using fingernail or a knife by gently scraping away a small section of the cortex, the thin layer of skin on the lichen. Red color flashing results into a C+ reaction; this indicates the presence of certain acids that will, after an extended soak in ammonia, transform into some shade of purple dye. This test is a good place to start, but of course with some exceptions. If there is no color change means lichen responds C-, indicating that boiling water extraction may work best for this lichen. Most of the well-known dyes come from commonly found lichens enlisted:

- *Letharia* spp. — neon yellow dye
- *Umbilicaria* spp. — purple dyes
- *Lasallia* spp. — purple dyes
- *Ochroleccia* spp. — purple dyes
- *Flavopunctеliaflavantior* — purple dyes
- *Evernia prunastri* — lilac dye
- *Xanthoria* spp. — solar induced color-changing dyes
- *Teloshistes* spp. — solar induced color-changing dyes

There are many lichens that produce bold reddish-brown dyes in boiling water [7–9, 26].

7.5 Advantages and Limitations of Lichens Extracted Dyes

The primary and undisputed advantage of lichens colors are eco-friendliness. It doesn't introduce any harmful chemicals in to the environment and also safe to use for one and all. Its antibacterial property also another significant feature of lichen dyes. It has been reported also that these dye are also pest and insect repellent, because lichen dyed fiber become tasteless for insects and pests. It also gives a musky smell to the fibers. Moreover, lichen dyed fiber is higher in quality than synthetic dyed one due to its natural substrate use. Lichen dyes do not require any moderator or intermediary agent for color fastening. From mixing of various lichens, we can produce more color options and change the color intensity and stability can be changed with the change of pH value.

However, lichen dyes are not readily available for all due to its high cost; this is because of its long saturation time 1–3 months as per the method requirements. The need for more work on dying technique to use all types of fibers is easily not limited to wool and silk only. Social awareness also plays a vital role in its popularity and market growth along with research funding. Color stability also sometimes act as a negative features for textile functionalization.

7.6 Conclusion and Future Scope of Lichen Dyes

Nowadays, many dyers prefer the natural dyes over synthetic one, due to gradual growth of market requirement of naturally dyed fibers. The use of natural dye made textile material exclusive and nature friendly. In addition to that lichen dyes are antibacterial and anti-inflammatory due to certain antibiotic compounds; the further study and extraction of such natural antibiotic compounds add a new dimension to textile industry and anti-inflammatory application creates a new possibility of naturopathy use such fibers. The uses of lichen dyes become more prominent after reduction in its extraction duration and cost. Maybe in future, we have vibrating color fast dye with lichens.

References

1. Perez-Llano, G.A., Lichens their biological and economic significance. *Bot. Rev.*, 10, 1, 1–65, 1944.
2. Yusuf, M., Shabbir, M., Mohammad, F., Natural colorants: Historical, processing and sustainable prospects. *Nat. Prod. Bioprospect.*, 7, 1, 123–145, 2017.
3. Molnár, K. and Farkas, E., Current results on biological activities of lichen secondary metabolites: A review. *Z. Naturforsch. C.*, 65, 3–4, 157–173, 2010.

4. Bačkorová, M., Jendželovský, R., Kello, M., Bačkor, M., Mikeš, J., Fedoročko, P., Lichen secondary metabolites are responsible for induction of apoptosis in HT-29 and A2780 human cancer cell lines. *Toxicol. In Vitro*, 26, 3, 462–468, 2012.
5. Muggia, L., Lichens as treasure chests of natural products. *Sim News*, 59, 3, 85–97, 2009.
6. Reddy, R.G., Veeraval, L., Maitra, S., Chollet-Krugler, M., Tomasi, S., Lohezic-Le Devehat, F., Chakravarty, S., Lichen-derived compounds show potential for central nervous system therapeutics. *Phytomedicine.*, 23, 12, 1527–1534, 2016.
7. Mitrović, T., Stamenković, S., Cvetković, V., Nikolić, M., Tošić, S., Stojičić, D., Lichens as source of versatile bioactive compounds. *Biol. Nyssana.*, 2, 1, 2011.
8. Australian National Botanic Gardens. https://www.anbg.gov.au/lichen/what-is-lichen.html (Retrieved on 08-07-2019).
9. Upreti, D.K., Divakar, P.K., Nayaka, S., Commercial and ethnic use of lichens in India. *Econom. Bot.*, 59, 3, 269–273, 2005.
10. Allen, A., Getting started with Lichens Dyes. https://www.fungimag.com/summer-2014-articles/LR2%20V7I2%2066-69%20Dies.pdf (Extracted on 01-08-2019).
11. Kok, A., A short history of the orchil dyes. *The Lichenologist*, 3, 2, 248–272, 1966.
12. Brisdelli, F., Perilli, M., Sellitri, D., Piovano, M., Garbarino, J.A., Nicoletti, M., Celenza, G., Cytotoxic activity and antioxidant capacity of purified lichen metabolites: An *in vitro* study. *Phytother. Res.*, 27, 3, 431–437, 2013.
13. Studzińska-Sroka, E. and Dubino, A., Lichens as a source of chemical compounds with anti-inflammatory activity. *Herba Pol.*, 64, 1, 56–64, 2018.
14. Comini, L.R., Vieyra, F.E.M., Mignone, R.A., Páez, P.L., Mugas, M.L., Konigheim, B.S., Cabrera, J.L., Montoya, S.C.N., Borsarelli, C.D., Parietin: An efficient photo-screening pigment *in vivo* with good photosensitizing and photodynamic antibacterial effects *in vitro*. Photochem. *Photobiol. Sci.*, 16, 2, 201–210, 2017.
15. Ristic, S., Rankovic, B., Stamenkovic, S., Biopharmaceutical potential of two Ramalina lichens and their metabolites. *Curr. Pharmaceut. Biotechnol.*, 17, 7, 651–658, 2016.
16. Melo, M.G.D., dos Santos, J.P.A., Serafini, M.R., Caregnato, F.F., de BittencourtPasquali, M.A., Rabelo, T.K., da Rocha, R.F., Quintans, L., Jr., de Souza Araújo, A.A., da Silva, F.A., Moreira, J.C.F., Redox properties and cytoprotective actions of atranorin, a lichen secondary metabolite. *Toxicol. In Vitro*, 25, 2, 462–468, 2011.
17. Odabasoglu, F., Cakir, A., Suleyman, H., Aslan, A., Bayir, Y., Halici, M., Kazaz, C., Gastroprotective and antioxidant effects of usnic acid on indomethacin-induced gastric ulcer in rats. *J. Ethnopharmacol.*, 103, 1, 59–65, 2006.
18. de Paz, G.A., Raggio, J., Gómez-Serranillos, M.P., Palomino, O.M., González-Burgos, E., Carretero, M.E., Crespo, A., HPLC isolation of antioxidant

constituents from Xanthoparmelia spp. *J. Pharmaceut. Biomed. Anal.*, 53, 2, 165–171, 2010.

19. Brisdelli, F., Perilli, M., Sellitri, D., Piovano, M., Garbarino, J.A., Nicoletti, M., Bozzi, A., Amicosante, G., Celenza, G., Cytotoxic activity and antioxidant capacity of purified lichen metabolites: An *in vitro* study. *Phytother. Res.*, 27, 3, 431–437, 2013.
20. Marante, F.T., Castellano, A.G., Rosas, F.E., Aguiar, J.Q., Barrera, J.B., Identification and quantitation of allelochemicals from the lichen Lethariellacanariensis: Phytotoxicity and antioxidative activity. *J. Chem. Ecol.*, 29, 9, 2049–2071, 2003.
21. Bačkorová, M., Bačkor, M., Mikeš, J., Jendželovský, R., Fedoročko, P., Variable responses of different human cancer cells to the lichen compounds parietin, atranorin, usnic acid and gyrophoric acid. *Toxicol. In Vitro*, 25, 1, 37–44, 2011.
22. Kosanić, M., Ranković, B., Stanojković, T., Rančić, A., Manojlović, N., Cladonia lichens and their major metabolites as possible natural antioxidant, antimicrobial and anticancer agents. *LWT-Food Sci. Technol.*, 59, 1, 518–525, 2014.
23. Barreto, R.S., Albuquerque-Júnior, R.L., Pereira-Filho, R.N., Quintans, J.S., Barreto, A.S., DeSantana, J.M., Santana-Filho, V.J., Santos, M.R., Bonjardim, L.R., Araújo, A.A., Quintans-Júnior, L.J., Evaluation of wound healing activity of atranorin, a lichen secondary metabolite, on rodents. *RevistaBrasileira de Farmacognosia*, 23, 2, 310–319, 2013.
24. Shaheen, S., Iqbal, Z., Hussain, M., First Report Of Dye Yielding Potential And Compounds Of Lichens; A Cultural Heritage Of Himalayan Communities, Pakistan. *Pak. J. Bot*, 51, 1, 341–360, 2019.
25. White, P., Oliveira, R., Oliveira, A., Serafini, M., Araújo, A., Gelain, D., Moreira, J., Almeida, J., Quintans, J., Quintans-Junior, L., Santos, M., Antioxidant activity and mechanisms of action of natural compounds isolated from lichens: A systematic review. *Molecules*, 19, 9, 14496–14527, 2014.
26. Allen, A., Getting started with lichen dyes. *Fungi*, 7, 2–3, 2014.

8

A Review on Trends and Opportunity in Edible Lichens

Mohd Yusuf

Department of Chemistry, YMD College, M. D. University, Nuh, Haryana, India

Abstract

Lichens are special composition in which symbiotic relation occurs, consisting of a solo fungal, usually an ascomycete, and an algal ally that take part in photosynthesis process. They have been well known for their role as traditional medicines due to active substances present within. From very long time, lichens have served in many kinds such as coloration of textiles, disease management, and a few species were recognized as food. These valuable natural resources are being used as medicines, dyes, perfume, food, fodder, spice, and for miscellaneous purposes. Besides, some lichens are choice of attention for use in food ingredient, mainly in spices and delicious dishes preparation. However, there are few studies have been conducted on the edibility of lichens; therefore, more work is needed for understanding. According to literature studies, the chapter highlights the utilization of edible lichens and their future dimension as well as opportunities.

Keywords: Lichen, edible lichen, spice, nutrition

8.1 Introduction

All living organisms required proper nutrition, one of the most important factors to survive. Ancient people around the world had feed mainly uncooked food and lived dependent on a raw food diet. They ate raw vegetables, fruits, roots, and meat in general way which were full of nutrition values. On-time lapsing, people have used the plan of obtaining additional nutrition from the natural environment onto many food options which

Email: yusuf1020@gmail.com

Mohd Yusuf (ed.) Lichen-Derived Products: Extraction and Applications, (189–202) © 2020 Scrivener Publishing LLC

can be easily available to them including cooked food and additional natural diversity of plants/insects/animals [1]. Generally, lichen-derived natural colorants, especially violet and purple coloring constituents have been widely utilized during the period of Middle Ages in European countries [2]. In addition, at that time, it went on to become one of the unique dyeing tradition for the Mediterranean and European cultures [2, 3]. In European countries, the regular crops were very badly destroyed by nature's drought condition during the mid of the eighteenth century. This situation discovered new alternative foods to survive. Lichen species made a better option to the quest of hunger by eating. Soon, the lichens for food have gain popularity among troubled peoples, due to their easy availability, cheapness, and high nutritive values. But, some species contain toxic substances which made severe health issues. Furthermore, the lack of conventional food during wars and nature's disasters emphasized in searching new opportunities for survival, most common are edible plant species. In the pre-war period, most people had almost no knowledge about the edible wild vegetation. Alternative ways were developed by them for nutrition and their survival. However, they learned fast during the war period [4].

Lichens, slow-growing species, structurally unique are traditionally considered as mutualisms between fungi and photoautotrophic algal species containing many phytoconstituents and secondary metabolites. Recent studies revealed to the significant biological potential of lichen-derived metabolites possessing active functionalities, for example, antifungal, antibacterial, antitumor, antiviral, antioxidant, and anti-inflammatory [4–10]. However, many lichens have much nutritional value and can be used as safe to eat, but some species contain significant quantities of certain toxic chemicals, for example, vulpinic acid (Figure 8.1) in *Letharia vulpine* and *Bryoria tortuosa*, associated with serious health effects to human [4–6].

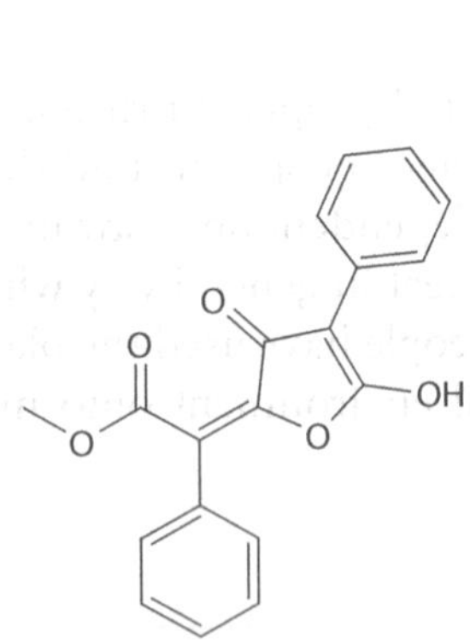

Figure 8.1 Vulpinic acid.

The chapter encompasses the literature basis exploration on edible perspectives of various lichen species with their future outlook.

8.2 Feed and Fodder

Many cultures across the world had been included lichens in their diet. Although several lichen species contain much nutritive value and were only used as emergency food mainly in times of famine. Also, certain species specified themselves as a staple food as well as a delicacy on the special occasion. There are rare documents showing lichen's feeding frequently. When we think about lichens feeding, two obstacles are often encountered: firstly, lichen have sufficient amount of polysaccharides (generally indigestible to humans), and secondly, they may contain mildly toxic phytochemicals/secondary metabolites that must be removed prior to feed [11]. According to some studies, lichens have a good source of nutrients like polysaccharides, dietetics, and medicines [12, 13]. Many yellow appearing lichen species are found to be toxic as they contain high percentages of toxic substances like vulpinic acid or usnic acid (Figure 8.2), and so, cannot be utilized as food [14]. There are certain cases where people would rather die from chronic hunger than eat a wild plant, animal, or lichen during the world war [15]. Thus, like edible parts of many plants and mushrooms, edible lichens could be regarded as suitable for use as feed and fodder [4, 14], only after necessary processing and detoxification. There are limited lichen genera containing certain species which are utilized as feed and fodder. Some of which are discussed under.

8.2.1 *Cetraria islandica*

Cetraria islandica, best recognized in *Parmeliaceae* family, commonly known as "*Iceland's moss*", is distributed frequently from Canada to Alaska; south to the Pacific Northwest, New England, Colorado, Wyoming, the Great Lakes states, and in the alpine regions from Appalachians to

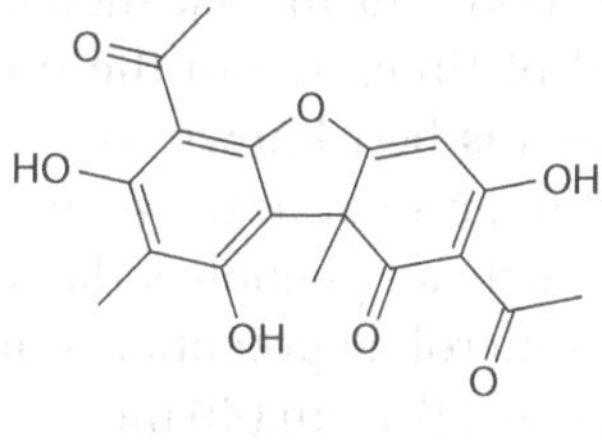

Figure 8.2 Usnic acid.

Tennessee [16]. *Cetraria islandica* is a small to medium having brown colored fruticose thallus type with tufted, lobed, and irregularly branched in appearance that grow loosely on the soil [17]. Lapps and Ice-landers were used *C. islandica* as an emergency food after getting rid of the harmful contents by boiling or by adding to the solution containing sodium carbonate base. Scandinavians have been used as a folk medicine to treat pulmonary disorders [18]. In Northern Europe, *C. islandica* was reported as an important source of food for humans. The people of Northern Europe had been cooked this lichen species as a porridge, pudding, bread, soup, or directly as salad. Additionally, the reindeer, also known as the caribou in North America, is a species of deer and native to the tundra, boreal, Arctic, sub-Arctic, and mountainous regions of Siberia, North America, and Northern Europe. *C. islandica* serves as food source in the winter diet of caribou [19, 20]. A number of biologically active constituents were isolated in *C. islandica* lichen, for example, phenolics, diterpenoids, depsides, and other benzenoid, as well as non-benzenoid derivatives [16–20]. However, it is used to make folk medicines and in a few traditional dishes in the seventeenth century [19]. According to the Finnish Food Authority, the Iceland moss (*C. islandica*) refers to be safe and may be used in Finland in tea [21]. Further, *C. islandica* was also noticed to use as spice as well as food decorative ingredient among few Finnish community [21].

8.2.2 *Roccella montagnei*

Roccella montagnei, a lichenized genus of fungi-algal symbiont, comes under the family *Roccellaceae*, containing fruticose-type thallus with gray-brownish and flat or terete branched. Generally, *Roccella* grows on vertical, overhanging rocks and cliffs or on trees and shrubs and primarily distributed along with the marine, arid, subtropical coastal areas with a Mediterranean type of climate, and widely found in the Mediterranean areas, the central Chilean coast, Mexican coast, Southern California, and Southern Africa [22]. It is most commonly distributed in Coromandel Coast, Tamil Nadu, and Southeast Coast of India [23]. A few lichens have been widely used in the food and in folk medicine in several countries over an extremely period of time, due to the presence of inherent toxic substances [17]. This lichen is best known for the production of derived dyestuffs like litmus, archil, and cudbear. Many active components were identified in this lichen such as phenolics, lactones, depsides, erythrin etc., found significant antimicrobial potential, along with the considerable quantity of carotenoids, chiefly β-form (40 mg per 100 gm approx.) made a good fodder with the adequate supply of vitamin-A [16, 20, 22].

8.2.3 *Cladonia rangiferina*

Cladonia rangiferina synonym to *C. rangiferinus*, commonly known as reindeer lichen/deer moss/caribou moss/gray reindeer moss, grows in both hot and cold climates well and is light-colored and fruticose in appearance belonging to the family *Cladoniaceae*. It has also been reported as an important food for reindeer (caribou) [18]. The lichen grows on the ground surface as well as rocks simultaneously from high mountains to low and widely distributed in southern Japan, Finland, North America, Siberia, Himalayas of India, and Nepal [24]. Biologically, active constituents were observed in *C. rangiferina* lichen, for example, phenolics, diterpenoids, depsides, labdane, abietane, and furan derivatives [16, 20–22].

Traditionally, peoples who lived in north in North America and Siberia were used to eat with great fervor the partially digested *C. rangiferina* obtained from the rumen of reindeer that had already killed [25]. Also, certain peoples in northern countries such as Finland, Norway, and Sweden ate the *C. rangiferina* lichen with a systematic recipe. For example, firstly, the lichen was boiled with rye grain, followed by the mixture rinsed with clean water, dried in oven or furnace, then the mixture ground into a fine flour to bake bread finally [20, 26]. Recently, certain species of lichens may be used safely in food supplements, beverages, and tea-like products as well as for seasoning and decoration [21].

8.2.4 *Bryoria fremontii*

The genus *Bryoria* contains almost 23 species along with *Bryoria fremontii*, commonly known as "*wila*", dark brown, hanging, hair-like lichen belongs to the family *Parmeliaceae*. All the species possess similar physical appearance, and so, it is very difficult to distinguish significantly among the different species of the genus *Bryoria*. The simplest characteristic that can be distinguished to some extent is "*wila*" from the other species of *Bryoria*. On the basis of the fact, the main branches of the genus grow to be thickened (usually <0.4 mm wide, somewhat flattened, twisted, and wrinkled in the older ones), while other species generally have narrower the chief branches [27, 28].

The frequent growing areas for *wila* are western North America, northern Europe, Russia, and the interior of British Columbia (Albertan Rockies, Montana, Idaho, and Wyoming, Washington, Oregon, and California). It is an important traditional food for many First Nations in North America [27, 29]. In Europe, literature dealing with the edible black tree lichen *Alectoria jubata*, ethnographic cites as *B. fremontii* by Indian natives was sometimes employed as an alternative fodder for cattle and other pet mammals. *B. fremontii* has been used as an important source of food for Indian natives of the western North America [30].

Interestingly, edible *B. fremontii* has been utilized for food, but the similar specimen *B. tortuosa* was found inedible by lichenologists depending on the concentration of vulpinic acid (a toxic content) [22, 30, 31]. Prof. T. Spribille, lichenologist and microbiologist, and Prof. J. P. McCutcheon, from University of Montana, analyzed DNA of both the edible *B. fremontii* and the *inedible B. tortuosa* specimens and confirmed that both species had the same algal and fungal components [9, 32]. In addition, both *B. fremontii* and *B. tortuosa* showed morphologically very distinct observation. At present, *wila* is not eaten in large scale, but some people communicate and serve this traditional food as a special feast. The people often eat cooked *B. fremontii* thallus with berries, fish eggs, or directly by making soup with other edible ingredients and also boiled in water to make porridge [33].

8.2.5 *Parmelia perlata*

Parmelia perlata, synonym to *Parmotrema perlatum*, locally known as black stone flower/include *shaileyam* in Sanskrit, *kalpasi* in Tamil, *Chadila*. etc., in India, belongs to the family *Parmeliaceae* and is a foliose species of lichen used as a spice in India that provides a distinct as well as unmistakable flavor [34]. *P. perlata* grows chiefly on trees/trunks and rocks and distributed in the temperate Himalayas and Northeast India, Eastern Nepal, West of Britain, and tropical regions. The thallus of this lichen is large, light-gray to pale-blue patch-shaped with rounded lobes and often have black hairs at the edges [3, 6, 34]. *Parmelia* lichens are important feed source for the certain caterpillars of *Lepidoptera*, for example, bagworm moth—*Taleporia tubulosa* [35, 36]. The lichen *P. perlata* was found to contain several unique phytochemicals in it such as, phenolics, phenolic acids, and others, i.e., atronin, lecaronic acid, salazinic acid, etc. [30–35] responsible for the traditional uses in several countries of Asia, US, and UK, Africa [34–38] as folk medicines prescribed in bronchitis, antimicrobial, antiviral, antioxidant, analgesic, antipyretic, vomiting, heart diseases, skin diseases, wound, stomach disorders, bleeding piles, scabies, leprosy, and many more [28, 33–38].

8.2.6 *Umbilicaria Species*

Rock tripe, which grows mainly on rocks, is the common name for various lichen species of the genus *Umbilicaria* of the family *Umbilicariaceae*.

These lichens are distributed throughout northern parts of North America like New England, Asia, Japan, and Korea [36, 39]. Many *Umbilicaria species* were documented to the use of folk medicine and food [31, 40]. For example, *Umbilicaria esculenta* is commonly used as a food and restorative medicine in different parts of India, Nepal, China, etc. Moreover, *U. esculenta* and *U. crustulosa* were also used to make in a variety of traditional as well as occasional food by the peoples of Asian, Korean, and Japanese origin [31, 39, 41]. The specimens within the genera *Umbilicaria* as well as *Actinogyra* were found a particular eventually food in emergency, known to Cree, Chipewyan, and other Indigenous Indians and to the Inuit peoples as a *fall back* in times of famine but considered as the secondary food system [42].

8.2.7 *Usnea florida*

Usnea genus endowed with numerous species of fruticose-type lichens is candidly hard to distinguish from one another. The genus is generally called as beard lichens due to their filamentous and tufted, appearance, and flaccid nature. The most abundant species of *Usnea* lichens is *U. florida* which is safe to eat and edible, distributed in Southern and western British Isles containing sufficient nutrients with high rated vitamin C [43]. *Usnea* lichens are found very sensitive to atmospheric pollution especially sulphur dioxide (SO_2) and, therefore, grow properly only in unpolluted humid areas, for example in the woodlands of hills and their outskirt zones. Many of *Usnea species* have been used to treat various diseases as traditional therapies that contain potent bioactive substances, best known is usnic acid ($C_{18}H_{16}O_7$), antibiotic, and antifungal agent [44, 45].

8.3 Lichens as Food Additives

8.3.1 Lichen-Based Tea

Usnea florida, which appears gray-green throughout, forms a bush on a tree with thin branches. The branches are flat and leaf-like that produce spore-discs with about 10–20 cm in length. It is commercially available in the form of raw, powder, capsule, or tincture extract. The diluted tincture in water may be taken orally to treat strong bacterial and fungal infections such as colds, influenza, respiratory infections, strep throat, sore throats, and gastrointestinal disorders [43–45]. In the current decade, *U. florida* is used to make a healthy and hygienic tea mixture [46, 47]. The dried tissues of *U. florida* is used for health benefits and bear a germ-killing effect and health-wise. Amazon is an online distribution company that sold lichen tea in recent time under the name "*Usnea Lichen Moss Tea*" [48].

8.3.2 Lichens as Spice

Parmelia perlata is locally known as black stone flower, belongs to the family *Parmeliaceae*, and is a foliose species of lichen used as spice in India that provides a distinct as well as unmistakable flavor [34, 36]. In southern and northern India, the special fragrance of *P. perlata* typically used not only in the preparation of several high-quality meat/fish dishes like nahari, Bombay/Hyderabadi biryani, and stews, but also is used in vegetarian dishes like shahi paneer, curry, etc. [49, 50]. In addition, Rai and Limbu communities of east Nepal have been used *P. perlata* as a spice, a dye, as well as bio-indicator for the inference of heavy metal pollutants [34, 51].

8.3.3 Food Coloring Aspects

Around the globe, Europeans and Romans have used lichens as an important source of dyeing for orchil, violet and purple pigments obtained chiefly from *Roccella spp*. Moreover, in the Canary Islands during the eighteenth century, the dyestuffs derived from lichens were become economically important [52]. Also, tribal people of Southwest United States and chemists still make use of litmus paper (dye from lichen extract soaked on paper) prepared from extracts of *Roccella spp*. which is an important indicator for pH value estimation [2, 22]. Shukla *et al.* performed a detailed study on several lichen species. They reported a number of Indian lichens chiefly counted from different localities and places such as Tamil Nadu, Maharashtra, Uttarakhand, and Jammu and Kashmir. They observed that various lichen species contain a range of colored compounds such as phenolics, xanthones, carotenoids, etc., bearing purple, orange, brown, pink, blue, green, and yellow colorants [53].

According to the Food and Drugs Administration of United States, nature-derived products possessing flavoring as well as adjuvants properties may be used safely as a food ingredient. The ague is acceptable only on the following two major criteria:

(a) if they follow all the minimum quantity standards as well as all the principles of good manufacturing practice to produce their intended physical/technical efficacy;

(b) if they consist of one or more combination, may be employed as food additive in the most convenient aspects such as various plant parts, oils and their extracts, concentrates, absolutes, resins, balsams, gums, waxes, etc., with flavoring and medicinal abilities, and so, generally granted as safe.

Evernia prunastri, commonly known as oak moss, is a species of lichen is proved as safe by USFDA (Food and Drugs Administration of United States) [54]. However, the use of lichen-derived colorants is being used for several substrates from natural like cotton, silk, wool to synthetic ones such as nylon [55].

8.4 Lichens as Snuffing Agents

Snuffing is a process known for intake a powdered substance through the nose. A best known example is smokeless tobacco powder made from ground or pulverized and dried tobacco leaves. Three species of *Parmelia* genus are found in India, *P. perlata*, *P. perofrata*, and *P. sanacti-angelii* which possess astringent, resolvent, laxative, and carminative properties. In this regard, these species are presumed to be effective in the treatment of various diseases and ailments such as amenorrhoea, dyspepsia, piles, scabies, leprosy, excessive salivation, spermatorrhoea, wounds, and throat, stomach, and blood circulatory disorders [6, 49, 52, 53]. The powdered species also observed to have cephalic snuff ability to treat headache [56]. Another lichen's (*Parmotrema nilgherrense*) in whole part is used as cephalic snuff as a liniment for headaches [49]. Moreover, the powdered *Pyrenocarpus* lichen that grows on trees possesses psychoactive nature and thus is used as an alternative to tobacco snuff [57].

8.5 Conclusion and Future Outlook

Lichens, a unique organism, offer the widest diversity of natural products responsible for the different pharmacological activities, such as antimicrobial, anticancer, antioxidant, and anti-infammatory, and many more biological activities. *Roccella* lichen is best known for the production of derived dyestuffs like litmus, archil, and cudbear. Fortunately, there are indigenous and tribal peoples practicing lichens as their food source particularly in countries of Asia, US, UK, and Europe. Anciently, many lichen species were used as folk medicine and emergency food options. However, many lichens have much nutritional value and can be used as safe to eat, but some species contain significant quantities of certain toxic chemicals, such as vulpinic acid in *Letharia vulpine* and *Bryoria tortuosa*, associated with serious health concerns to human. In 2019, USFDA made directive on the use of flavoring and medicinal ingredients of natural origin in which these are referred to be safe products, which may be utilized in food as additives in accordance with safety regulations [54]. And thus, *Evernia prunastri*, a species of lichen is proved as safe. Therefore, based on the literature documented, the safety of lichens when eaten is a major topic still needed to be explored. Also, more R&D is required onto slow-growth rate of suitable lichens so that they can be made economically viable for industrial-scale exploitation.

References

1. Hanley, N., Willis, K., Powe, N., Anderson, M., Valuing the benefits of biodiversity in forests, in: *Report to the Forestry Commission*, Centre for Research in Environmental Appraisal and Management (CREAM), University of Newcastle, UK, 2002.
2. Raisanen, R., Dyes from lichens and mushrooms, in: *Handbook of natural colorants*, T. Bechtold and R. Mussak (Eds.), pp. 183–200, Wiley, UK, 2009.
3. Rather, L.J., Jameel, S., Ganie, S.A., Bhat, K.A., Lichen Derived Natural Colorants: History, Extraction, and Applications, in: *Handbook of Renewable Materials for Coloration and Finishing*, M. Yusuf (Ed.), pp. 103–114, Scrivener Publishing LLC, Beverly, USA, 2018.
4. Redzic, S., The Nutrition of People in the War Using Wild Edible Plants. *Coll. Antropol.*, 34, 2, 551–570, 2010.
5. Huneck, S., New Results on the Chemistry of Lichen Substances, in: *Progress in the chemistry of organic natural products*, W. Herz, H. Falk, G.W. Kirby, R.E. Moore (Eds.), pp. 81–95, Springer, Wien, New York, 2001.
6. Baron, G., *Understanding Lichens: A Beginner's Guide Lichens*, Richmond Publishing Co., Slough, UK, 1999.
7. Ekman, S. and Tønsberg, T., Most species of Lepraria and Leproloma form a monophyletic group closely related to Stereocaulon. *Mycol. Res.*, 106, 1262–1276, 2002.
8. Smith, C.W., Aptroot, A., Coppins, B.J., Fletcher, A., Gilbert, O.L., James, P.W., Wolseley, P.A. (Eds.), *The lichens of Great Britain and Ireland*, British Lichen Society, London, 2009.
9. Spribille, T., Tuovinen, V., Resl, P., Vanderpool, D., Wolinski, H., Aime, M.C., Schneider, K., Stabentheiner, E., Toome-Heller, M., Thor, G., Mayrhofer, H.,

Basidiomycete yeasts in the cortex of ascomycete macrolichens. *Sci.*, 353, 6298, 488–492, 2016.
10. Oksanen, I., Ecological and biotechnological aspects of lichens. *Appl. Microbiol. Biotechnol.*, 73, 723–734, 2006.
11. Redzic, S., Wild edible plants and their traditional use in the human nutrition in Bosnia and Herzegovina. *Ecol. Food Nutr.*, 45, 189–232, 2006.
12. Gorin, P.A.J. and Iacomini, M., Polysaccharides of the lichens *Cetraria islandica* and *Ramalina usnea*. *Carbohydrate Res.*, 128, 119–32, 1984.
13. Gulcin, I., Oktay, M., Kufrevioglu, O.I., Aslan, A., Determination of antioxidant activity of lichen *Cetraria islandica* (L.). *Ach. J. Ethnopharmacol.*, 79, 325–329, 2002.
14. Watson, F., Kulenovic, I., Vespa, J., Nutritional status and food security: Winter nutrition monitoring in Sarajevo 1993-1994. *Eur. J Clinic. Nutr.*, 49, 2, S23–32, 1995.
15. Redzic, S., Barudanovic, S., Pilipovic, S., Wild mushrooms and lichens used as human food for survival in war conditions: Podrinje-Zepa Region (Bosnia and Herzegovina, W. Balkan). *Human Ecol. Rev.*, pp.175–187, 2010.
16. Hale, M.E., Jr., *Lichen handbook: A guide to the lichens of eastern North America*, p. 178, Smithsonian Institution Press, Washington, DC, 1961.
17. Lichens. https://www.fs.fed.us/database/feis/lichens/cetisl/all.html (Retrieved on 22.12.2019)
18. Chopra, R.N. and Chopra, I.C., *Indigenous drugs of India*, II Ed., Academic publishers, Kolkata, India, 1994.
19. Vors, L.S. and Boyce, M.S., Global declines of caribou and reindeer. *Global Change Biol.*, 15, 11, 2626–2633, 2009.
20. Ahti, T. and Hepburn, T.L., Preliminary studies on woodland caribou range, especially on lichen stands, in: *Ontario. Res. Rep. (Wildlife) No. 74. Toronto*, ON: Ontario Department of Lands and Forests, Research Branch, p. 134, 1967.
21. Novel Foods. Finnish Food Authority https://www.ruokavirasto.fi/en/companies/food-sector/production/common-requirements-for-composition/novel-foods (Retrieved on 22.12.2019)
22. Tehler, A., Dahlkild, Å., Eldenäs, P., Feige, G.B., The phylogeny and taxonomy of Macaronesian, European and Mediterranean *Roccella* (*Roccellaceae, Arthoniales*). *Symbolae Botanicae Upsalienses*, 34, 1, 405–428, 2004.
23. Logesh, A.R., Upreti, D.K., Kalaiselvam, M., Nayaka, S., Kathiresan, K., Lichen flora of Pichavaram and Muthupet mangroves (Southeast coast of India). *Mycosphere*, 3, 6, 884–888, 2012.
24. Upreti, D.K., Key to the species of lichen genus Cladonia from India and Nepal. *Feddes Repertorium*, 98, 7-8, 469–473, 1987.
25. Crawford, S.D., Lichens used in traditional medicine, in: *Lichen secondary metabolites*, pp. 31–97, Springer, Cham, 2019.
26. Airaksinen, M.M., Peura, P., Ala-Fossi-Salokangas, L., Antere, S., Lukkarinen, J., Saikkonen, M., Stenbäck, F., Toxicity of plant material used as emergency food during famines in Finland. *J. Ethnopharmacol.*, 18, 3, 273–296, 1986.

27. Kravchenko, A.V., Records of the protected species Bryoria fremontii (Parmeliaceae, Ascomycotina) in Arkhangelsk and Vologda regions. *Botanicheskii Zhurnal*, 88, 102–104, 2003.
28. Gómez-Serranillos, M.P., Fernández-Moriano, C., González-Burgos, E., Divakar, P.K., Crespo, A., Parmeliaceae family: phytochemistry, pharmacological potential and phylogenetic features. *RSC Adv.*, 4, 103, 59017-59047, 2014.
29. Velmala, S., Myllys, L., Halonen, P., Goward, T., Teuvo, A.H.T.I., Molecular data show that *Bryoria fremontii* and *B. tortuosa* (*Parmeliaceae*) are conspecific. *Lichenol.*, 41, 3, 231–242, 2009.
30. Turner, N.J., Economic importance of black tree lichen (*Bryoria fremontii*) to the Indians of western North America. *Economic Bot.*, 31, 4, 461–470, 1977.
31. Devkota, S., Chaudhary, R.P., Werth, S., Scheidegger, C., Indigenous knowledge and use of lichens by the lichenophilic communities of the Nepal Himalaya. *J. Ethnobiol. Ethnomed.*, 13, 1, 15–22, 2017.
32. Goward, T., Twelve readings on the lichen thallus VI-Reassembly. *Evansia*, 26, 3, 91–97, 2009.
33. Kirkpatrick, R.C., Zou, R.J., Dierenfeld, E.S., Zhou, H., Digestion of selected foods by Yunnan snub-nosed monkey Rhinopithecus bieti (Colobinae). *Am. J. Phys. Anthropol.*, 114, 2, 156–162, 2001.
34. Bhattarai, T., Subba, D., Subba, R., Nutritional value of some edible lichens of east Nepal. *J. Appl. Bot.*, 73, 11–14, 1999.
35. Pöykkö, H. and Tammaru, T., Countergradient vs. cogradient variation in growth and diapause in a lichen-feeding moth, Eilema depressum (Lepidoptera: Arctiidae). *J. Evolut. Biol.*, 23, 6, 1278–1285, 2010.
36. Pescott, O.L., Simkin, J.M., August, T.A., Randle, Z., Dore, A.J., Botham, M.S., Air pollution and its effects on lichens, bryophytes, and lichen-feeding Lepidoptera: review and evidence from biological records. *Biol. J. Linnean Soc.*, 115, 3, 611–635, 2015.
37. Aarland, R.C., Peralta-Gómez, S., Sanchéz, C.M., Parra-Bustamante, F., Villa-Hernández, J.M., León-Sánchez, F.D.D., Perez-Flores, L.J., Rivera-Cabrera, F., Mendoza-Espinoza, J.A., A pharmacological and phytochemical study of medicinal plants used in Mexican folk medicine. *Indian J. Tradit. Know.*, 14, 4, 550–557, 2015.
38. Susana, P.G., Aarland, R.C., Lara, M.G.C., Lara, M.J., Alberto, M.J., Toxicity analysis, phytochemical and pharmacological study of the plant known as Mora herb, collected at the environmental education center of yautlica (CEA-Yautlica). *Asian J. Plant Sci.*, 12, 4, 159–164, 2013.
39. Gerson, H., Cudmore, B., Mandrak, N.E., Coote, L.D., Farr, K., Baillargeon, G., Monitoring international wildlife trade with coded species data. *Conserv. Biol.*, 22, 1, 4–7, 2008.
40. Fragoso, G. and Ferriss, S., Monitoring international wildlife trade with coded species data: response to Gerson et al. *Conserv. Biol.*, 22, 6, 1648–1650, 2008.

41. Ranković, B., Mišić, M., Sukdolak, S., Evaluation of antimicrobial activity of the lichens *Lasallia pustulata, Parmelia sulcata, Umbilicaria crustulosa*, and *Umbilicaria cylindrica. Microbiol.*, 76, 6, 723–727, 2007.
42. Turner, N.J. and Aderkas, P.V., Sustained by First Nations: European newcomers use of Indigenous plant foods in tempterate North America. *Acta Soc. Bot. Poloniae*, 81, 4, 295–315, 2012.
43. Wirtz, N., Printzen, C., Sancho, L.G., Lumbsch, T.H., The phylogeny and classification of Neuropogon and Usnea (Parmeliaceae, Ascomycota) revisited. *Taxon*, 55, 2, 367–376, 2006.
44. Alexandria, V.A., *In The Alternative Advisor: The Complete Guide to Natural Therapies and Alternative Treatments*, Time-Life Inc., Flairfax, UK, 1997.
45. Behera, B.C., Verma, N., Sonone, A., Makhija, U., Optimization of culture conditions for lichen Usnea ghattensis G. Awasthi to increase biomass and antioxidant metabolite production. *Food Technol. Biotechnol.*, 47, 1, 7–12, 2009.
46. Usnea. https://www.encyclopedia.com/plants-and-animals/plants/plants/usnea (Retrieved on 22.12.2019)
47. Usnea. https://www.mountainroseherbs.com/products/usnea/profile (Retrieved on 22.12.2019)
48. Usnea-Lichen-Moss-Tea. https://www.amazon.com/Usnea-Lichen-Moss-Tea-Loose/dp/B008X8NCP6 (Retrieved on 22.12.2019)
49. Upreti, D.K., Bajpai, R., Nayaka, S., Singh, B.N., Jain, A.K., *Ethnolichenological studies in India: future prospects. Indian ethnobotany: emerging trends*, Scientific Publisher, Jodhpur, India, 2016.
50. Ravindran, P.N., *The encyclopedia of herbs and spices*, pp. 195–233, CAB Int., Wallingford, UK, 2017.
51. Goyal, P.K., Verma, S.K., Sharma, A.K., Biological and phytochemical aspects of lichen Parmelia perlata: A review. *Int. J. Res. Ayurveda Pharm.*, 7, 1, 102–107, 2016.
52. Muggia, L., Lichens as treasure chests of natural products. *Sim News*, 59, 3, 85–97, 2009.
53. Shukla, P., Upreti, D.K., Tiwari, P., Assessment of dye yielding potential of Indian lichens. *Ind. J. Plant Sci.*, 3, 1, 57–63, 2014.
54. USFDA 2019 directive no. 21CFR172.510 (Revised). *Food for Human Consumption*, Title 21, Vol. 3.
55. Rawat, H., Sharma, E., Fatima, N., Value-addition of silk using natural dye extracted from Lichen (*Evernia cirrhatum*). *J. Appl. Nat. Sci.*, 10, 2, 627–632, 2018.
56. Lakshmi, V., Ameta, K., Mishra, V., Srivastava, A., Agarwal, S.K., Palit, G., Mahdi, A.A., Gastroprotective effect of ethanolic extract of Parmelia perlata in rats. *J. Phytopharmacol.*, 2, 6, 19–25, 2013.
57. Prance, G.T., Ethnobotanical notes from Amazonian Brazil. *Economic Bot.*, 26, 3, 221–237, 1972.

9

Biotechnological Applications of Lichens

Md. Aftab Alam[1]*, Rizwana Khatoon[1], Shamsul Huda[1], Niyaz Ahmad[2] and Pramod Kumar Sharma[1]

[1]*Department of Pharmacy, School of Medical and Allied Science, Galgotias University, Greater Noida, India*

[2]*Department of Pharmaceutics, College of Clinical Pharmacy, Imam Abdul Rahman Bin Faisal University, Dammam, Kingdom of Saudi Arabia*

Abstract

Secondary lichens-related metabolites consisting of aliphatic, cycloaliphatic, aromatic, and terpenic compounds are special and indicate exciting biological and pharmacological activity compared to higher plants. Lichens are an ideal model for research into their bioactive metabolites. The green algae or cyanobacteria form lichen, often together, a symbiotic relationship between the fungus and a partner in the chlorophyll. For several reasons, lichens are valuable organisms. It can be used in the area of natural coloring. Lichens are photobiont and mycobiont composite organisms. Traditionally, across the globe, lichens are used and several examinations have demonstrated the encouraging pharmacology of the lichens, as well as insecticidal action. Many important medicinal values such as antiviral, antibacterial, antifungal, antioxidant, cytotoxic outcome, and others are derived from the special and biologically effective compounds presented in lichen thalli. Many medical approaches to treating such illnesses or disorders like diabetes, Alzheimer's disease, malignancy, and overweightness tend to include the suppression of certain enzymes. The therapeutic potential of lichen, which has an important effect in the Pharma industry, is reflected in all these multiple lichen metabolite activities. This article discusses the biotechnological applications of lichens.

Keywords: Lichens, biotechnology, antimicrobial, bio-indicators, anticancer

**Corresponding author*: aftab.alam@galgotiasuniversity.edu.in

Mohd Yusuf (ed.) Lichen-Derived Products: Extraction and Applications, (203–220) © 2020 Scrivener Publishing LLC

9.1 Introduction

Lichens are a symbiotic link between a photobiont and a mycobiont that supports itself. The fungus mycobiont is unique, dominating the combination, and the photobiont is called algae or cyanobacteria. Most lichens may tolerate intense environmental conditions that are harsh to each partner's survival [1–3]. Lichen is a Greek term that means surface growth on olive bark. Mycobiont loses its true identity and the hybrid organism is known as lichen which acts as one entity both morphologically and physiologically while combining the algal component, phycobiont or photobiont, and fungus element. Crustose, foliose, and fruticose are the three main forms of lichen production. Leprosy, placodioid, squamulose, and dimorphic are some intermediate types of development. In dissimilaratmospheric circumstances and on various substrates, lichen can mature. On other plants, the lichens are classified as epiphytic. On rocks underwater and historical sites and residences, lichens can also grow [4]. They can grow on stones, outcrops, tree trunks, or shrubs like scattered parts. Lichens reveal mainly their vegetative sections to the sun to allow the photobiont to capture solar radiation power. Thalli or lichenized stroma containing feature secondary compounds are these vegetative organisms [5]. It has been used for decades in traditional medicines. To date, approximately 200 components from 13,500 species of lichen-forming fungi described in the world have been reported [6, 7].

Of the over 800 recognized secondary metabolites, 80% are confined to the lichenized form. They contain specific secondary metabolites including depsides, depsidones, dibenzofuran, or xanthones. Aromatic polyketides, especially depsides, depsidones, diphenyl ethers, and dibenzofurans, are the more popular lichen compounds, giving lichens a rich origin of phenolic compounds [8]. They take many biological steps including antibiotics, antifungal, antiproliferative, antitumor, cytotoxic effects, antimicrobial, antivirals, antimycobacterial, antiprotozoal, anti-inflammatory, antipyretic, and analgesic [9–15]. As a source of dyes and sacrificial fuel, lichen has potential use as a spice, medicine, food, and fragrance. Across dissimilarportions of the creation, lichens are used medically. Two types of lichen treatments, like tuberculosis (TB), diarrhoea, nausea, cold, bleeding, skin disorders, bronchial irritation, diabetes, and injury cures, are used in both traditional and certain indigenous medicine systems. Lichen is also used for the treatment of wound healing [16–22].

9.2 Biotechnological Applications of Lichens

9.2.1 Antioxidant Activity

Because of their phenolic content, the extracts of some Lichens have been documented for antioxidant activities. Antioxidants inhibit and prevent species of reactive oxygen that may reasonof degenerative illnesses. Natural antioxidants are favored for medical applications over many artificial antioxidants, which can be toxic [23]. The antioxidant possessions of methyl orsellinate, atranorin, osellinic acid, and lecanoric acid were investigated by Jayaprakasha and Rao (2000) [24]. In extracts from Antarctic lichens, Bhattarai *et al.* (2008) [25] recorded greater antioxidant activity than lichens from temperate or tropical regions. *Parmotremastupeum (Nyl.) Hale* (Parmeliaceae) phenolic components from the Lichen as well as methyl orsenillate, orsenillic acid, atranorin, and lecanoric acid exposed strong antioxidant action [24]. The *in vivo* research of lichen *Cetraria islandica* recorded antioxidant activities [26].

Aoussar *et al.* [27] introduced the identification of phenolic compounds and testing of dichloromethane, acetone and methanol extracts of *Evernia prunastri L.* and *Pseudevernia furfuracea L.* antioxidant function [27]. Clauda *et al.* (2014) assessed the antioxidant action, total phenol contentment, and flavonoid content of four lichen fungal taxa in the genera *Cladonia*, *Peltigera*, *Thamnolia*, and *Cora* together in Mérida (Venezuela) and statistically assessed the association among total phenolic and flavonoid contentment and antioxidant action of the extracts of lichen studied [28]. The four species *Parmotrema grayana*, *Heterodermia obscurata*, *Cladonia sp.*, and *Roccella montagnei* were evaluated by Thadhani *et al.* in 2011. And by SOR scavenging assay, NOR scavenging assay and radical scavenging assay by DPPH [29].

9.2.2 Antimicrobial Activity

The antimicrobial activity of Lichens *Cladonia furcata*, *Lecanora atra*, and *Lecanora muralis* was calculated via the broth microdilution process with 96-well micro-titer plates to assess the minimum inhibitory concentration against six species of bacteria *Bacillus mycoides*, *Bacillus subtilis*, and *Staphylococcus aureus* (Gram positive bacteria); *Enterobacter cloaceae*, *Escherichia coli*, and *Klebsiella pneumoniae*, (Gram-negative bacteria); and 10 species of fungi *Aspergillus flavus*, *Aspergillus fumigatus*, *Botrytis cinerea*, *Candida albicans*, *Fusarium oxysporum*, *Mucor mucedo*, *Paecilomyces variotii, Penicillium purpurescens, Penicillium verrucosum*, and *Trichoderma harsianum*. The extract of *Cladoniafurcata* was the high effective antimicrobial agent with a minimum concentration of 0.78 to 25mg/ml [30].

Agar disc diffusion and broth microdilution approaches were used to exam antimicrobial action of lichens. Methods for measuring lichens' (*Caloplacapusilla, Protoparmeliopsismuralis*, and *Xanthoria parietina*) antimicrobial actionin contradiction of the bacteria's species *Bacillus subtilis, Escherichia coli, Enterococcus faecalis, Klebsiella pneumoniae, Pseudomonas aeruginosa, Staphylococcus aureus*, and *Staphylococcus epidermidis* were the diffusion of agar disc and broth microdilution. The development of Gram-positive bacteria was strongly inhibited by *P. muralis, Bacillus subtilis, Enterococcus faecalis, Staphylococcus aureus*, and *Staphylococcus epidermidis* (MICs from 6.67 to 100.00 μg/ml) [31].

In acetone, methanol, and ethanol, the lichen *Usnea ghattensis* were removed. *In vitro* antimicrobial action was initially verified using Kirby-Bauer disc diffusion process and verified by a lowest concentration of inhibitors with the Broth microdilution technique against six bacteria, three-gram positive (*Staphylococcus aureus, Streptococcus faecalis*, and *Bacilluscereus*), and three-gram negative (*Escherichia coli, Pseudomonas aeruginosa*, and *Salmonella typhimurium*). Ethanol extract with an inhibition area of 29.8 ± 0.6 mm and 12.3 ± 0.5 mm diameters at a concentration of 0.2 mg/ml was most effective against *Bacillus cereus* and *Pseudomonas aeruginosa*. Extract of acetone and methanol showed nearly comparable action counter to *Staphylococcus aureus* and the inhibition zone was 24.6 ± 0.5 and 24.7 ± 0.4 mm. Only methanol extract with a 13.5 ± 0.8 mm zone showed activity against *Streptococcus faecalis*. The MIC value for *Staphylococcus aureus* and *Streptococcus faecalis* was 6.25 μg/ml and 25 μg/ml, compared to 3.125 μg/ml and 200 μg/ml, respectively, for *Bacillus cereus* and *Pseudomonas aeruginosa* [32]. Maximum of the antibacterial actions were verified on *Bacillus, E. coli, Pseudomonas, Kleibsiella, Staphylococcus aureus, Salmonella, Candida, Yersinia*, and *Proteus sp.* [33–44].

9.2.3 As Bioindicators

Bioindicators or biomonitors are organisms that respond with their life functions to environmental pollution and provide quantitative data on the surrounding environment. Lichens are among the most susceptible species to environmental changes at the ecosystem level and can be used as a biomonitoring system in addition to chemical and physical monitoring methods. Some lichens consume many toxic substances from the environment, such as sulfur dioxide, fluorides, and heavy metals. They easily accumulate pollutants through their entire surface, such as heavy metals and other components from the ambient air [45]. Lichens are used to recognize PAHs (polycyclic aromatic hydrocarbons), phenols, and trace metals as bioindicators and/or bioaccumulation [46]. The main responsibility for the production of PAHs in the combustion procedures from road traffic, manufacturing, and home-based heating [47, 48].

Lichens can be used in two different ways as bioindicators and/or biomonitors [49–51]:

1. Viaplotting all species in a precisezone (Process A); and
2. By discreteselection of the species of lichens and measuring contaminants that gather in the thallus; or viaresettling lichens from an unadulterated region to a polluted area, then computing morphological variations in lichen thallus and/or assessing physiological constraints and/or assessing pollutant bioaccumulation (Process B).

Usnea longissima is one of the most susceptible species of lichen to air pollution. Its well-being has been recognized as an air pollution measure in some European countries. The existence of *U. Longissima* shows pure water, although its absenteeism shows contaminated air. It was used to reinforce hair and too used in women's hygienic products [52].

9.2.4 Anticancer Activity

Extracts of *fruticulous* or *foliaceous* lichen species have a cytotoxic outcome on four cell lines of human malignancy. The four malignancy cell lines are cK-562: chronic myelogenous leukaemia, the U251: glioblastoma, the DU145: Prostate carcinoma, and the MCF7: breast Adenocarcinoma. Among these, the extracts from *Cladonia convoluta*, *Cladoniarangiformis*, *Parmelia caperata*, *Platismatia glauca*, and *Ramalina cuspidata* showed interesting activities particularly on the cell lines of human cancer [53].

For the first time in human U87MG-GBM cell lines and main rat cerebral cortex (PRCC) cells, the antitumorlatent of three lichen secondary metabolites as well asolivetoric acid (OLA) and physodic acid (PHA) remoted from *Pseudevernia furfuracea (L.) Zopf* (Parmeliaceae) and psoromic acid (PSA) remoted from *Rhizoplaca melanophthalma (DC.) Leuckert* (Lecanoraceae) was examined. OLA, PHA, and PSA values of median inhibitory concentration (IC50) were 125.71, 698.19, and 79.40 mg/L for PRCC cells and 17.55, 410.72 and 56.22 mg/L for U87 MG cells individually. The cytotoxic effects of these metabolites have been shown to have a

Table 9.1 Various activities of Lichens.

Antioxidant activity of lichens		
Lichen extracts/compounds	**Assay bioactive type**	**Ref.**
Ramalina pollinaria	DPPH & ALP	[54]
R. pacifica	DPPH, ALP, SOI, NOS, & TEAC	[55]
Parmotrema	DPPH	[56]
Antimicrobial activity of lichens		
Lichens	**Microorganisms**	
Usnea barbata	*Micrococcus viradans*	[57]
Everniastrumcirrhatum	*Listeria monocytogenes*	[58]
Usnea ghattensis	*B. licheniformis, B. megaterium*	[59]
Cetrariaaculeata	*Escherichia coli*	[58]
Usnea subfloridans	*A. niger*	[60]
Parmotrema tinctorum	*Cladosporium sphaerospermum*	[61]
Evernia prunastri, Hypogymnia physodes, Cladonia portentosa	*Phytophthora infestans*	[62]
Anticancer activity of lichens		
Lichens	**Lichen metabolites**	
Xanthoria parietina	Parietin	[63]
Parmotrema reticulatum	Catechin, purpurin, tannic acid and reserpine	[64]
Xanthoparmelia chlorochroa	Usnic, salazinic, constictic, and norstictic acids	[65]
Anti-insecticidal activity of lichens		
Lichens	**Targeted insects**	
Cladonia substellata	*A. aegypti larvae*	[66]

(*Continued*)

Table 9.1 Various activities of Lichens. (*Continued*)

Anti-insecticidal activity of lichens		
Lichens	**Targeted insects**	
Letharia vulpina and Peltigera rufescens	*Sitophilus zeamais adults*	[67]
Usnea longissima	*S. granarius*	[68]
Letharia vulpina	Larvae of yellow-striped armyworm *Spodoptera ornithogalli*	[69]
Enzyme inhibition activity of lichens		
Lichens	**Enzyme inhibitors**	
Usnea complanata	Inhibition of angiotensin converting enzyme (ACE)	[70, 71]
Pseudevernia furfuracea	Trypsin	[72]
Umbilicaria esculenta	Telomerase	[73]
Anti-diabetic action of lichens		
Lichens	**Action**	
Ramalina sinensis	α-Amylase	[74]
Parmotrema cooperi	Antiglycation	[75]
Punctelia borreri	α-Amylase	[76]

positive connection with concentration, LDH action, and oxidative DNA harm [77].

The tenuiorin methyl orsellinate and orcinol derivatives, present in *Peltigera leucophlebia (Nyl.) Gyeln* (Peltigeraceae) extracts, showed *in vitro* prohibited action counter to soybean15-lipoxygenase. Tenuiorin and methyl orsellinate have been addedverified on cultured human breast, pancreatic, and colon cancer cell lines for antiproliferative action. Bianthraquinone glycosides, colleflaccinosides remoted from *Collema flaccidum (Ach.) Ach.* (Collemataceae), were confirmed to have antitumor activity in Israel and Russia [78]. *Cladoniafurcata*, *Lecanoraatra*, and *Lecanoramuralis* lichens anticancer action was established in contrast to cell lines FemX (human melanoma) and LS174 (human colon carcinoma) using MTT system. Both extracts are originated to have high anticancer action with IC_{50} values ranging from 8.51 to 40.22 μg/ml for both cell lines [30].

9.2.5 Anti-Insecticidal Activity

Killing mosquito larvae is a good way to minimize mosquito populations in breeding grounds until they reach adulthood [79]. Compounds of lichen, *viz.*, (+) and (−)-usnic acid, vulpinic acid, and stictic acid were screened by poisonousness and antifeedant assays for insecticidal actioncounter to larvae of polyphagous *Spodoptera littoralis.* (+) and (−)-usual acid and vulpinic acid showed strong larvae mortality though stictic acid did not reason larvae death [80]. The insecticidal efficacy of three lichens extracts, *viz.*, was evaluated by Emsen *et al.* (2015). *Lecanora muralis*, *Letharia vulpina*, and *Peltigera rufescens* against *Sitophilus granarius* adults. The treatment with the extract instigated in dose-dependent insect mortality with lower action after the extended exposure period. The values of the LC50 for *L. muralis*, extracts from *L. vulpina*, and *P. rufescens* are 0.666, 0.505, and 0.328 mg/ml, individually [81].

Silva *et al.* (2009) investigated possible insecticide effects on *Nasutitermescorniger Motschulsky* termite of lectin isolated from *Cladonia verticillaris (Raddi)* Fries. It was mentioned that as a result of their work, the preparations of *C. verticillaris* can regulate termites (or other insects) that are economically important to agriculture and the wood industry [82]. The effects of methanol, chloroform, and *L. pulmonaria*'s water extracts on the life span of *Drosophila melanogaster Meigen* (vinegar fly) have been explored in another review. The methanol extract was found to be more active than chloroform extracts and the extract of water was relatively weaker than methanol and chloroform extracts [83].

Extracts from lichen *Letharia vulpine* have shown powerful insecticidal activity counter to *Spodoptera ornithogalli* and *S. littoralis* [84]. The study conducted by Martins *et al.* (2018) [85] exposed dose-dependent insecticide action of usnic, fumarprotocetrary, and barbatic acids extracted from *Cladonia substellata* lichens, *C. Verticillaris*, and *Cladia aggregata* counter to the termite *Nasutitermescorniger*, respectively.

9.2.6 Enzyme Inhibition Activity

Lichens are active enzyme prohibitors of ornithine decarboxylase and arginine decarboxylase, such as usnic acids, resorcinol derivatives, and atranorin which affect the polyamine metabolism [86]. *Umbilicaria esculenta* extracts have severely inhibited mold and mammalian disaccharide hydrolytic enzymes. Inhibitory activity of glucosidase was also reported from extracts of *Parmelia austrosinensis* and *Parmelia praesorediosa* [87]. Different lichens solvent extracts, *viz.*, *Ramalina sinensis*, *leucomelosheteroderma*, *sp. Herpothallon*, and α-amylase were tested for *Parmatorema reticulatum*. The extracts showed concentration-dependent amylase inhibitory activity [88].

Similar *Parmelia perlata* solvent extracts are tested for inhibitory activity of salivary amylase. In the case of methanol extract (94.74% prohibition) and aqueous extract (40.80% prohibition), the maximum and least inhibitory activity was observed [89]. *Parmotrema tinctorum* ethyl acetate extract demonstrated the α-glucosidase inhibitory activity with an IC50 value of 58.45 ± 1.24 μg/ml [90]. Romagni *et al.* (2000) [91] showed usnic acid phytotoxic activity isolated from *Alectoria sarmentosa (Ach.)* with irreversible inhibition of phydroxyphenylpyruvate dioxygenase activity (through adeceptive IC_{50} value of 70 nM). Medication with composite (–)-usnic acid resulted in concentration-dependent bleaching of cotyledonary tissues along with chlorophyll and carotenoid decline. Furthermore, in the case of (+) enantiomer of usnic acid, the least bleaching effect was observed.

In the lichen extracts, *Umbilicaria esculenta*, prohibitor action toward β-glucosidase was noticed. The extract exposed significant prohibition of mould and mammalian disaccharide hydrolytic enzymes, but a feeble or no inhibition of hydrolytic polysaccharide enzymes excluding glucoamylase and laminarinase. The extract prohibitor was very constant and retained further than 95% of its action once treated with water, acid, alkali, and certain hydrolytic enzymes [92].

9.2.7 Antidiabetic Activity

Several studies have established a positive potential for lichens to be explored as powerful antidiabetic mediators. Their hypoglycemic function has been tested in various approaches, as well as by their constrain behavior of carbohydrate hydrolyzing enzymes (α-amylase and α-glucosidase) or proteins tyrosine phosphatase 1B (PTP1B), which is documented as the main negative controller in insulin signalling, or by their antioxidant role in restoring insulin-secreting pancreatic cells [93].

To investigate for 42 days, the *in vitro* antioxidant and antihyperglycemic effect of *Parmotrema hababianum* ethanol extract in streptozocin-induced diabetic rats. The extract exposed with maximum presence almost all antioxidant activities. Extract treatment in diabetic rats at a dose level of 100 and 200 mg/kg, associated to diabetic switch and glibenclamide at a dose level of 4 mg/kg, and biochemical limitslike blood glucose, total cholesterol, triglycerides, HDL, LDL, insulin, total protein, urea, and creatinine were evaluated. The extract showed a positive correlation ($p < 0.001$) relative to the control increasing blood glucose level. Also, there was a substantial decrease ($p < 0.01$) in levels of TC, TG, LDL, SGPT, SGOT, urea, and creatinine and a substantial increase ($p < 0.01$) in levels of HDL and insulin. The results of this study show significant effectiveness in the treatment of diabetes with powerful antioxidant and antihyperglycemic activity [94].

In another study, selected lichens extract of methanol and ethyl acetate were investigated for the order to inhibit α-amylase activity for their antidiabetic function. The lichen extracts' amylase inhibitory activity has been established against fungal diastase (amylase). The extract resulted in a dose-dependent amylase activity inhibition. The inhibition of α-amylase activity was tested separately for three dissimilar concentrations: 5, 10, and 15 mg/ml of methanol and ethyl acetate extracts of lichens *Flavoparmelia caperata* and *Physcia aipolia*. The maximum inhibition of *F. caperata* was 49% at 15 mg/ml, tracked by 46% *P. aipolia*. Whereas, the extract of ethyl acetate showed 43% in *F. caperata* and 32% at a concentration of 15 mg/ml in *P. aipolia*. Hence, the potential of these extracts for managing hyperglycemia is indicated. To control diabetes, lichens could be used in the development of active agents to regulate the postprandial glucose level [95].

9.3 Conclusion and Future Dimension

Lichens generate secondary metabolites for medicinal, nutritional, food, perfumery, and dyeing purposes. Lichens have long been underestimated by mycologists and agrochemists, given their wide range of biological activities, primarily because of its slow development of the essence and problems of artificial cultivation. The stage has therefore not yet reached for the extensive industrial manufacturing of lichen metabolites. Further investigation and progress are needed to advance, refine, and expand hopeful, extremely industrial and national lichen-based innovations. Biotechnologies for the development of natural antioxidants, antimicrobial, anti-insecticidal, antipyretic, and anticancer agents would benefit the biopharmaceutical industry by commercializing them. The capacity that lichens has for human health and our environment needs to be properly examined and exploited. The literature reviewed indicates that lichen is important pointers for the lasting surveillance of environment and air contamination and shows antidiabetic, enzyme inhibition, anticancer, antitumor, antimicrobial activity, antioxidant activity, etc. As air pollution is rapidly increasing and global warming issues are growing, the world faces many challenges that only environmentally friendly lichens can solve. So, it is our responsibility to save our mother nature and grow valuable lichens.

References

1. Denton, G.H. and Karlén, W., Lichenometry: Its application to Holocene moraine studies in southern Alaska and Swedish Lapland. *Arct. Antarct. Alp. Res.*, 5, 347–372, 1973.
2. Stocker-Wörgötter, E., Metabolic diversity of lichen-forming ascomycetous fungi: Culturing, polyketide and shikimate metabolite production, and PKS genes. *Nat. Prod. Rep.*, 25, 188–200, 2008.
3. Kekuda, T.R., Vinayaka, K.S., Swathi, D., Suchitha, Y., Venugopal, T.M., Mallikarjun, N., Mineral composition, total phenol content and antioxidant activity of a *MacrolicheneverniastrumCirrhatum (Fr.) Hale* (Parmeliaceae). *J. Chem.*, 8, 1886–1894, 2011.
4. Nayaka, S., Methods and techniques in collection, preservation and identification of lichens, in: *Plant Toxonomy and Biosystematics: Classical and Modern Methods*, pp. 101–105, 2014.
5. Ahmadjian, V. and Reynolds, J.T., Production of biologically active compounds by isolated lichenized fungi. *Sci.*, 133, 700–701, 1961.
6. Manojlovic, N.T., Solujic, S., Sukdolak, S., Krstic, L., Isolation and antimicrobial activity of anthraquinones from some species of the lichen genus Xanthoria. *J. Serb. Chem. Soc.*, 65, 555–560, 2000.
7. Tay, T., Türk, A.Ö., Yılmaz, M., Türk, H., Kıvanç, M., Evaluation of the antimicrobial activity of the acetone extract of the lichen *Ramalinafarinacea* and its (+)-usnic acid, norstictic acid, and protocetraric acid constituents. *ZeitschriftfürNaturforschung C.*, 59, 384–388, 2004.
8. Huneck, S. and Yoshimura, I., Identification of lichen substances, in: *Identification of lichen substances*, pp. 11–123, 1996.
9. Pandey, A., Lichens: A resource chest of herbal antimicrobial compounds. *Int. J. Theoretic. Appl. Sci.*, 9, 137–146, 2017.
10. Verma, N., Behera, B.C., Makhija, U., Antioxidant and hepatoprotective activity of a lichen *Usnea ghattensis in vitro. Appl. Biochem. Biotechnol.*, 151, 167–181, 2008.
11. Manojlović, N.T., Vasiljević, P., Jusković, M., Najman, S., Janković, S., Milenković-Andjelković, A., HPLC analysis and cytotoxic potential of extracts from the lichen, *Thamnolia vermicularis* var. *subuliformis*. *J. Med. Plants Res.*, 4, 817–823, 2010.
12. Illana Esteban, C., Líquenesusadosenmedicinatradicional. *Bol. Soc. Micol. Madrid.*, 36, 163–174, 2012.

13. Behera, B.C., Verma, N., Sonone, A., Makhija, U., Antioxidant and antibacterial activities of lichen *Usnea ghattensis in vitro. Biotechnol. Lett.*, 27, 991–995, 2005.
14. Halama, P. and Van Haluwin, C., Antifungal activity of lichen extracts and lichenic acids. *Bio. Control.*, 49, 95–107, 2004.
15. Wei, X., Jeon, H.S., Han, K.S., Koh, Y.J., Hur, J.S., Antifungal activity of lichen-forming fungi against *Colletotrichum acutatum* on hot pepper. *Plant Pathol. J.*, 24, 202–206, 2008.
16. Rogers, R.D., Medicinal lichens. *Fungi*, 7, 56–58, 2014.
17. Upreti, D.K., Divakar, P.K., Nayaka, S., Commercial and ethnic use of lichens in India. *Econ. Bot.*, 59, 269, 2005.
18. Shukla, P., Upreti, D.K., Nayaka, S., Tiwari, P., Natural dyes from Himalayan lichens. *Indian. J. Trad. Knowl.*, 13, 195–201, 2014.
19. Singh, S., Upreti, D.K., Lehri, A., Paliwal, A.K., Quantification of lichens commercially used in traditional perfumery industries of Uttar Pradesh, India. *Indian J. Plant Sci.*, 4, 29–33, 2015.
20. Weissbuch, B.K., Medicinal lichens: The final frontier. *J. Am. Herbal. Guild.*, 12, 23–28, 2016.
21. Londoño-Castañeda, P.A., Buril, M.L.L., Rego-Cunha, I.P., Silva, N.H., Honda, N.K., Pereira, E.C., Andrade, L.H.C., Lichens used in the traditional medicine by the Pankararu Indigenous Community, Pernambuco-Brazil. *Global J. Sci. Front. Res.: C Biol. Sci.*, 17, 15–22, 2017.
22. Huneck, S., The significance of lichens and their metabolites. *Die Naturwissenschaften*, 86, 559–570, 1999.
23. Zambare, V.P. and Christopher, L.P., Biopharmaceutical potential of lichens. *Pharm. Biol.*, 50, 778–798, 2012.
24. Jayaprakasha, G.K. and Rao, L.J., Phenolic constituents from the lichen *Parmotremastuppeum (Nyl.)* Hale and their antioxidant activity. *Zeits. Für. Naturfor. C.*, 55, 1018–1022, 2000.
25. Bhattarai, H.D., Paudel, B., Hong, S.G., Lee, H.K., Yim, J.H., Thin layer chromatography analysis of antioxidant constituents of lichens from Antarctica. *J. Nat. Med.*, 62, 481, 2008.
26. Gülçin, İ., Oktay, M., Küfrevioğlu, Ö.İ., Aslan, A., Determination of antioxidant activity of lichen *Cetrariaislandica (L) Ach. J. Ethnopharmacol.*, 79, 325–329, 2002.
27. Aoussar, N., Manzali, R., Nattah, I., Rhallabi, N., Vasiljevic, P., Bouksaim, M., Douira, A., Manojlovic, N., Mellouki, F., Chemical composition and antioxidant activity of two lichens species (*Pseudeverniafurfuracea L.* and *Evernia prunastri L.*) collected from Morocco. *J. Mat. Environ. Sci.*, 8, 1968–1976, 2017.
28. Plaza, C.M., de Torres, L.E.D., Lücking, R.K., Vizcaya, M., Medina, G.E., Antioxidant activity, total phenols and flavonoids of lichens from Venezuelan Andes. *J. Phar. Pharmacol. Res.*, 2, 138–147, 2014.
29. Thadhani, V.M., Choudhary, M.I., Ali, S., Omar, I., Siddique, H., Karunaratne, V., Antioxidant activity of some lichen metabolites. *Natur. Prod. Rese.*, 25, 1827–1837, 2011.

30. Ranković, B.R., Kosanić, M.M., Stanojković, T.P., Antioxidant, antimicrobial and anticancer activity of the lichens *Cladoniafurcata*, *Lecanoraatra* and *Lecanoramuralis*. *BMC. Complem. Alter. Med.*, 11, 97, 2011.
31. Felczykowska, A., Pastuszak-Skrzypczak, A., Pawlik, A., Bogucka, K., Herman-Antosiewicz, A., Guzow-Krzemińska, B., Antibacterial and anticancer activities of acetone extracts from *in vitro* cultured lichen-forming fungi. *BMC. Complem. Altern. Med.*, 17, 300, 2017.
32. Srivastava, P., Upreti, D.K., Dhole, T.N., Srivastava, A.K., Nayak, M.T., Antimicrobial property of extracts of Indian lichen against human pathogenic bacteria. *Interdisc. Perspec. Infec. Dis.*, 2013, 1–7, 2013.
33. Ingolfsdottir, K., Bloomfield, S.F., Hylands, P.J., *In vitro* evaluation of the antimicrobial activity of lichen metabolites as potential preservatives. *Antimicrob. Agents Chemother.*, 28, 289–292, 1985.
34. Yılmaz, M., Türk, A.Ö., Tay, T., Kıvanç, M., The Antimicrobial Activity of Extracts of the Lichen *Cladoniafoliacea* and Its (–)-Usnic Acid, Atranorin, and Fumarprotocetraric Acid Constituents. *Zeitsc. Für. Naturfor.*, 59, 249–254, 2004.
35. Ranković, B. and Mišić, M., The antimicrobial activity of the lichen substances of the lichens *Cladoniafurcata*, *Ochrolechiaandrogyna*, *Parmeliacaperata* and *Parmeliaconspresa*. *Biotech. Biotech. Equi.*, 22, 1013–1016, 2008.
36. Karthikaidevi, G., Thirumaran, G., Manivannan, K., Anantharaman, P., Kathiresan, K., Balasubaramanian, T., Screening of the antibacterial properties of lichen *Roccella belangeriana (Awasthi)* from *Pichavaram mangrove (Rhizophora* sp.*)*. *Adv. Biol. Res.*, 3, 127–131, 2009.
37. Karagoz, A., Dogruoz, N., Zeybek, Z., Aslan, A., Antibacterial activity of some lichen extracts. *J. Med. Plants Res.*, 3, 1034–1039, 2009.
38. Tay, T., Türk, A.Ö., Yılmaz, M., Türk, H., Kıvanç, M., Evaluation of the antimicrobial activity of the acetone extract of the lichen *Ramalinafarinacea* and its (+)-usnic acid, norstictic acid, and protocetraric acid constituents. *Zeits. Für. Naturfor.*, 59, 384–388, 2004.
39. Martins, M.C.B., Lima, M.J.G.D., Silva, F.P., Azevedo-Ximenes, E., Silva, N.H.D., Pereira, E.C., *Cladiaaggregata* (lichen) from Brazilian northeast: chemical characterization and antimicrobial activity. *Brazi. Arch. Biol. Tech.*, 53, 115–122, 2010.
40. Manojlovic, N.T., Vasiljevic, P.J., Markovic, Z.S., Antimicrobial activity of extracts and various fractions of chloroform extract from the lichen *Laurerabenguelensis*. *J. Biolog. Res.*, 13, 27, 2010.
41. Ranković, B., Kosanić, M., Sukdolak, S., Antimicrobial activity of some lichens and their components. *Rec. Adv. Clin. Med.*, 37, 279–286, 2010.
42. Santiago, K., Borricano, J., Canal, J.N., Marcelo, D.M.A., Perez, M.C.P., Dela Cruz, T.E.E., Antibacterial activities of fruticose lichens collected from selected sites in Luzon Island, Philippines. *Phili. Sci. Lett.*, 3, 18–29, 2010.
43. Swathi, D., Suchitha, Y., Kekuda, P.T.R., Venugopal, T.M., Vinayaka, K.S., Mallikarjun, N., Raghavendra, H.L., Antimicrobial, anthelmintic and insecticidal activity of a *Macrolicheneverniastrumcirrhatum (Fr.) Hale*. *Int. J. Drug Develop. Res.*, 2, 780–789, 2010.

44. Zambare, V.P., Zambare, A.V., Christopher, L.P., Antioxidant and antibacterial activity of extracts from lichen *Xanthoparmeliasomloensis*, native to the black hills, South Dakota, USA. *Int. J. Med. Sci. Technol.*, 3, 46–51, 2010.
45. Conti, M.E. and Cecchetti, G., Biological monitoring: Lichens as bioindicators of air pollution assessment-a review. *Enviro. Pollu.*, 114, 471–492, 2001.
46. Migaszewski, Z.M., Gałuszka, A., Pasławski, P., Polynuclear aromatic hydrocarbons, phenols, and trace metals in selected soil profiles and plant bioindicators in the Holy Cross Mountains, South-Central Poland. *Environ. Int.*, 28, 303–313, 2002.
47. Samanta, S.K., Singh, O.V., Jain, R.K., Polycyclic aromatic hydrocarbons: environmental pollution and bioremediation. *Trends Biotechnol.*, 20, 243–248, 2002.
48. Grynkiewicz, M., Polkowska, Ż., Namieśnik, J., Determination of polycyclic aromatic hydrocarbons in bulk precipitation and runoff waters in an urban region (Poland). *Atmos. Environ.*, 36, 361–369, 2002.
49. Richardson, D.H.S., Lichens as biological indicators- Recent developments, in: *Bioindicators and Environmental Management*, Jeffrey, D.W. & Madden, B. (eds.), pp. 263–272, Academic Press, London, 1991.
50. Seaward, M.R.D., Lichens and sulphur dioxide air pollution: field studies. *Environ. Rev.*, 1, 73–91, 1993.
51. Gries, C., Lichens as indicators of air pollution, in: *Lichen Biology*, T.H. Nash III (Ed.), pp. 240–254, Camb. Univ. Press, Camb, 1996.
52. Brodo, M.I., Sharnoff, S.D., Sharnoff, S., *Lichens of North America*, pp. 721–723, Yale. Univ. Press, New Haven and London, 1999.
53. Bézivin, C., Tomasi, S., Lohézic-Le Dévéhat, F., Boustie, J., Cytotoxic activity of some lichen extracts on murine and human cancer cell lines. *Phytomed.*, 10, 499–503, 2003.
54. Gulluce, M., Aslan, A., Sokmen, M., Sahin, F., Adiguzel, A., Agar, G., Sokmen, A., Screening the antioxidant and antimicrobial properties of the lichens *Parmeliasaxatilis, Platismatiaglauca, Ramalinapollinaria, Ramalinapolymorpha* and *Umbilicarianylanderiana. Phytomed.*, 13, 515–521, 2006.
55. Verma, N., Behera, B.C., Sharma, B.O., Glucosidase inhibitory and radical scavenging properties of lichen metabolites salazinic acid, sekikaic acid and usnic acid. *Hace. J. Bio. Chem.*, 40, 7–21, 2012.
56. Rajan, V.P., Gunasekaran, S., Ramanathan, S., Murugaiyah, V., Samsudin, M.W., Din, L.B., Biological activities of four *Parmotrema* species of Malaysian origin and their chemical constituents. *J. App. Pharma. Sci.*, 6, 36–43, 2016.
57. Madamombe, I.T. and Afolayan, A.J., Evaluation of antimicrobial activity of extracts from South African *Usnea barbata. Pharma. Biol.*, 41, 199–202, 2003.
58. Türk, A.Ö., Yılmaz, M., Kıvanç, M., Türk, H., The antimicrobial activity of extracts of the lichen *Cetrariaaculeata* and its protolichesterinic acid constituent. *Zeits. Für. Naturfo.*, 58, 850–854, 2003.

59. Behera, B.C., Verma, N., Sonone, A., Makhija, U., Evaluation of antioxidant potential of the cultured mycobiont of a lichen *Usnea ghattensis. Phytother. Res. Int. J. Dev. Pharma. Toxico. Eva. Nat. Prod. Deriv.*, 19, 58–64, 2005.
60. Ekong, U.S., Mgbii, A.I., Adikwu, M.U., Evaluation of the *in vitro* combination effect of colloidal silver concentrate on the antifungal activity of ethanolic extract of the lichen *Usnea subfloridans. Niger. Anna. Nat. Sci.*, 8, 1–5, 2008.
61. Gomes, A.T., Honda, N.K., Roese, F.M., Muzzi, R.M., Marques, M.R., Bioactive derivatives obtained from lecanoric acid, a constituent of the lichen *Parmotrematinctorum (Nyl.) Hale* (Parmeliaceae). *Rev. Brasi. de Farmaco.*, 12, 74–75, 2002.
62. Halama, P. and Van Haluwin, C., Antifungal activity of lichen extracts and lichenic acids. *BioControl.*, 49, 95–107, 2004.
63. Bačkorová, M., Bačkor, M., Mikeš, J., Jendželovský, R., Fedoročko, P., Variable responses of different human cancer cells to the lichen compounds parietin, atranorin, usnic acid and gyrophoric acid. *Toxicol. In Vitro*, 25, 37–44, 2011.
64. Ghate, N.B., Chaudhuri, D., Sarkar, R., Sajem, A.L., Panja, S., Rout, J., Mandal, N., An antioxidant extract of tropical lichen, *Parmotremareticulatum*, induces cell cycle arrest and apoptosis in breast carcinoma cell line MCF-7. *PloS One*, 8, e82293, 2013.
65. Shrestha, G., El-Naggar, A.M., St. Clair, L.L., O'Neill, K.L., Anticancer activities of selected species of North American lichen extracts. *Phytother. Res.*, 29, 100–107, 2015.
66. Bomfim, R.R., Araujo, A.A., Cuadros-Orellana, S., Melo, M.G., Quintans-Junior, L.J., Cavalcanti, S.C., de Bioprocessos, L.D.E., Larvicidal activity of *Cladoniasubstellata* extract and usnic acid against Aedes aegypti and Artemia salina. *Latin. Am. J. Phar.*, 28, 580–584, 2009.
67. Yildirim, E., Emsen, B., Aslan, A., Bulak, Y., Ercisli, S., Insecticidal activity of lichens against the maize weevil, *Sitophilus zeamaisMotschulsky* (Coleoptera: Curculionidae). *Egypt. J. Biol. Pest. Control.*, 22, 151–6, 2012.
68. Yildirim, E., Aslan, A., Emsen, B., Cakir, A., Ercisli, S., Insecticidal effect of *Usnea longissima* (Parmeliaceae) extract against *Sitophilus granarius* (Coleoptera: Curculionidae). *Int. J. Agric. Biol.*, 14, 303–6, 2012.
69. Slansky, F., Jr., Effect of the lichen chemicals atranorin and vulpinic acid upon feeding and growth of larvae of the yellow-striped armyworm, *Spodopteraornithogalli. Environ. Ento.*, 8, 865–868, 1979.
70. Behera, B.C., Mahadik, N., Morey, M., Antioxidative and cardiovascular-protective activities of metabolite usnic acid and psoromic acid produced by lichen species *Usnea complanata* under submerged fermentation. *Pharma. Biol.*, 50, 968–979, 2012.
71. Mahadik, N.D., Morey, M.V., Behera, B.C., Makhija, U.V., Naik, D.G., Cardiovascular-protective, antioxidative, and antimicrobial properties of natural thallus of lichen *Usnea complanata. Lat. Am. J. Pharm.*, 30, 220–228, 2011.

72. Proksa, B., Adamcova, J., Sturdikova, M., Fuska, J., Metabolites of *Pseudeverniafurfuracea (L.) Zopf.* and their inhibition potential of proteolytic enzymes. *Die. Pharm.*, 49, 282–283, 1994.
73. Xu, B., Li, C., Sung, C., Telomerase inhibitory effects of medicinal mushrooms and lichens, and their anticancer activity. *Int. J. Med. Mushrooms*, 16, 7–28, 2014.
74. Hengameh, P., Rashmi, S., Rajkumar, H.G., *In vitro* inhibitory activity of some lichen extracts against α-amylase enzyme. *Eur. J. Biomed. Pharma. Sci.*, 3, 315–318, 2016.
75. Choudhary, M.I., Ali, M., Wahab, A.T., Khan, A., Rasheed, S., Shyaula, S.L., Rahman, A.U., New antiglycation and enzyme inhibitors from *Parmotremacooperi. Sci. China Chem.*, 54, 1926–1931, 2011.
76. Valadbeigi, T. and Shaddel, M., Amylase inhibitory activity of some macrolichens in Mazandaran province, Iran. *Physi. Pharma.*, 20, 215–219, 2016.
77. Emsen, B., Aslan, A., Togar, B., Turkez, H., *In vitro* antitumor activities of the lichen compounds olivetoric, physodic and psoromic acid in rat neuron and glioblastoma cells. *Pharmaceut. Bio.*, 54, 1748–1762, 2016.
78. Řezanka, T. and Dembitsky, V.M., The colleflaccinosides, two chiral bianthraquinone glycosides with antitumor activity from the lichen *Collemaflaccidum* collected in Israel and Russia. *Nat. Prod. Res.*, 20, 969–980, 2006.
79. Vinayaka, K.S., Kumar, P.S.V., Kekuda, P.T.R., Krishnamurthy, Y.L., Mallikarjun, N., Swathi, D., Proximate composition, antioxidant, anthelmintic and insecticidal activity of a *Macrolichenramalina Conduplicans Vain.* (Ramalinaceae). *Eur. J. App. Sci.*, 1, 40–46, 2009.
80. Emmerich, R., Giez, I., Lange, O.L., Proksch, P., Toxicity and antifeedant activity of lichen compounds against the polyphagous herbivorous insect *Spodopteralittoralis. Phytochem.*, 33, 1389–1394, 1993.
81. Emsen, B., Yildirim, E., Aslan, A., Insecticidal activities of extracts of three lichen species on *Sitophilus granarius (L.)*(Coleoptera: Curculionidae). *Plant Protec. Sci.*, 51, 155–161, 2015.
82. Silva, M.D., Sa, R.A., Napoleao, T.H., Gomes, F.S., Santos, N.D., Albuquerque, A.C., Xavier, H.S., Paiva, P.M., Correia, M.T., Coelho, L.C., Purified *Cladoniaverticillaris* lichen lectin: Insecticidal activity on *Nasutitermescorniger* (Isoptera: Termitidae). *Int. Biodet. Biodegrad.*, 63, 334–340, 2009.
83. Uysal, H., Altun, D., Aslan, A., Drosophila Melanogaster'deLobaria Pulmonaria (l.) Hoffm. LikenininÖmürUzunluğuÜzerineEtkisi. *TÜBAV. Bilim. Der.*, 2, 271–276, 2009.
84. Khanuja, S.P.S., Tiruppadiripuliyur, R.S.K., Gupta, V.K., Chand, P., Garg, A., Srivastava, S.K., Verma, S.C., Saikia, D., Darokar, M.P., Shasany, A.K., Pal, A., Antimicrobial and anticancer properties of methyl-beta-orcinolcarboxylate from lichen. US20070099993, 2007.
85. Martins, M.C., Lopes, R.S., Barbosa, P.S., Santiago, R., Rodrigues, B.R., de Albuquerque, A.C., Flacão, E.P.S., Lima, V.L.D.M., da Silva, N.H., Pereira,

E.C., Efects of Usnic, Barbatic and Fumarprotocetraric acids on Survival of *Nasutitermescorniger* (Isoptera: Termitidae: Nasutitermitinae). *Sociol.*, 65, 79–87, 2018.

86. Boustie, J. and Grube, M., Lichens-a promising source of bioactive secondary metabolites. *Plant. Gene. Res.*, 3, 273–287, 2005.
87. Lee, K.A. and Kim, M.S., Glucosidase inhibitor from *Umbilicaria esculenta. Cana. J. Microbiol.*, 46, 1077–1081, 2000.
88. Hengameh, P., Rashmi, S., Rajkumar, H.G., *In vitro* inhibitory activity of some lichen extracts against α-amylase enzyme. *Eur. J. Biomed. Pharm. Sci.*, 3, 315–318, 2016.
89. Tekale, S.S., The A-amylase inhibitory and dye degradation potential of the lichen (*Parmeliaperlata*). *Intern. J. Basi. App. Res.*, 8, 256–260, 2018.
90. Salin Raj, P., Prathapan, A., Sebastian, J., Antony, A.K., Riya, M.P., Rani, M.P., Biju, H., Priya, S., Raghu, K.G., *Parmotrematinctorum* exhibits antioxidant, antiglycation and inhibitory activities against aldose reductase and carbohydrate digestive enzymes: an *in vitro* study. *Nat. Prod. Res.*, 28, 1480–1484, 2014.
91. Romagni, J.G., Meazza, G., Nanayakkara, N.D., Dayan, F.E., The phytotoxic lichen metabolite, usnic acid, is a potent inhibitor of plant p-hydroxyphenylpyruvate dioxygenase. *FEBS. Lett.*, 480, 301–305, 2000.
92. Lee, K.A. and Kim, M.S., Glucosidase inhibitor from *Umbilicaria esculenta. Cana. J. Microbiol.*, 46, 1077–1081, 2000.
93. Umeno, A., Horie, M., Murotomi, K., Nakajima, Y., Yoshida, Y., Antioxidative and antidiabetic effects of natural polyphenols and isoflavones. *Mole.*, 21, 708, 2016.
94. Ganesan, A., Mahesh, A., Sundararaj, J.P., Mani, K., Ponnusamy, P., Antihyperglycemic and anti-oxidant activity of various fraction of *Parmotremahababianum*in streptozotocin-induced diabetic rat. *Bang. J. Pharm.*, 11, 935–939, 2016.
95. Shivanna, R., Parizadeh, H., Garampalli, R.H., Screening of lichen extracts for *in vitro* antidiabetic activity using alpha amylase inhibitory assay. *Int. J. Bio. Pharma. Res.*, 6, 364–367, 2015.

10

Potential of Lichen's Extract as Natural Photosensitizer for Dye-Sensitized Solar Cells

Asim Mansha[1]*, Sana[1], Muhammad Usman[1], Sadia Asim[2] and Matloob Ahmad[1]

[1]*Department of Chemistry, Govt. College University Faisalabad, Faisalabad, Punjab, Pakistan*
[2]*Department of Chemistry, Govt. College Women University Faisalabad, Faisalabad, Punjab, Pakistan*

Abstract
During earlier few decades, dye-sensitized solar cells (DSSCs) have gained widespread attention of scientific communities due to their valuable characteristics with respect to silicon solar cells such as low cost, environmental impact, and ease of fabrication. Dyes based on natural origin related to different types of chemical classes have been successfully utilized in DSSCs mechanisms, like- anthocyanin, chlorophyll, carotenoid, coumarins, anthraquinoes etc. In this chapter, the dimension of lichen-based dyes/extracts as DSCCs is highlighted on the basis of latest literature exploration.

Keywords: Lichen, DSCCs, dyes, anthocyanin, carotenoid, coumarins, anthraquinoes

10.1 Introduction

In preceding two decades on account of the scientific advancements along with primary interest in transformation of energy field, the dye-sensitized solar cells (DSSC) emerged as most striking research subject. Basically, DSSC is constituted by a porous layer of conducting substance (like

**Corresponding author*: asimmansha@gcuf.edu.pk

Mohd Yusuf (ed.) Lichen-Derived Products: Extraction and Applications, (221–244) © 2020 Scrivener Publishing LLC

titanium dioxide), covered with a molecular dye (i.e., as in photosynthesis phenomenon in chlorophylls of green plant parts) that absorbs sunlight to make ATPs. The platinum-based catalyst is located above the electrolyte solution under which a titanium dioxide was mixed. As in a conventional alkaline battery, an anode (TiO_2) and a cathode (Pt) are placed on either side of a liquid conductor [1]. In 1991, O'Regan and Grätzel introduced modern DSSC contrivance depending on 10-μm thickness; TiO_2 nanoparticles are utilized here as an optically clear film having high surface area and monolayer of a charge transfer dye are made for coating on it with ideal spectral characteristics which then in term are used for sensitizing a film for light harvesting. The overall incident photon-to-current conversion efficiency (IPCE) yield obtained by the transforming of electric current from incident photon in simulated solar light was 7.1–7.9% while in diffuse daylight it was 12%. A large/short circuit current density J_{SC} (greater than 12 $mAcm^{-2}$) and exceptional stability (sustaining at least five million turnovers without decomposition) and low cost are utilized to make the practical application feasible. Remarkably effective efficiencies are observed when incident solar energy flux having high ratio of 46% are harvested by device, and efficiencies greater than 80% are utilized for electric current transformation from incident photons [1, 2].

As renewable energy producer devices, DSSC have become the most favorable suitor among low worth solar cells including their research because of the extensive accessible coupled materials and their process of among sound efficiencies. They gave an acute strength and an excellent efficiency when in DSSC synthetic dyes are utilized as sensitizers [3]. Among these advantages synthetic dyes have a variety of restrictions like movement to experience a degradation operation of lethal materials and an acute worth [4]. Natural sensitizers which are bio compatible are used to render these restrictions. Phenomenon of light absorption and in TiO_2 conduction band the addition of charges are resulting from a chemical reaction that take place because of the plant pigments like flavonoid, anthocyanin, chlorophyll and carotenoid present in the natural sensitizers [4, 5].

DSCCs convert visible radiation into electricity by using sensitization of the cells. Performances of DSCC are mainly due to the activation ability of dye used as a sensitizer and the electrode. There are many studies have been performed to assess the activation ability of natural origin dyes and pigments from natural resources [5–11]. Generally, in DSSC, the chemical constituents are used found in flowers, seeds, fruits, barks, leaves, etc., as sensitizer and possess excellent stability, higher absorption in the visible range of solar spectrum, excellent electron injection, and efficient metal-to-ligand charge transfer characteristics [12]. In the area of this technology, natural dyes applications have promising development. High cost of metal complex sensitizers is cutting down by the natural dyes and also expensive chemical synthesis process is replaced through simple extraction process. Natural dyes are considered as abundant, easily extractable, eco-safe and environmentally-friendly materials [5, 13].

Lichen dyes in industrial revolution serves an important function. Until the discovery for synthetic dyes in 1950s, lichens were utilized as dyeing of textile materials, and hence, they had high economically viable for many centuries. Importantly lichens were used for purple and violet colors dyeing rather than most lichen species also contain yellowish, brownish, and reddish brown colored components (Figure 10.1) [14–16].

Operation and components of the DSSC based upon the natural dye's illustration, are the important feature of this chapter. Here, in this chapter, the mostly used natural dyes sensitizers along with various plant pigments are described too. It also includes an overview of the research done on the field of natural dues as sensitizers. The formation of electrolyte solution and electrodes is discussed in this chapter along with the advantages and restrictions faced in DSSC.

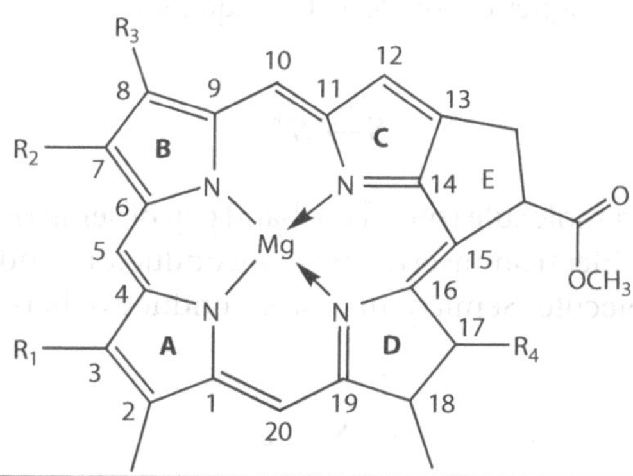

Chlorophylls	R_1	R_2	R_3	R_4
Chlorophyll *a*	CH=CH_2	CH3	CH_2CH_3	$(CH_2)_2COOC_{20}H_{39}$
Chlorophyll *b*	CH=CH_2	CHO	CH_2CH_3	$(CH_2)_2COOC_{20}H_{39}$
*Chlorophyll c_1	CH=CH_2	CH3	CH_2CH_3	CH_2=CHCOOH
*Chlorophyll c_2	CH=CH_2	CH3	CH=CH_2	CH_2=CHCOOH
*Chlorophyll c_3	CH=CH_2	$COOCH_3$	CH=CH_2	CH_2=CHCOOH
Chlorophyll *d*	CHO	CH_3	CH_2CH_3	$(CH_2)_2COOC_{20}H_{39}$

#double bond occur in all chlorophylls c (c1,c2,c3) between C-17 and C-18

Figure 10.1 Molecular structures of major chlorophylls.

10.2 DSSC Operations and Components

10.2.1 Operating Principle

Depending upon the phenomenon of solar radiation transformation to electrical energy, a semiconductor device operates DSSC. It based upon many constituents given as:

a) Fluorine doped tin oxide (FTO)/Indium doped tin oxide (ITO) are generally contained by the clear conducting oxide (TCO). Graphene, aluminum doped TiO_2, and zinc oxide are reported as some cheaper materials utilized as an alternative.
b) Mesoporous metal oxide layer is normally obtained from TiO_2 nanoparticles and can perform as photo anode.
c) Photoanode attaches the sensitizer (dye molecules).
d) Redox reaction is caused by the iodide-tri iodide electrolyte, electrolyte.
e) Platinum is smeared on a glass, counter electrode.

Initially, dye molecule is at ground state (S). Here, at this level of energy, the semiconductor material of anode lies near the valance band (non-conductive). Dye molecules excited when light falls on them and move from their ground energy level to their higher energy level (S*) equation.

$$S \xrightarrow{hv} S^*$$

Now, an excited molecule routed the band gap of semiconductor because of higher energy. Electron inserted in semiconductor conduction band by oxidizing dye molecule. Semiconductor is conductive here so electron can move freely.

$$S \rightarrow S^+ + e^-$$

By using the process of diffusion, electrons moved to the current collector of an anode. If connection is made, then electrical load will be powered. In electrolyte the iodide gave an electron to regenerate the oxidized dye molecule (S^+).

$$S^+ + \frac{3}{2}I^- \rightarrow S + \frac{1}{2}I_3^-$$

On the cathode, the reduction of triiodide can regenerate the iodide in turn.

$$\frac{1}{2}I_3^- + e^- \rightarrow \frac{3}{2}I^-$$

10.3 Constituents

DSCC has four fundamental parameters which are:

1. Semi-conductor electrode
2. Electrolyte
3. Counter electrode
4. Dye molecules (Sensitizers)

DSSC is fabricated by soaking porous layer of TiO_2 with dye molecule (sensitizer) which can absorb sunlight such as chlorophyll in green leaves sealed it to counter electrode (CE) soaked with electrolyte solution, to prevent the leakage, hot melt tape is used.

10.3.1 Working Electrode/Semi-Conductor Electrode

In DSSC system, the working electrode was fabricated by using clear glass plate because they have comparatively low cost as a substrate made of FTO or ITO and settling conductive coating in the form of oxide semiconducting materials having thin film TiO_2, Nb_2O_5, ZnO, SnO_2 (n-type), and NiO (p-type) on substrate. Among these oxide materials, the cheapest is TiO_2 also non-toxic and have good stability. To yield open circuit voltage (VOC), short circuit current (ISC), efficiency (η), and IPCE TiO_2 are used as semiconductor electrode. Under UV radiations, it exhibits photo catalytic degradation of organic compounds. In region of UV, 4% of incident radiation absorbs.

By using spin coating technique Chou *et al.* (2011) [17] stated the formation of composite electrode. When contrasted with prevalent DSSC which demonstrate a 3.27% of conversion energy while 3.8% of the TiO_2/NiO composite particles doped DSSC was recorded. The efficiency enhancement was ascribed to the effects of the NiO. Consequently, Chou *et al.* (2012) developed electrodes of TiO_2 coated with ZnO utilizing dip coating methods along with assess their application in DSCCs, and found satisfactory results [18]. Yan *et al.* (2011) [19] used a double layered photo-anode made up of highly crystalline TiO^2 nanooctahedra and agglutinated mesoporous TiO^2 microspheres as high efficiency DSCCs. Using a fine film of semiconductor or insulator for TiO_2 coating to make a variation of TiO_2 can enhance the efficiency and photo electrical properties of it by producing energetic properties which in turn avoid the back reaction. The reformation of the TiO_2 layer with gadolinium was made by Zalas *et al.* (2011) [20], and it adsorbed the lesser quantity of dye. Under 400 W/m^2 tungsten-halogen lamp irradiation the efficiency exhibited by the modified TiO_2 was 0.74%, while contrasted with the non-modified electrode, it shows a 0.55% efficiency.

10.3.2 Electrolyte

The main part through which conductivity of an internal electric ion produced *via* spreading within film of TiO_2 having mesoporous surface provided to DSSC including a critical point for determining the working of cell is an electrolyte. Because of this study, the various types of electrolytes multifold attempts have been performed such as room temperature ionic liquid electrolytes (RTIL), redox couple, volatile organic solvent (VOS), polymer electrolytes, and so on.

10.3.2.1 Volatile Organic Solvent

IPEC peak values are exhibitedranging from 74 to 78%; the most frequently used electrolyte in DSSC is VOS. They have restrictions such as the requirement of complex scaling process and the poor long span chemical, thermal, and electrochemical stability of cells, but they have high efficiency. VOS have restricted use over practical applications because they can cause a loss of supplements and degradation due to their low boiling point [2-picoline (128°C) and TBP (197°C)].

10.3.2.2 Room Temperature Ionic Liquid Electrolytes (RTIL)

At room temperature because of the ionic conductivity of high concentration, non-volatility, non-inflammability and their chemical stability, they have much more concern as a substitute to VOS. When compared to VOS, non-volatile solvents have higher viscosity which leads to higher energy conversion. They should have to enhance the conductivity and the connection between working and CEs. The problems such as volatilization and leakage produced by using RTIL prevent its utilization in DSSC despite their benefits. They also organized the threat of evaporation, solvent leakage and cause limitation to the shape of device. Polymer electrolytes have been introduced to get over these problems.

10.3.2.3 Polymer Electrolytes

Polymer electrolytes are formed by dissolving metal salts having low lattice energy in polymer matrices like ether, amide, or ester linkage. They not just have an excellent thermal stability but also have an excellent ionic conductivity. They overcome dispute of solvent leakage including sealing. There is a powerful Lewis-type acid-base interactions present between the cations and the donor atom present in structure and combined with rapid segmental motion of these electrolytes which, in turn, developed an ionic conductivity. They gave the easier assembling of cells as compared to the liquid electrolytes. With DSSC components, they are non-corrosive. When contrasted to other electrolytes by snaring liquid solvents, inorganic salts like ethylene carbonate, and an organic solvent, the gel polymer electrolytes (GPE) are formed. When compared to formal polymer electrolytes, they gave an excellent relationship between electrodes and have low vapor pressure and higher ionic conductivity.

To set liquid electrolytes sol gel silica with nanocomposites organic-inorganic gel electrolytes are used by quasi solid state DSSC which have many benefits are as follows: (1) As they carry out uniform environment for ionic movement with liquid electrolytes exhibiting higher ambient ionic conductivity. (2) As liquid electrolytes carry inside gel nets the appliance instability is routed which is produced through liquid electrolytes that were evaporated. (3) Electrical conductivity is verified between Titania and the electrolyte when liquid sol can move into Titania mesoporous structure.

10.3.2.4 Redox Couple

To calculate the ionic conductivity of DSSC iodide salts are an effectively known apart from all electrolytes. But some flaws have been occurred in I^-/I^{3-} redox couple as partial absorption of visible light near 430nm by I^3-species and corrosion of metallic grids. They are useful to regenerate the oxidized dye efficiency. In 2003, for the improvement of solar cell working, Kusama and Arakawa (2004) [21] have been recommended the additives to I^-/I^{3-} redox couple. In acetonitrile, I^-/I^{3-} redox electrolyte was crammed with bis (tetrabutyl ammonium) cis-bis (thiocyanato) bis (2,2'-bipyridine-4-carboxylic acid, 4'-carboxylate) ruthenium II as sensitizer for DSSC to know strength of alkylpyridine additive. These additives enhance the solar cell fill factor including voltage of open circuit and decreased current of short circuit. The exhibited efficiency of DSSC under solar radiation of am 1.5 was 7.6%. They have the redox mediator having highly positive potential. This type of electrolyte also has some drawbacks of energy loss during dye regeneration even though they have shown the high efficiency. Therefore, current researches have been made to overcome the drawbacks.

In addition, another group of Min *et al.* (2011) [22] utilized an organic mediators as 2,2,6,6-tetramethyl-1-piperidyloxy and different redox mediators as $SCN/(SCN)_3$, $SeCN/(SeCN)_3$, (CO^{+2}/CO^{+3}), and (CO^{+}/CO^{+2}). To decrease the losses occurred due to voltage because of Nernst potential of iodine-based redox couple and to hinder the photo current leakage as caused by the light absorptionby trioxide ions, various redox couples synthesized to overcome the drawback of efficiency and $I-/I^{3}-$ is well-known till now. TBP on the surface of TiO_2 decreased the recombination caused by the back transfer to an electrolyte. These solid-state electrolytes are manufactured for encountering the problem related to ionic liquid.Sometimes an electrolyte has fifth components known as cations where dye and the spectra which are obtained by absorption don't intermix.

10.3.3 Counter Electrode

The energetic difference between the nanocrystalline counter electrodes and the Fermi level to nanocrystalline TiO_2 films are used to calculate the open circuit voltage in DSSC. In DSSC utilizing carbon (C) or platinum (Pt) counter electrodes are fabricated. We can also say that the counter electrodes are the triodide ions which produced by minimizing iodide ions in electrolyte, Pt is commonly used as counter electrodes. Being stable and clear electrical resistance of less value and electro catalytic action of higher value for redox reaction of triodide can show an ideal CE property.

When platinum used as CE, it gives some limitations such as high temperature sintering, instability in redox electrolyte and resource scarcity with high energy conversion efficiency. Because of these limitations caused by the platinum (Pt) quest of new CE having lesser worth, CE of excellent catalytic property which has chemical stability increases. As alternative materials, various carbonaceous materials were endeavored. Less catalytic action demonstrated by the various carbon materials as CE is requited through porous structure for catalytic film to enhance the surface area. To get efficiencies equivalent to the Pt CE the CEs of carbon are required having some μm thickness greater than Pt. For DSSC among carbon, graphene appeared as a potential catalyst. Preparations of compounds which are related to graphene through spray deposition technology, electrophoretic deposition, chemical oxidation and doctor blading technique are most effective ways. The stand-alone graphene films have been synthesized by the Cruz *et al.* using graphene oxide (GO) [23]. Reference equivalent efficiencies have been shown by the graphene films made by spray deposition technology. Currently most effective alternative carbon nanotubes counter has been adopted as electrode because of its low cost, high efficiency, and large surface area. Furthermore, the most adopted technique is direct growth of vertically aligned carbon nano tube (VACNT) [21, 22].

10.4 Natural Dye Sensitizers

Acton *et al.* (2014) studied the lichenized fungi extract as dye sensitizer for harnessing energy from the sun and observed the convert up to 15% of solar energy into electrical energy, which are cheaper to manufacture than traditional photovoltaic system [24]. Several pigments and various colors are exhibited by naturally available fruits, bacteria, leaves, and flowers etc., that can be obtained and utilized in DSSC. Because of their absorption in region of UV, ease of formation, relative abundance including environmental affability natural dyes have a meaningful importance in the DSSC as photosensitizers. Worth of DSSC has been an efficient when based upon the formation route for natural dyes because it does not have any noble metals such as Ru. Transmitted or reflected wavelength of plant tissues are changed by the interaction of sunlight with the exhibited electronic structure of plant pigments and leads toward the process of pigmentation. Wavelength of highest absorbance (λ_{max}) and the color noticed by humans explain the pigments. When contrasted with formatted dyes and pigments of natural origin, these easily exhibit sensitizing abilities [20–24] such as chlorophyll, carotenoid, flavonoids, anthocyanin etc., as shown in Table 10.1.

10.4.1 Chlorophylls

Plant green color given by natural photosynthetic pigments contains chlorophyll. The main role of chlorophyll is an effective harvesting and transduction of light energy for photosynthesis which is mainly obtained by their spectral properties. Chlorophyll "a" and chlorophyll "b" are the two main kinds of chlorophyll. Chlorophylls and its derivatives have an ability to absorb red and blue light because of which they can be utilized as sensitizers in DSSC. Methyl trans-32-carboxy -pyrophero-phorbide is most effective derivative of chlorophyll "a". When contrast with chlorophyll "a"

Table 10.1 Dyes and pigments of natural origin with classes used as dye sensitizers.

Color	Pigment type	Pigment group	Examples
Green	Delphinidin Cyanidin	Flavonoid Flavonoid	Mostly blue fruits and flowers Rarely occurred blue such as Ipomoea flower
Yellow	Chalcone Betaxanthin Flavonol Aurone	Flavonoid Betalain Flavonoid Flavonoid	Dianthus Portulaca flowers Cotton flowers Antirrhinum majus flower
Black	Delphinidin	Flavonoid and Carotenoid mix	Viola flowers
Cream	Flavones	Flavonoid Carotenoid	Cream flowers Some red flower
Pink to red	Pelargonidin and/or cyanidin Betacyanin Anthocyanin and Carotenoid	Flavonoid Betalain Carotenoid Flavonoid and Carotenoid	Red and Pink flower and fruits such as Eustoma grandiflorum flower and Apple fruit Bougainvillea flowers Marigold flowers Tulipa Flowers

the chlorophyll "b" has exhibited an absorbance spectrum which describe characteristic blue tinge and redshift. Chlorophyll a and chlorophyll b are the most abundant types in higher plants and generally in ratios between 3:1 and 1:1 [25]. The molecular structures of major chlorophylls are shown in Figure 10.1.

10.4.2 Anthocyanins

Besides chlorophyll, the usual pigments groups which clear to naked eye of human beings are anthocyanin. Chemically anthocyanin depends upon the frame of C_{15} with a chromane ring having second aromatic ring

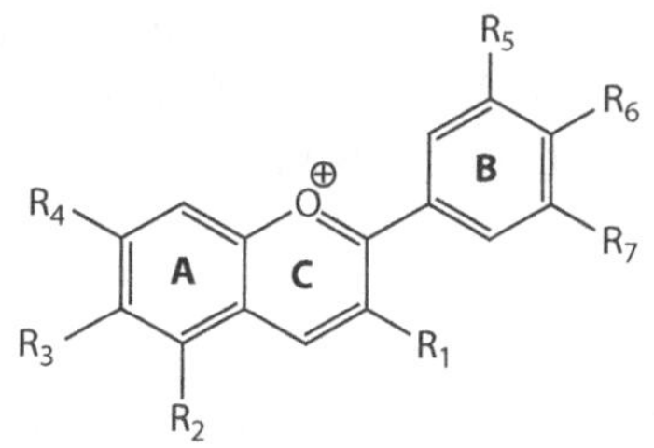

Figure 10.2 Chemical structure of anthocyanins.

B in position 2 (C_6-C_3-C_6). Figure 10.2 shows the chemical structure of anthocyanins.

10.4.3 Anthocyanidins

As per compounds such as monosides, bioside, trioside etc. are formed from the sugar molecules, anthocyanin has been classified into 17 kinds. By considering sugar diversity along with the feasible structural points of glycosylation the types increased. It takes in mind the pleasant color of the flowers, fruits and leaves from scarlet to blue and for the condition of an incident light. In plants, anthocyanin varies from plant to plant that adoring plants such as dianthus and petunia have one main type, while rosa and verbena have mixtures. Fruits are also main cause of anthocyanins. Grapes contain almost all kinds of anthocyanin. Cherry, fi apple, and peach contain cyanidin while pomegranate and eggplant contain delphinidin. In flower mostly epidermas cell and sometimes mesophyll have anthocyanin. Excitation and transfer of an electron toward the porous TiO_2 film conduction band from the anthocyanin molecules are encouraged by the bond of

Table 10.2 Positions of hydroxyl groups in anthocyanidins.

Name	Position	Produced colors
Cyanidin	3,5,7,3',4'	Crimson and magenta
Luteolinidin	5,7,3',4',	Orange
Aurantindin	3,5,6,7,4'	Orange
Triacetidin	5,7,3',4',5'	Red
Delphinidin	3,5,7,3',4',5'	Blue and purple

carbonyl and hydroxyl to TiO_2 semiconductor filmthat is present in anthocyanin molecules (Figure 10.2). Position of subsituated hydroxyl groups in basic anthocyanin structure is encapsulated in Table 10.2.

10.4.4 Flavonoids

The most significant group of natural ingredients that provide colors to flowers is flavonoids. Gessiman and Hinreiner first used the word "flavonoid" for all the compound that depends upon the flavones having C_6-C_3-C_6 skeleton. Flavone depends upon the two benzene rings attached in the way that are easy to differentiate the one flavonoid compound from other. In the visible region only, some flavonoids have an ability to absorb light. Figure 10.3 shows the commonly used chemical structures of flavonoids.

10.4.5 Carotenoids

Many fruits and flowers with different colors such as red, yellow, and orange are provided by the isoprenoids which have the large family of carotenoids (over 600 members). Backbone chain of C_{40} hydrocarbons required to differentiate the carotenoids that induced an oxygen and structural modification. Chemical structures of several carotenoids are shown in Figure 10.4.

The absorbance of light developed by the chlorophyll pigments is the main structural property of carotenoids. In carotenoids, photo induced

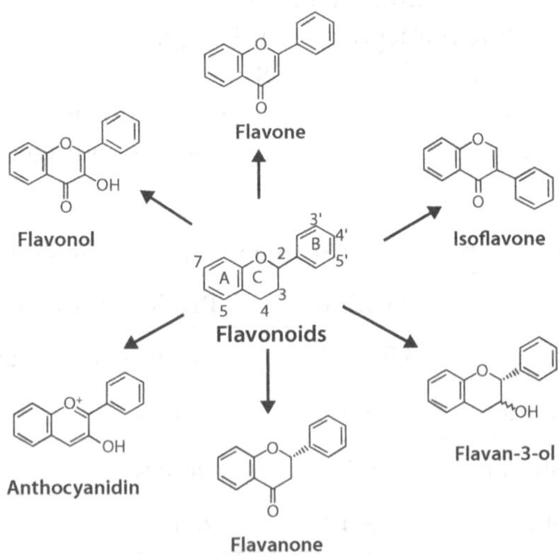

Figure 10.3 Chemical structure of flavonoids.

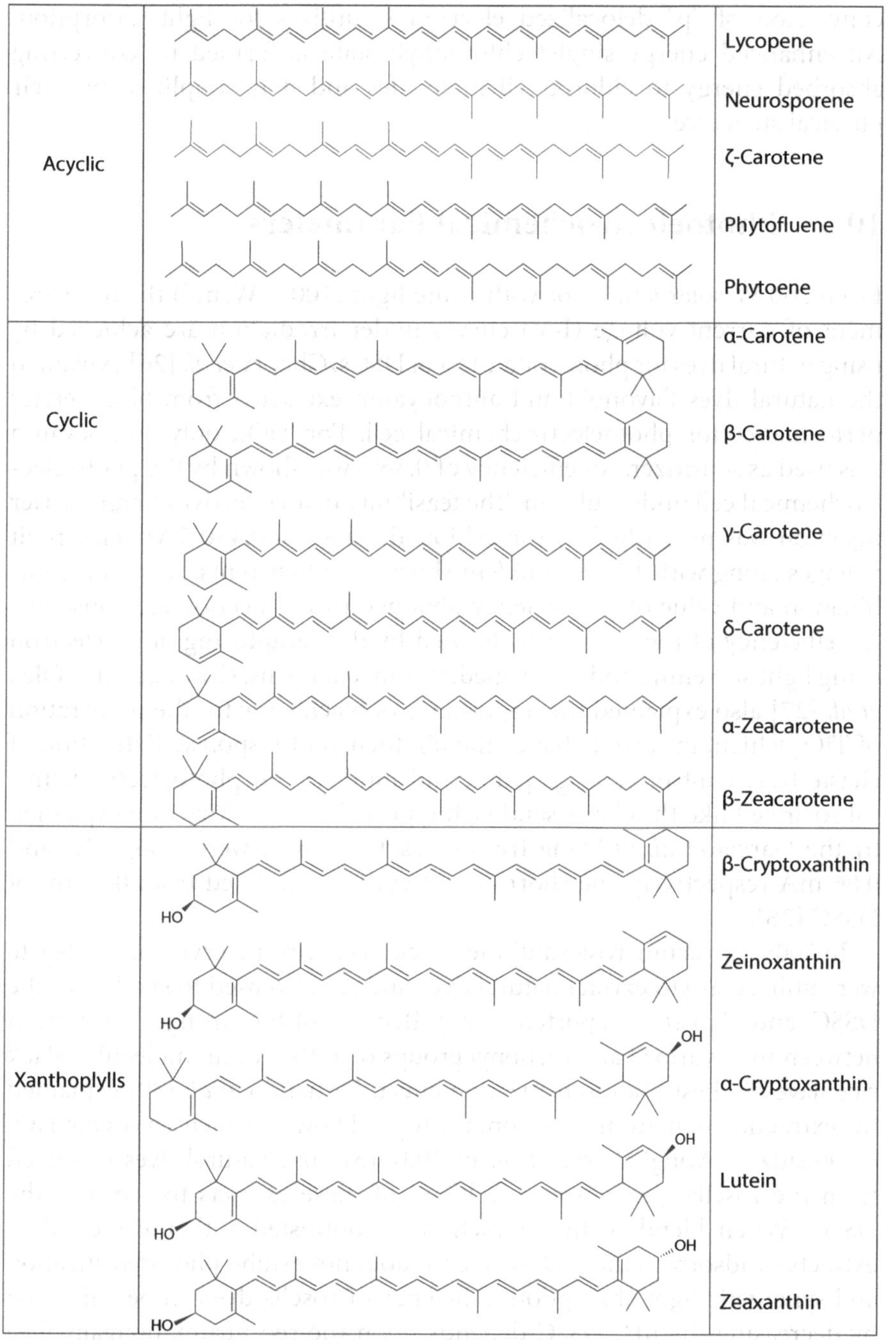

Figure 10.4 Chemical structures of several carotenoids.

conversion of "p" delocalized electron is utilizes for light absorption. An enhanced energy singlet chlorophyll state is formed by converting absorbed energy to chlorophyll molecules and it is simplified by their physical structure.

10.5 Photoelectrochemical Parameters

From 300 W solar simulator with white light (100 $mWcm^{-2}$) the measurement of current-voltage (I-V) curves under irradiation are achieved by using natural dyes for photovoltaic test of DSCs. Chepry *et al.* [26] explained the natural dyes flavonoid and anthocyanin extracted from blackberries performance for photoelectrochemical cell. For TiO_2, only anthocyanin was used as sensitizer. An efficiency of 0.56% was shown by the photo electrochemical cell under full sun. The feasibility of an effective charge carrier injection was prescribed by the exhibited values of 0.4–0.5 Vopen circuit voltages along with 1.5–2.2 mA/cm^2short circuit currents. Rate of recombination and value of an efficiency obtained from injection are constants. An efficiency of the cell can be limited by the recapturing of an electron using light screening and redox mediator mechanisms. Consequently, Olea *et al.* [27] also explained the importance of blackberry for the preparation of TiO_2 which, in turn, enhance the photocurrent response. Extraction of chaste tree, fruit of cabbage-palm, and mulberry in photoelectrochemical solar cell like thin layer sandwich type as TiO_2 sensitize rare explained by the Garcia *et al.* utilizing fresh extracts of these were 1.06, 0.37, and 0.86 mA respectively, the short-circuit current exhibited from 0.5 cm^2 of DSSC [28].

In 2006, capsicum, rosa xanthine, black rice, kelp and erythrina variegate were utilized as yto extract natural dyes and are also used to synthesize the DSSC, and Hao *et al.* reported it [29]. Because of the stronger interaction between the hydroxyl and carbonyl groups of anthocyanin molecules black rice have the best photosensitization effect. Fernando *et al.* [30] explained the extraction of natural dyes from the topical flowers which serves for TiO_2 as sensitizer. Wonghcharee *et al.* in 2007 [31] use natural dyes extracted from the rosella, pea flower and blend of these extracts to sensitize the DSSC. When blend of the extraction is contrasted with the individual extracts to adsorb on the TiO_2 surface, it does not exhibit photosensitization and synergetic light absorption. Efficiency of rosella dye can be enhanced by decreasing the pH. DSSC depends upon the red turnip, bougainvillea and the purple wild sicilian prickly pear fruit juice which has been synthesized and characterized by the Calogero *et al.* (2010) [32]. DSSC shows a best interpretation resulting a 9.5 mA/cm^2 density and an 1.70% efficiency. Large amount of betaxantins (betalain pigments) are found in the red turnip having ball-shaped roots that is the major cause of its high efficiency. Anchoring on the surface of TiO_2 could be enhanced by the occupancy of the carboxylic groups in betalains. Bazargan studied the performance of nanostructured DSCC utilizing natural sensitizer operated with platinum and carbon coated CEs digestion [33].

Pigments of shisho plants are also utilized as sensitizer. Kumara *et al.* [34] illustrated the 0.53 open circuit voltage, 1.30% efficiency, and 4.80 mA current density. For the preparation of DSSC, almost all the natural dyes are extracted from the natural things that are used such as flowers, leaves, seeds, fruits, and beverages, and traditional Chinese medicines are utilized. Natural pigments of various types having anthocyanins are extracted from the tropical flowers are reported [35].

Additionally, an achievable dye has been reported for the enhancement of 13.4% efficieny of power transformation as good sensitizing ability in near IR photons having 1.48 eV (840 nm absorption onset) optical band gap [36]. Seven species of *Cortinarius* fungi group collected in the East Poland were investigated to be photosensitizer for DSCCs by Zalas *et al.* [37]. In the study, it is concluded that the abilities to absorb visible light by working electrodesas well as the absorption efficiency of dyes both were strictly depends upon the photoconversion process of the particular cell. As a result, *Cortinarius* fungi were found to be effective source of sensitizers for DSSCs. In a study, dyes which wasextracted from the fungi *Monascu* shaving 2.3% total efficiency was observed by Ito *et al.* [38]. Thus, the theoretical studies showed that similar monascus dyes are worked as potential sensitizer which was observed by studying four selected monascus dyes [39]. In a similar way of same group, small scale insects (cochineal and lac) are utilized to extract a group of dyes which can practically be used as photosensitizer in DSSC and enhanced photochemical properties are observed [40].

Lichen-derived dyes in industrial revolution have significant function. Dyeing of textile materials is commercially most valued application of lichens and lichen dyes are commercially more valued for dyeing textile materials and for many centurieshad high economic value until the discovery for synthetic dyes in 1950s. Due to political and economic pressure, European dye industries lagged the British dye industries to develop natural dyes from indigenous plants [41]. Orcein, a blend of various compounds/natural dyes, is the chief constituent of many lichen species, most abundant is *Roccella* sp., but here it was formed by utilizing hydrogen peroxide to carry oxidation reaction when ammonia was present [39]. Moreover, the

Table 10.3 Photoelectrochemical parameters of some natural sensitizers based on the literature [2–12, 22–26, 38–42].

Dye solution	Semiconductor electrode	Electrolyte	Pigments	λ_{max} (nm)	Jsc ($mAcm^{-2}$)	Voc (V)	FF	η (%)
Fructus lucii	TiO_2	I^-/I_3^-	Carotene	447,425	0.53	0.689	46.6	0.37
Rosella sp.	TiO_2	I^-/I_3^-	Anthocyanin	520	1.63	0.40	0.57	0.37
Rosella sp.	TiO_2	I^-/I_3^-	β-Orcein/γ-Orcein	460/462	1.62	1.71/ 0.860	0.742/ 0.741	–
Erythrina variegate	TiO_2	I^-/I_3^-	Chlorophyll	451,492	0.78	0.48	0.55	–
Blue pea flower	TiO_2	I^-/I_3^-	Anthocyanin	580	0.37	0.37	0.33	0.05
Raspberries	TiO_2	I^-/I_3^-	Betacyanin	540	0.26	0.42	64.8	1.50
Grapes	TiO_2	I^-/I_3^-	Betacyanin	560	0.09	0.34	61.1	0.38
Cheries	TiO_2	I^-/I_3^-	Betacyanin	500	0.46	0.30	38.3	0.18
Capsicum	TiO_2	I^-/I_3^-	Carotenoid	455	0.23	0.41	0.63	–

(Continued)

Table 10.3 Photoelectrochemical parameters of some natural sensitizers based on the literature [2–12, 22–26, 38–42]. (*Continued*)

Dye solution	Semicon-ductor electrode	Electrolyte	Pigments	λ_{max} (nm)	Jsc ($mAcm^{-2}$)	Voc (V)	FF	η (%)
Wild Sicilian prickly pear	TiO_2	I^-/I_3^-	Betalin	465	8,20	0.38	0.38	1.19
Chaste tree fruit	TiO_2	I^-/I_3^-	Anthocyanin	548	1.06	0.39	0.48	
Mulberry	TiO_2	I^-/I_3^-	Anthocyanin	543	0.86	0.42	0.43	
Spinach	TiO_2	I^-/I_3^-	Chlorophyll	437	0.47	0.55	0.51	0.13
Arugula	TiO_2	I^-/I_3^-	Chlorophyll	666	0.78	0.59	42.0	0.20
Brassica olercea	TiO_2	I^-/I_3^-	Anthocyanin	537	0.50	0.37	0.54	0.13
Shico	TiO_2	I^-/I_3^-	Chlorophyll	440,600	3.56	0.55	0.51	1.01

(*Continued*)

Table 10.3 Photoelectrochemical parameters of some natural sensitizers based on the literature [2–12, 22–26, 38–42]. (*Continued*)

Dye solution	Semicon-ductor electrode	Electrolyte	Pigments	λ_{max} (nm)	Jsc ($mAcm^{-2}$)	Voc (V)	FF	η (%)
Green Algae	ZnO, TiO_2	I^-/I_3^-	Chlorophyll	666	0.08 0.13	0.32 0.41	0.37 21.0	0.01 0.01
Walnuts	ZnO	I^-/I_3^-	Carotenoid	510	0.73	0.304	0.39	0.0104
Madder	TiO_2	I^-/I_3^-	Anthraquinone	540	0.540	0.389	0.69	0.10
Turmeric	TiO_2	I^-/I_3^-	Carotenoids	535	0.268	0.372	0.46	0.03
Tagetserecta	TiO_2	I^-/I_3^-	Xanthrophyll	465	2.891	0.475	0.606	0.8
Gardenia blue	TiO_2	I^-/I_3^-	Carotenoid	600	0.53	0.44	0.69	0.16
Cosmos sulphureus	TiO_2	I^-/I_3^-	Flavonoids	505,590	1.041	0.447	0.61	0.54
Black Carrot	TiO_2	I^-/I_3^-	Anthocyanin	540	1.302	0.4	0.47	0.25

same research group of Sang-aroon studied α, β and γ-orceins by density functional theory (DFT) and time-dependent TDDFT at the B3LYP/6-31+G(d) level to observe the function of DSCCs, and the integral equation formalism-polarizable continuum model (IEFPCM) employed to access the solvation effect [42]. In the study, TiO_2 was used as electrode with I–/ I_3– electrolyte oxidation potential and it was concluded that β and γ-orcein dyes possess the best computed parameters related to photochemical properties such as excited state oxidation potential, electron injection force, and photoinduced electron transfer and also open-circuit voltage. Natural dyes have been obtained from the seeds, and flowers. Fruits, leaves, trees barks and other parts with their photoelectrochemical parameters based on the literature [2–12, 22–26, 38–42] are given in Table 10.3.

10.6 Preparation of Dye Solution and Electrodes

Novel survey illustrates that the conversion efficiency higher than 12.0% are shown by the DSSC. With a catalytic CE, porous semiconductor working electrode and an electrolyte between these two electrodes DSSC demonstrated the sandwich like structure. Utilizing screen printing is one of the most stable and important methods of synthesizing DSSC working electrode.

10.6.1 Preparation of Working Electrode

Because of low electronic transfer and increased charge recombination of TiO_2, it has been recognized as the standard of semiconductor electrode from 1991. In 2004 DSSC having 10.2% efficiency was synthesized by Hočevar *et al.* [43] with distinct TiO_2 paste which was gained by disintegrating Ti-isopropoxide into ethylene glycol and citric acid blend.

The following steps are performed to synthesize the working electrode of DSSC with TiO_2 particles.

i. Preparation and homogenization of TiO_2 particles colloid.
ii. On the surface of FTO glass, the TiO_2 particles was accumulated by using spin coating and then for 1h it was sintered in a high furnace temperature at 500°C.
iii. At 70°C, the film of TiO_2 articles with the FTO glass substrate was dissolved in N-17 dye and ethyl alcohol solution for 6 h. Here, for the DSSC calculation the TiO_2 electrode have an arear of 0.25 cm^2.

To get the suitable working electrode film thickness on glass substrate surface of FTO anα-step surface profiler can be used.

10.6.2 Preparation of CE

Utilizing film of Pt sandwiched between the FTO glass substrate and NiO film the process for the preparation of CE of DSSC is as follows

i. In a furnace of high temperature at 500°C process of annealing the Ni powder was performed to get the P-type NiO.
ii. Utilizing an ultrasonic homogenizer NiO colloid which was made by blending 1g of NiO with solvents such as acetylacetone (0.1 ml), water (20ml),Triton X-100 (0.1 ml) andethanol (1ml) are homogenized for 30 min.
iii. Utilizing an E-beam evaporator (Ion-sputter coater), on FTO glass substrate a Pt film had been accumulated which was made by depositing a 2 to 3 ml NiO colloid using spin coating.
iv. Process of sintering was performed to this substrate for 1 h in a furnace of high temperature at 500°C.

CE FTO was accumulated with the Pt film the ion sputter coater having 10-2 Torr vacuum level an E-beam evaporator with 4×10^{-7} Torr were used to demonstrate the performance of DSSC. Here, 4.0 cm^2 area is used for the Pt film. In this study two stage spin coating was used (1) to eliminate an unwanted NiO colloid from substrate the time of 5 s and 1000 rpm rotation speed and spin coating of two stage was utilized; (2) to homogenize the NiO film on substrate the time duration of 15 s and 1500 rpm was used. Utilizing the digital camera and scanning electron microscope the picture and micrograph of CE respectively are achieved.

10.6.3 Preparation of Dye Solution

Different natural dyes are extracted from the various natural materials and different type of low dye concentration in solution phase nearly 500-800 μM (depending upon the dye sample) are prepared from them because dyes show some aggregation properties. Synthesis of some such dyes solutions is as:

At different temperatures 1g of blue pea flower or fresh rosella was extracted in two different solvents of 100ml for 30 min. To get the pure dye solution solid remains are filtered out from the solution. At the ratio of 1:1, the two dyes solutions are blended to get the mixed dye solution by volume. To observe the effect of extracting water was utilized as solvent of extraction at varying temperature from 25 to 100°C. By contrasting water and ethanol extracted dyes extracting solvent effect was observed. By blending 0.5 mM Ru-dye (N-749 or N-719) with ethanol was also used to synthesize the dye solution for DSSC.

10.7 Conclusion and Future Outlook

From the beginning of DSSC like semiconductor electrodes to the deep in DSSC like CEs and natural dye sensitizers all are overviewed in this chapter. Among the limitations that are faced in DSSC based on natural dyes, each part of them has been formed by using variety of materials. The need of eco-friendliness and environmental concerns has developed the replacement of the synthetic dyes with natural dyes to make cleaner alternatives. Solar cells manufactured by using natural dyes have not only tendency to exhibit less amount of light electricity conversion efficiency but also have an advantages as they are less expensive, friendly to the environment, nontoxic, and easily accessible as contrasted with the solar cells manufactured using synthetic dyes. Because of this, more concern are developed toward the natural sensitizers. As of now, pigments of plants that are present in higher amount having greater stability and better results as compared to their counterparts. Main progresses achieved, nowadays, are natural dyes and pigments in DSSC as important photosensitizers. However, lichen-derived dyes and pigments have shown a tendency to be used as photosensitizers in DSSC. But, a few researches were carried out so far. In this regard, more Research & Development is required to explore enhancement in DSSCs with sustainable and eco-friendly approaches.

References

1. Sharma, K., Sharma, V., Sharma, S.S., Dye-sensitized solar cells: Fundamentals and current status. *Nanoscale Res. Lett.*, 13, 1, 381, 2018. https://doi.org/10.1186/s11671-018-2760-6
2. O'Regan, B. and Grätzel, M., A low-cost, high-efficiency solar cell based on dye-sensitized colloidal TiO_2 films. *Nature*, 353, 6346, 737–740, 1991.
3. Calogero, G., Yum, J.H., Sinopoli, A., Di Marco, G., Grätzel, M., Nazeeruddin, M.K., Anthocyanins and betalains as light-harvesting pigments for dye-sensitized solar cells. *Sol. Energy*, 86, 5, 1563–1575, 2012.

4. Ye, M., Wen, X., Wang, M., Iocozzia, J., Zhang, N., Lin, C., Lin, Z., Recent advances in dye-sensitized solar cells: From photoanodes, sensitizers and electrolytes to counter electrodes. *Mater. Today*, 18, 3, 155–162, 2015.
5. Yusuf, M., Shabbir, M., Mohammad, F., Natural colorants: Historical, processing and sustainable prospects. *Nat. Prod. Bioprospect.*, 7, 1, 123–145, 2017.
6. Gong, J., Liang, J., Sumathy, K., Review on dye sensitized solar cells (DSSCs): Fundamental concepts and novel materials. *Renew. Sustain. Energy Rev.*, 16, 5848–5860, 2012.
7. Phinjaturus, K., Maiaugree, W., Suriharn, B., Pimanpaeng, S., Amornkitbamrung, V., Swatsitang, E., Dye-sensitized solar cells based on purple corn sensitizers. *Appl. Surface Sci.*, 380, 101–107, 2016.
8. Lim, A., Ekanayake, P., Lim, L.B.L., Bandara, J.M.R.S., Co-dominant effect of selected natural dye sensitizers in DSSC performance. *Spectrochim Acta A: Mol. Biomol. Spectros.*, 167, 26–31, 2016.
9. Aung, S.H., Hao, Y., Oo, T.Z., Boschloo, G., Kinetic study of carminic acid and santalin natural dyes in dye-sensitized solar cells. *J. Photochem. Photoiol. A: Chem.*, 325, 1–8, 2016.
10. Ho, C. and Rong, L., Fabrication of natural sensitizer extracted from mixture of purple cabbage, roselle, wormwood and seaweed with high conversion efficiency for DSSC. *J. Nanosci, Nanotechnol.*, 16, 4, 2072–2075, 2016.
11. Eli, D., Ahmad, M., Maxwell, I., Ezra, D., Francis, A., Sarki, S., Dye sensitized solar cells using natural dyes extracted from Roselle (*Hibiscus Sabdariffa*) flowers and Pawpaw (*Carica Papaya*) leaves as sensitizers. *J. Energy Nat. Resour.*, 5, 1, 11–15, 2016.
12. Al-Ghamdi, A.A., Gupta, R.K., Kahol, P.K., Wageh, S., El. Shirbeeny, W., Improved solar efficiency by introducing grapheme oxide in purple cabbage dye sensitized TiO_2 based solar cell. *Solid State Commun.*, 183, 56–59, 2014.
13. Yusuf, M., Mohammad, F., Shabbir, M., Eco-friendly and effective dyeing of wool with anthraquinone colorants extracted from Rubia cordifolia roots: Optimization, colorimetric and fastness assay. *J. King Saud Univ.- Sci.*, 29, 2, 137–144, 2017.
14. Cardon, D., *Natural Dyes: Sources, Tradition, Technology and Science*, Archetype Publications, London, 2007.
15. Casselman, K.D., *Lichen Dyes and Dyeing: The New Source Book*, Courier Corp., Dover Publications, Mineola, New York, USA, 2001.
16. Casselman, K.D., Magic, mystery and mayhem: Lichen dyes old and new, in: *Publication of the Natural Dyes Product Research and Development Project*, EVTEK Institute of Art and Design, Vantaa, 2002.
17. Chou, C.S., Lin, Y.J., Yang, R.Y., Liu, K.H., Preparation of TiO_2/NiO composite particles and their applications in dye-sensitized solar cells. *Adv. Powder Technol.*, 22, 1, 31–42, 2011.
18. Chou, C.S., Chou, F.C., Kang, J.Y., Preparation of ZnO-coated TiO_2 electrodes using dip coating and their applications in dye-sensitized solar cells. *Adv. Powder Technol.*, 215, 38–45, 2012.

19. Yan, K., Qiu, Y., Chen, W., Zhang, M., Yang, S., A double layered photoanode made of highly crystalline TiO_2 nanooctahedra and agglutinated mesoporous TiO_2 microspheres for high efficiency dye sensitized solar cells. *Energy Environ. Sci.*, 4, 6, 2168–2176, 2011.
20. Zalas, M., Walkowiak, M., Schroeder, G., Increase in efficiency of dye-sensitized solar cells by porous TiO_2 layer modification with gadolinium-containing thin layer. *J. Rare Earths*, 29, 8, 783–786, 2011.
21. Kusama, H. and Arakawa, H., Influence of alkylaminopyridine additives in electrolytes on dye-sensitized solar cell performance. *Sol. Energy Mater. Sol. Cells*, 81, 1, 87–99, 2004.
22. Min, J., Won, J., Kang, Y.S., Nagase, S., Benzimidazole derivatives in the electrolyte of new-generation organic dye-sensitized solar cells with an iodine-free redox mediator. *J. Photochem. Photobiol. A: Chem.*, 219, 1, 148–153, 2011.
23. Cruz, R., Tanaka, D.A.P., Mendes, A., Reduced graphene oxide films as transparent counter-electrodes for dye-sensitized solar cells. *Sol. Energy*, 86, 2, 716–724, 2012.
24. Acton, N., Bernazani, M., Hill, J., Hinton, M., Vazquez, A., Gipson, K., Nagel, J., Design of a Climate Adaptable Solar Energy system using biomimetic inspiration from a lichen symbiosis. In: *Systems and Information Engineering Design Symposium (SIEDS)*, April 25, IEEE, Charlottesville, VA, USA, pp. 177–182, 2014.
25. Khan, S.A., Yusuf, M., Agarwal, P., Prasad, L., Chlorophylls as Pigment: A Contemporary Approach, in: *Handbook of Renewable Materials for Coloration and Finishing*, M. Yusuf (Ed.), pp. 115–123, Scrivener Publishing LLC, Beverly, USA, 2018.
26. Cherepy, N.J., Smestad, G.P., Grätzel, M., Zhang, J.Z., Ultrafast electron injection: Implications for a photoelectrochemical cell utilizing an anthocyanin dye-sensitized TiO_2 nanocrystalline electrode. *The J. Phys. Chem. B*, 101, 45, 9342–9351, 1997.
27. Olea, A., Ponce, G., Sebastian, P.J., Electron transfer *via* organic dyes for solar conversion. *Sol. Energy Mater. Sol. Cells*, 59, 1–2, 137–143, 1999.
28. Garcia, C.G., Polo, A.S., Iha, N.Y.M., Fruit extracts and ruthenium polypyridinic dyes for sensitization of TiO_2 in photoelectrochemical solar cells. *J. Photochem. Photobiol. A: Chem.*, 160, 1–2, 87–91, 2003.
29. Hao, S.C., Wu, J.H., Huang, Y.F., Lin, J.M., Natural dyes as photosensitizers for dye-sensitized solar cell. *Sol. Energy*, 80, 2, 209–214, 2006.
30. Fernando, J.M.R.C. and Senadeera, G.K.R., Natural Anthocyanins As A Photosensitizes For Dye Sensitized Solar Devices. *J. Curr. Sci.*, 95, 5, 663–666, 2008.
31. Wongcharee, K., Meeyoo, V., Chavadej, S., Dye-sensitized solar cell using natural dyes extracted from rosella and blue pea flowers. *Sol. Energy Mater. Sol. Cells*, 91, 7, 566–571, 2007.
32. Calogero, G., Marco, G.D., Cazzanti, S., Caramori, S., Argazzi, R., Carlo, A.D., Efficient Dye Sensitized Solar Cells Using Red Turnip And Purple Wild Sicilian Prickly Pear Fruits. *Int. J. Mol. Sci.*, 11, 254–267, 2010.

33. Bazargan, M.H., Performance Of Nanostructured Dye-Sensitized Solar Cell Utilizing Natural Sensitizer Operated With Platinum And Carbon Coated Counter Electrodes Digest. *J. Nanomater. Biostruct.*, 4, 723–727, 2009.
34. Kumara, N.T.R.N., Lim, A., Lim, C.M., Petra, M.I., Ekanayake, P., Recent progress and utilization of natural pigments in dye sensitized solar cells: A review. *Renew. Sust. Energy Rev.*, 78, 301–317, 2017.
35. Kazmerski, L.L., Photovoltaics: A review of cell and module technologies. *Renew. Sust. Energy Rev.*, 1, 1–2, 71–170, 1997.
36. Snaith, H.J., Estimating the maximum attainable efficiency in dye-sensitized solar cells. *Adv. Function. Mater.*, 20, 1, 13–19, 2010.
37. Zalas, M., Gierczyk, B., Bogacki, H., Schroeder, G., The Cortinarius fungi dyes as sensitizers in dye-sensitized solar cells. *Int. J. Photoenergy*, 1–6, Article ID 653740, 2015. http://dx.doi.org/10.1155/2015/653740.
38. Ito, S., Saitou, T., Imahori, H., Uehara, H., Hasegawa, N., Fabrication of dye-sensitized solar cells using natural dye for food pigment: Monascus yellow. *Energy Environ. Sci.*, 3, 7, 905–909, 2010.
39. Sang-aroon, W., Saekow, S., Amornkitbamrung, V., Density functional theory study on the electronic structure of Monascus dyes as photosensitizer for dye-sensitized solar cells. *J. Photochem. Photobiol. A: Chem.*, 236, 35–40, 2012.
40. Sang-aroon, W., Laopha, S., Chaiamornnugool, P., Tontapha, S., Saekow, S., Amornkitbamrung, V., DFT and TDDFT study on the electronic structure and photoelectrochemical properties of dyes derived from cochineal and lac insects as photosensitizer for dye-sensitized solar cells. *J. Mol. Model.*, 19, 3, 1407–1415, 2013.
41. Fahselt, D., Review article Secondary Biochemistry of Lichens. *Symbiosis*, 16, 117–165, 1994.
42. Sang-aroon, W., Kunmuak, K., Tontapha, S., Chaiamornnugool, P., Saekow, S., Amornkitbamrung, V., Theoretical insight into electronic and photoelectrochemical properties of orcein dyes relevant to dye-sensitized solar cells. *Monatshefte für Chemie-Chemical Monthly*, 145, 10, 1529–1537, 2014.
43. Hočevar, M., Berginc, M., Krašovec, U.O., Dye-Sensitized Solar Cells, in: *Sol-Gel Processing for Conventional and Alternative Energy*, M. Aparicio, A. Jitianu, L.C. Klein (Eds.), pp. 147–175, Springer, New York, USA, 2012.

11

Lichen-Derived Products as Sustainable Source of Natural Dyes

Shahid Adeel[1]*, Aqsa Majeed[2], Fazal-ur-Rehman[2], Muhammad Azeem[3], Naeem Iqbal[3] and Nimra Amin[2]

[1]*Department of Chemistry, Gov. College University Faisalabad, Faisalabad, Pakistan*
[2]*Department of Applied Chemistry, Gov. College University Faisalabad, Faisalabad, Pakistan*
[3]*Department of Botany Gov. College University Faisalabad, Faisalabad, Pakistan*

Abstract
Current environmental scenario about using synthetic products has led global community to take information about the toxic effects. Due to fatal diseases, eco-imbalances, and bad health conditions of the globe created by the effluent loads of synthetic dyes, the observer agencies have spread awareness about introduction of green products in all walks of life. The search for green products, researchers have found lichen as one of the alternative source of colorants for textiles. This chapter is dealing with the lichen as suitable source of the colorants in different fields, their classification, and future prospects of lichen in various fields. Hopefully, the chapter will add new information in the knowledge hub of researchers, dyers, as well as traders and academician to explore the new source of colorants using modern tools and will make the ease of their application in all fields.

Keywords: Corticolous, fungal, hemicolous lichen, radiation methods, natural dyes, sustainability terricolous

11.1 Introduction

To present the most fascinating and widely present organisms, the lichen was the name given by the Father of Botany, the great Greek philosopher, Theophrastus, about 200 years ago [1]. Lichens belong to Kingdom Protista and fungi, as these organisms possess a symbiotic relationship among a fungal and algal partner or both [2]. This symbiotic relationship was first

**Corresponding author*: shahidadeel@gcuf.edu.pk

Mohd Yusuf (ed.) Lichen-Derived Products: Extraction and Applications, (245–262) © 2020 Scrivener Publishing LLC

described by Frank in 1876 when he investigated the anatomy of five crustose lichens and assigned this relationship a name as "*Symbiotimus*" [2, 3]. Today, more than 20,000 lichen species have been known and have secondary metabolites that are produced by mycobionts. The upper cortex of lichen is the outer protective layer with fungal hyphae that are gelatinized and cemented together. The second layer is the algal layer and the third layer is the lichen body (thallus) which is made up of medulla with hyphae that are loosely interwoven into each other [3, 4].

Secondary metabolites in the lichen are responsible for the adoption of lichens to the harsh environmental conditions, either at polar latitudes or at extreme altitudes (up to 7,400 m) [3, 5] and can be grown on any substratum that may be soil, humus, stones, rocks, brick, lime plaster, leaves, tree trunk, decaying wood, etc., and other man-made substratum like iron pipes, asbestos sheet, lime or cement plaster, and glass panes and are able to shift from the metabolically active to the inactive state during periods of unfavorable conditions [3]. Due to the diversity of their nature, lichens are non-toxic, eco-friendly, and sustainable in nature which has functional metabolites that can be used in many segments of life, particularly in textiles [6–8]. Herein this chapter, we will discuss lichen as a source of natural dyes.

11.2 Classification of Lichen

Lichens can be classified, as given in Figure 11.1, into different classes based on

- Habitat
- Fungal partner
- Thallus structure

Based on habitat, they can be [3]

- Terricolous
- Humicolous
- Corticolous
- Lignicolous

Based on the fungal partner, they can be [9]

- Ascolichens (belonging to Ascomycota)
- Basicdolichens (belonging to Basidomyceta)

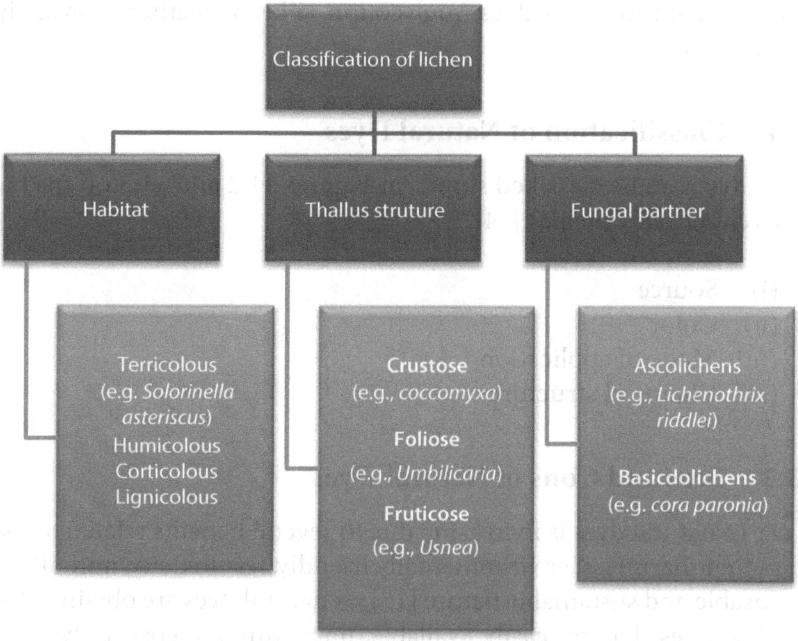

Figure 11.1 Flow chart presentation of classification of lichens.

Depending upon the morphology of thallus, they can be classified as [10]

- Crustose lichens being extremely flat
- Foliose lichen having a leafy thallus
- Fruticose lichens contain thallus with three-dimensional projections called podetia

11.3 Natural Dyes

Natural dyes are used since time immemorial, but synthetic dyes replace them due to their improved properties, wide availability, and economical sustainability [11]. As the environmental and health concerns are arising among people, the natural dyes are substituting the synthetic dyes. These dyes are obtained from natural sources and are gaining importance due to their biodegradable and bio-compatible nature. These dyes are known for dyeing many substrates such as wool, cotton, silk, and leather from ancient times [12, 13].

11.3.1 Classification of Natural Dyes

Natural dyes can be classified shown in Figures 11.2 and 11.3 on the basis of different criteria such as [14]

(i) Source
(ii) Color
(iii) Mode of application
(iv) Chemical structure

11.3.2 Pros and Cons of Natural Dyes

The use of natural dyes is increasing due to several benefits related to them such as their harmless, environmentally friendly, less toxicity, non-allergic, and reusable and sustainable nature [15], as natural dyes are obtained from natural sources that are easily available, impacting the cost of dying and making it the first choice of consumers, especially in the textile industry for dyeing [16]. Nowadays, many skins and other health diseases are caused by the dangerous UV radiation, 6% of UV-A region and 94% from the UV-b

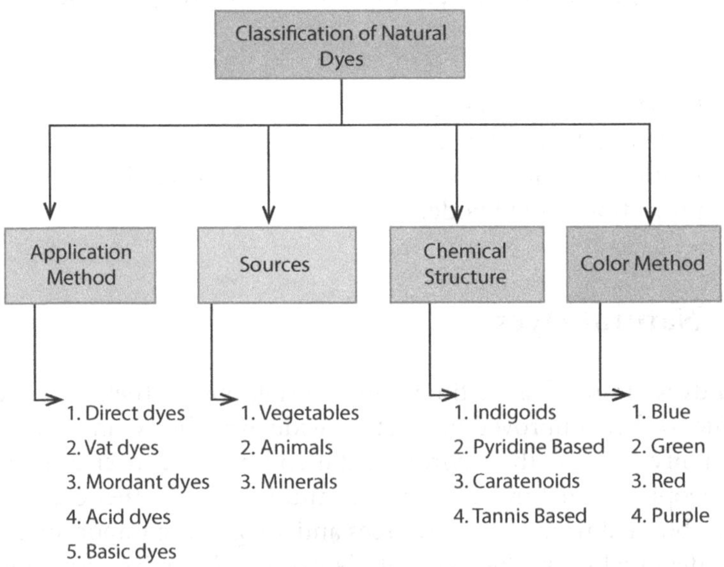

Figure 11.2 Classification of natural dyes.

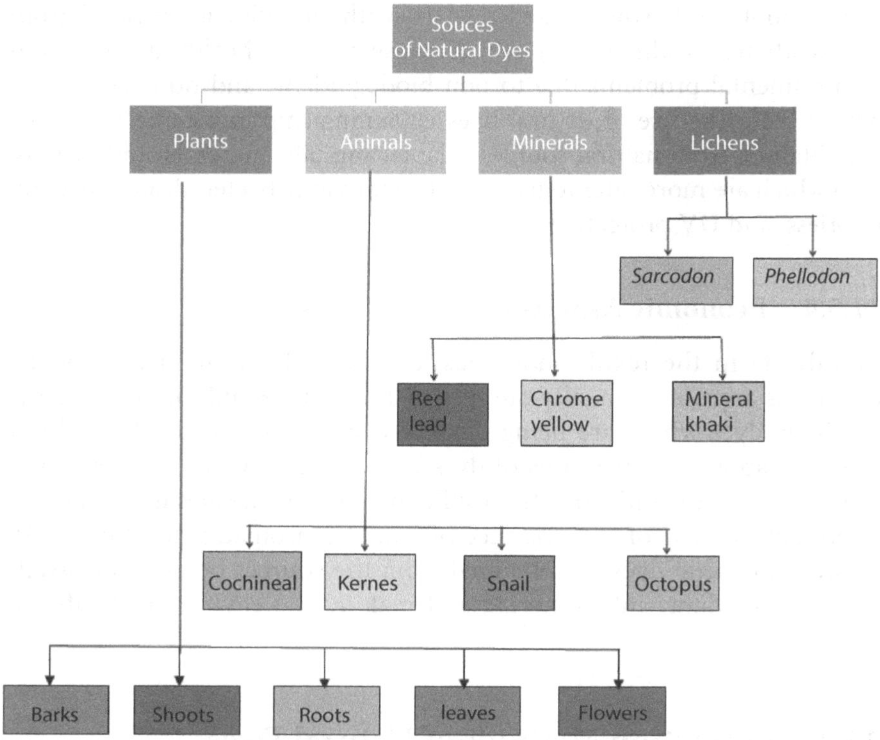

Figure 11.3 Sources of natural dyes.

region, reaching the earth's surface. Hence, cloths dyed with natural dyes are resistant to ultraviolet rays and protecting humans [17].

Despite having several benefits, natural dyes have several disadvantages that caused the increase in the use of synthetic dyes, as natural dyes are obtained from different plant sources that contain varying amounts of the same colorants depending upon their source of origin. Moreover, different dyes obtained from different plants which can be found only in a particular season, causing hindrance in natural dyes production. As dyeing with natural dyes consumes many chemicals and mordants which causes environmental pollution [18, 19]. Dyes also have problems related to colorfastness and maintenance of color strength causing hindrance in its use in the textile industry [20].

11.3.3 Environmental and Health Aspects of Natural Dyes

Developed countries are shifting from synthetic products to natural products, as the changing environmental conditions have forced the world to think about the harms created due to synthetic effluents released from the synthetic products [21] which cause various health diseases and environmental problems due to non-biodegradable and non-ecofriendly nature [22]. The use of natural dyes is gaining importance because they are obtained from natural sources (plants, animals, minerals, and lichens, etc.) which are more safer to use, eco- friendly, anti-bacterial, anti-oxidant, odorless, and UV protective [14, 23].

11.3.4 Economic Aspects of Natural Dyes

Worldwide in the textile industries, estimated 1.7 million tons of the chemicals are used annually out of which about 0.8 million tons are the synthetic dyes which are being employed in various fields [16, 23] out of which about 200, 000 tons of dyes were lost as a waste in dyeing and finishing in textile industry. These effluents causing hurdles in wastewater treatment because of the presence of complex aromatic structures [24]. However, natural dyes are obtained from the sources being easily available, anti-oxidant, anti-bacterial, and making the environment safer to use [25].

11.4 Extraction Methods of Natural Dyes

To overcome the problem related to the extraction of the natural dyes, conventional methods were used but these methods are costly, time-taking, and require a lot of labor. Conventional methods include [26, 27].

- Heating
- Stirring
- Overnight soaking

So, by keeping in view the limitations of the conventional methods, scientists are now exploring modern methods for the extraction of natural dyes because of their effectiveness regarding cost, labor, and time. These methods include [25, 28, 29].

- Microwave method
- Ultrasonic method
- Gamma radiations
- Enzyme analysis
- Plasma membrane method

11.5 Lichen as a Source of Natural Dye

The use of lichens as a source of natural dyes has been employed since ancient times [1]. As lichens and fungi contain symbiotic relationships due to which both of them contain some type of colorants [30]. Lichen dyes require not mordants to fix them to fiber and known as substantive dyes [31, 32]. Lichens contain a class is known as paranoid lichens, that have compounds depsides and depsidones (structures displayed in Figure 11.4), formed by acetate-polymelonate pathway by joining phenolics rings that are the main sources for the production of dyes in lichens [1]. They impart different colors depending upon the species such as red, brown, yellow, and blue [32]. Different species imparting different colors are presented in Figure 11.5.

In history, literature present evidence of lichen used as a source of dyes, as lichens belong to a peculiar group of natural colorants and has a long history of use as a source of natural dyes in Europe, Britain, and other parts of the world [32]. They are important candidates for dyeing wool and silk, especially used for imparting purple and violet colors collectively known as orchil lichens. Nowadays, orchil is used to color embracing fluids into a product called cudbear, signifying the point that how history influences future practices [33].

Many dyestuffs in lichen and fungi are benzoquinone derivatives, specifically terphenylquinone compounds. As mentioned earlier, due to symbiotic relationship fungi and lichens produces different dyes and colorants as intermediate metabolites like *A. alternata*, *A. gillusniger*, *H. lopiusnidulans*, *B. psisgrisea*, *P. schweinitzii*, *S. fuscoindicus*, *H. ceslactiflourum*, *Emericella nidulans*, *D. sanguinea*, *F. verticillioides*, and *P. purpurogenum* [34].

11.5.1 Extraction Methods of Colorants From Lichen

Lichens give different colors depending upon different species, but these species sometimes give a range of colors depending upon the extraction

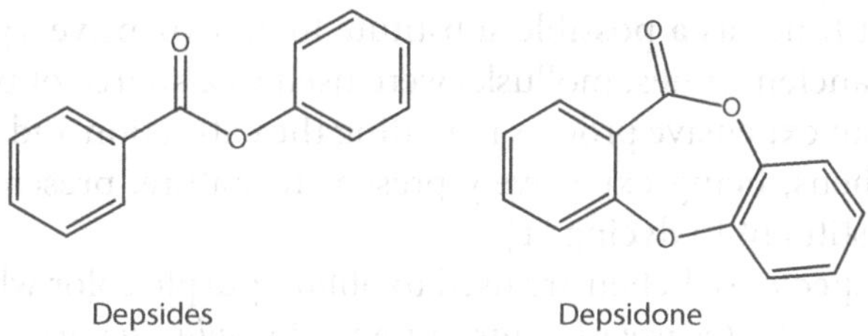

Figure 11.4 Functional components of lichens bearing different groups.

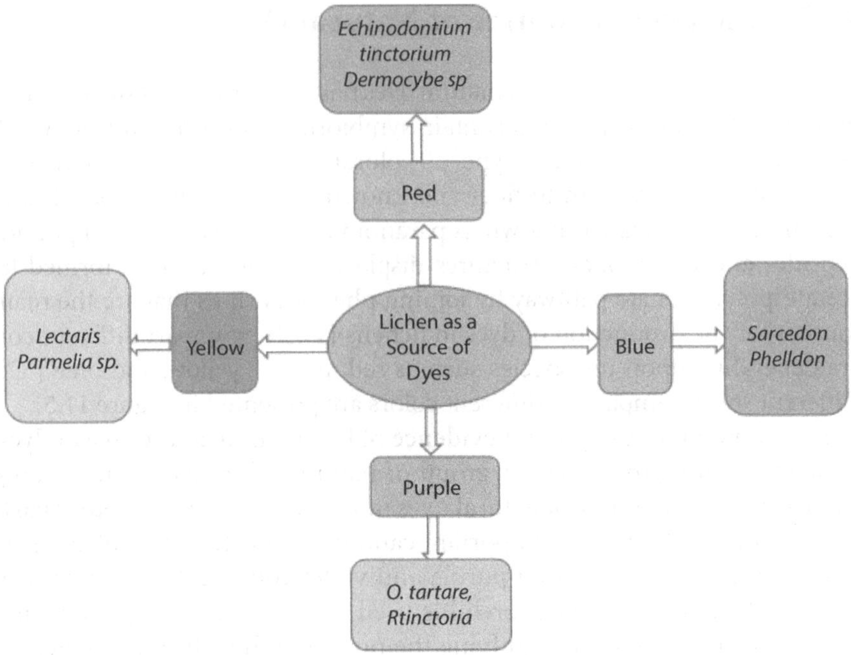

Figure 11.5 Lichen as a source of natural dyes depending upon different species.

methods. There are three extraction methods commonly used for extraction as shown in Figure 11.6 [1, 39].

1. Ammonia Fermentation Method (AFM)
2. Boiling Water Method (BWM)
3. Dimethyl sulphoxide Extraction Method (DEM)

11.5.2 Lichen as a Source of Purple Dye

Orchil is a purple dye from lichens, extensively used in industry for dying from ancient times as a possible substitute to an expensive Tyrian purple dye [35]. In ancient times, mollusks were used as a source of purple color, but this was an expensive process regarding the extraction and purification process. Lichens, being extensively present in nature, present them as a possible constituent to dyeing [1].

Different species of lichen are used to obtain purple color which include *Lasallian pustulata*, *Ochrolechia tartare*, and *Roccella tinctoria* by using different methods [36] and used to dye wool, silk, marbles, feathers, wood,

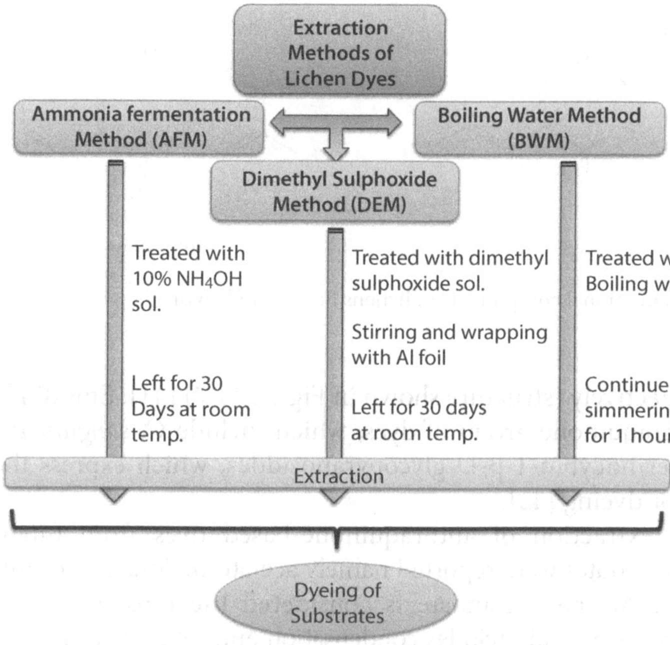

Figure 11.6 Schematic presentation of extraction methods of lichen dyes.

and leather by using litmus (a water-soluble pH indicator dye mixture, extracted from *Roccella sp.*) as a reagent to react with alkalis and acids [33].

Usually, this dye is extracted by a process in which ammonia (mostly Urine) is used as a soaking medium for lichens to allow chemical precursors transformations, but this process is time taking as the process of conversion of orsellinic acid to orcinol and production of phenoxazonic compounds (portions) is slow [37]. However, each dye could have its specific chemical precursors, orceins are the final precursors of each species [38, 39] but some authors have chosen the name "orsallia" when it is obtained from the genus *Umbilicaria*of lichens. These lichens from *Rocella* sp. Contain 3–4% of lacaronic acid of dry weight [40]. However, the colorants from *Rocella* and litmus are *pH*-dependent as they can impart red color in an acidic environment and bluish violent in a basic environment.

11.5.3 Lichen as a Source of Red Dye

Different species of lichen are specific for imparting red color on dyeing. These species include *Boletales sp.*, *Echinodontium tinctorium*, *Cortinarius*, *Dermocybe sp.* Containing pulvinic acid, emodin, and dermocybin coloring

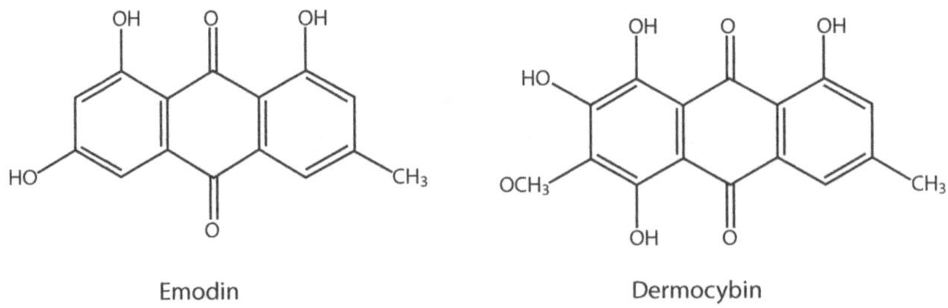

Figure 11.7 Functional components of lichens bearing red color.

agents, respectively (structure shown in Figure 11.7) [41]. Emodin belongs to the anthraquinone group of dyes, which include C. *sanguineus*, emodin, and dermocybin-1-β-D-glycopyranosiddes, which express the great potential for dyeing [42].

For the extraction of anthraquinone-based dyes from fungus two bio-synthetic routes were reported namely acetate-malonate and shikimate-mevalonate. Acetate-melanoate is considered the most important bio-synthetic pathway and includes condensation and folding of the polyketide chain, derived from acetate units. Acetate-melanocyte pathway showed that emodin, physcion, and endrocrocin produced yellow color are generated at the starting of the pathway and, red color generating compounds, i.e., dermocybin, dermorubin, and 5-chloro-dermorubin were produced at the end of the pathway. In this whole pathway, emodin is the most important coloring agent and most abundantly found anthraquinone. There are several synthetic methods most of them based on biogenetic schemes [39, 41].

Lichen as a source of red dye is mostly used to dye wool, cotton and silk fabrics. As a mordant, they are used to dye wool and polyamides, as a disperse dye to polyester and polyamides [31, 32].

11.5.4 Lichen as a Source of Yellow Dye

Lichens being diverse are the source of yellow dyes. Different species of lichens such as *Polyporus mori*, *Fomes fomentarius*, *Laricifomes officinalis*, *Lactarius sp.*, *Parmelia sp.*, *Xanthoria*, *Parietina*, and *Sticta aurata* impart yellow color in dyeing [39, 32]. These species contain telephonic acid, atranorin, physcion, and pulvinc acid as a coloring reagent. However, atranorin is only found in lichen species *Parmelia* and *Xanthoria parietina* and gyrophic acid in *Ochrolechia tartarea* and *Lasallia pustulata sp.* The derivative of atranorin such as salizinic acid (structure given in Figure 11.8) also imparts color ranging from yellowish to brownish in color [40, 43–46].

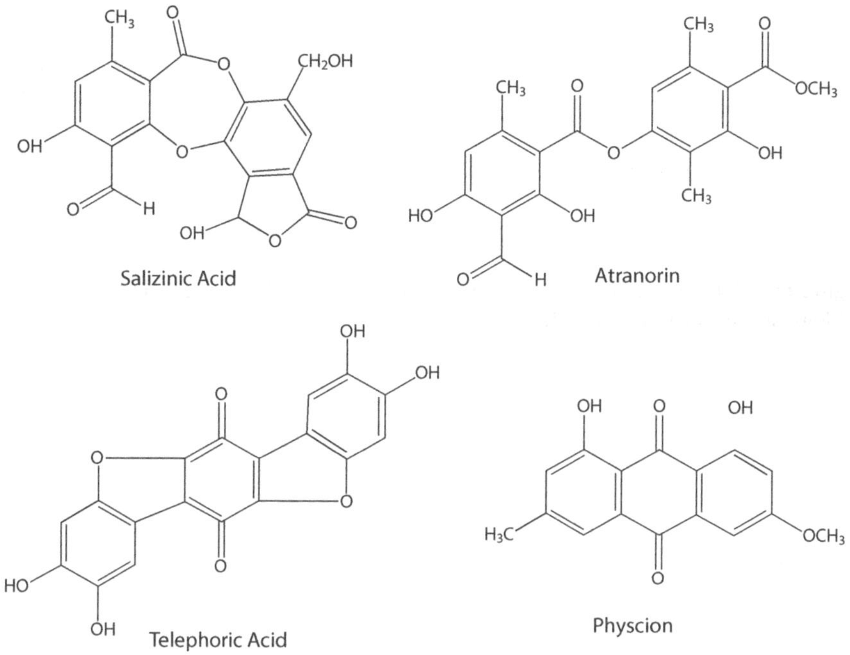

Figure 11.8 Functional derivativesof lichens bearing yellow color.

Two types of methods were reported to obtain colorants from lichen. First is the processing with ammonia, which includes enzymatic hydrolyzation, decarboxylation, and oxidation processing. The second method is boiling with water, as *Letharia vulpine* produces a yellow color in the boiling water method [44, 46].

11.5.5 Lichen as a Source of Blue Dye

In lichens, *Sarcodon, Phellodon, Hydnellum, Telephora, H. nidulans,* and *P. atrotomentosus* species are used to impart blue colors. These all species contain terphenylquinone compounds which are the main colorants for producing blue hues in dyeing. These species contain coloring components such as polyphoric acid, atromentin, telephofic acid, and telephoric acid (structure displayed in Figure 11.9). These compounds are used to dye wool fabric [41].

11.5.6 Isolation Medium Dependant Colorants from Lichens

As *Echinodontium tinctorium* gives red color and lichens having usnic acid (structure given in Figure 11.10) to provide clear yellow color, but

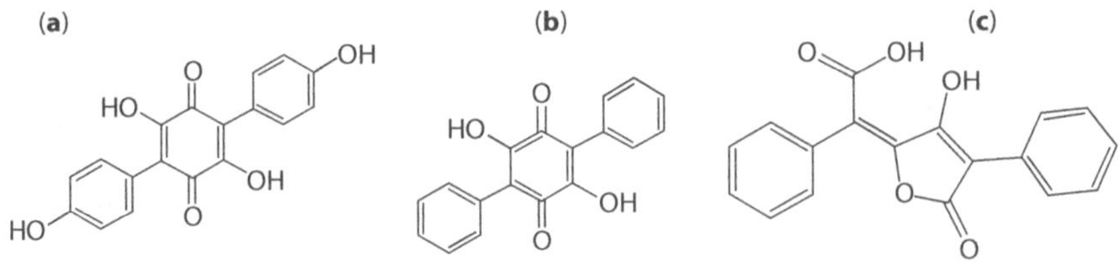

Atromentin Polymorphic acid Pulvinic Acid

Figure 11.9 Functional derivative of Blue colored Lichens (a) Atromentin (b) Polymorphic acid (c) Pulvinic Acid.

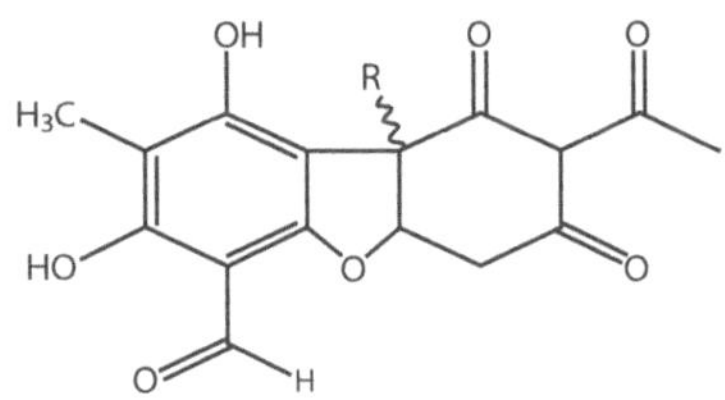

Usnic Acid

Figure 11.10 Functional component of lichen bearing red color.

the presence of other acids in species can provide other colors. A species of lichen, *Hypogymnia physodes*, have unsic acid, but in the hot water extraction method, it gives yellow-brown color [47]. *Evernia prunastri* produces orchil dye from evernic acid when produced through ammonia fermentation, with the BWM method, it produces a pale yellow color. Similarly, a species *Parmotrema reticulate* that contains salazinic acid, atranorin, and consalazinic acid produce light brown color in AFM method, deep olive buff in BWM, and sayal brown color in DEM [33].

However, some lichen species produce different color depending upon geographical areas like a species *Evernia furfuracea* (in America known as *Parmelia furfuraceain*) contain lecaronic acid which produce orchil dye, but in Britain, it contains olivetoric acid and presents itself as a non-orchil dye [48].

B. satschwensis, *E. cirrhatum*, *E. nepalenses*, *P. wallichiana*, *P. reticulatum*, *U. undulate*, and *X. somloensis* have Salzinic acid which produces orange dye through Ammonia fermentation method and brown dye through boiling water method, and if fermentation period is increased, it produces dark orange shade [5].

11.5.7 Future Prospectus of Lichen Dyeing

Colors are the most important and fascinating part of human life, and the textile industry is considered as the hub of coloration and dyeing. Natural dyes are used as the source of dyes since bronze times, but synthetic dyes replaced them due to their various improved properties over natural dyes. Now, the increased awareness among people has forced the people and industrialists to shift to the natural dyes again due to their non-toxic and environment-friendly nature.

Lichens are considered as important sources of natural dyes from ancient times. Tyrian purple is the oldest dye obtained from the lichen in WW-II, as dyeing with lichen is less costly due to its wide availability in every substratum. But dyeing with lichen was not explored much due to their poor colorfastness properties, hurdles in their collection as the whole family is required to collect the lichens, and poor extraction methods.

Now, many advances have been made for exploring the potential of lichens as a source of natural dyes due to their wide availability and medicinal properties, and now, these medicinal properties are an important factor known for textile and health sector applications. Lichens contain many lichen acids that are the sources of colors such as lacaronic acid, emodin, salzanic acid, etc., but one lichen species may contain multiple lichen acids that can be explored through a focus on dyeing with natural dyes. Emodin, most widely present lichen colorant, has many medicinal properties that can be explored in the textile field. Earth is damaging due to climate change, ozone is reaching the earth, and natural dyes being UV protective are important for dyeing. However, many works are done and much more is needed to explore the potential of natural dyes for dyeing with these natural resources. So, it is important to check the natural dyes for their anti-microbial, anti-toxic, and coloring properties and to explore new advanced techniques for their extraction and improving their colorfastness properties.

11.6 Conclusion

Lichens are an important source of natural dyes and provide a wide range of colors from red to brownish depending upon different species and extraction methods. Lichen acid is the source of colors in lichen species, presenting lichen as a major candidate for its dyeing properties due to harmless, odorless, anti-oxidant, anti-bacterial, and UV protective nature of the lichens. So, dyeing with lichen is a major source for the textile industry due to its advantages. Lichens as novel source of colorants should be explored and introduced in all fields particularly textiles, flavor, perfumes, etc., due to having good ayurvedic characteristics.

References

1. Shukla, P., Upreti, D.K., Nayaka, S., Tiwari, P., Natural dyes from Himalayan lichens. *Ind. J. Trad. Knowl.*, 13, 1, 195–201, 2014.
2. Hawksworth, D.L., *Lichenization: The origins of a fungal life-style:Recent Advances in Lichenology*, vol. 2, D. Upreti, P. Divakar, V. Shukla, R. Bajpai (Eds.), pp. 1–10, Springer, New Delhi, 2015.
3. Lumbsch, H.T. and Rikkinen, J., *Evolution of Lichens: The fungal community: Its organization and role in the ecosystem*, 4th Ed., J. Dighton and J.F. White (Eds.), pp. 53–64, CRC Press, Taylor and Francis Group, USA, 2017.
4. Upreti, D.K., Bajpai, R., Nayaka, S., *Lichenology: Current research in India: Plant Biology and Biotechnology*, 1st Ed., vol. 1, B. Bahadur, M. VenkatRajam, L. Sahijram, K.V. Krishnamurthy (Eds.), pp. 263–280, Springer, New Delhi, India, 2015.
5. Shukla, P. and Upreti, D.K., *Lichen dyes: Current scenario and future prospects: Recent Advances in Lichenology*, 1st Ed, vol. 2, D.K. Upreti, P.K. Divakar, V. Shukla, R. Bajpai (Eds.), pp. 209–229, Springer, New Delhi, India, 2015.
6. Varol, M., Anti-breast cancer and anti-angiogenic potential of a lichen-derived small-molecule: Barbatolic acid. *Cytotechnol*, 70, 6, 1565–1573, 2018.
7. Honegger, R., Edwards, D., Axe, L., The earliest records of internally stratified cyanobacterial and algal lichens from the lower Devonian of the Welsh Borderland. *New Phytologist.*, 197, 264–275, 2013.
8. Osyczka, P., Boroń, P., Lenart-Boroń, A., Rola, K., Modifications in the structure of the lichen *Cladonia* thallus in the aftermath of habitat contamination and implications for its heavy-metal accumulation capacity. *Environ. Sci. Poll. Res.*, 25, 2, 1950–1961, 2018.
9. Khan, A., Hussain, M.T., Jiang, H., Dyeing of silk fabric with natural dye from camphor (*Cinnamomumcamphora*) plant leaf extract. *Color. Technol.*, 134, 4, 266–270, 2018.
10. Adeel, S., Khan Khosa, K., Salman, M., Rafi, S., Revitalization of carotenoid-based natural colorants in applied field: A Short Review. *Text. Cloth.*, 1, 45–78, 2019.
11. Cooksey, C.J., Tyrian purple: 6,60-dibromoindigo and related compounds. *Molecules*, 6, 736–769, 2001.
12. Yusuf, M., Shabbir, M., Mohammad, F., Natural colorants: Historical, processing and sustainable prospects. *Nat. Prod. Bioprospect*, 7, 1, 123–145, 2017.
13. Amin, N., Adeel, S., Ahamd, T., Muneer, M., Haji, A., Sustainable application of cochineal-based anthraquinone dye for the coloration of bio-mordanted silk fabric. *Environ. Sci. Pollut. Res.*, 27, 7, 6851–6860, 2020.

14. Ratnapandian, S., *Natural colorants and its recent developments: Sustainable Technologies for Fashion and Textiles*, 2nd Ed, R. Nayak (Ed.), pp. 189–208, Woodhead Publishing, Duxford UK, 2019.
15. Nam, C. and Xiang, C., Natural dyeing application of used coffee grounds as a potential resource. *Int. J. Fashion Design, Technol.*, 12, 3, 335–345, 2019.
16. Adeel, S., Zia, K.M., Abdullah, M., Rehman, F.U., Salman, M., Zuber, M., Ultrasonic assisted improved extraction and dyeing of mordanted silk fabric using neem bark as source of natural colourant. *Nat. Prod. Res.*, 33, 14, 2060–2072, 2019.
17. Islam, S.U., Rather, L.J., Shabbir, M., Sheikh, J., Bukhari, M.N., Khan, M.A., Mohammad, F., Exploiting the potential of polyphenolic biomordants in environmentally friendly coloration of wool with natural dye from *Butea monosperma* flower extract. *J. Nat. Fibers*, 16, 4, 512–523, 2019.
18. Samanta, A.K. and Agarwal, P., Application of natural dyes on textiles. *Ind. J. Fibre Text. Res.*, 34, 384–399, 2009.
19. Rehman, F., Adeel, S., Hanif, R., Munner, M., Zia, K.M., Zuber, M., Jamal, M.A., Khosa, M.K., Modulation of marigold based lutein dye and its dyeing behaviour using UV radiation. *J. Nat. Fibers*, 14, 1, 63–70, 2009.
20. Adeel, S., Rafi, S., Azeem, M., Salman, M., Iqbal, N., Zuber, M., *Resurgence of natural dyes: Ideas and technologies for textile dyeing, In: Handbook of Textile Coloration and Finishing*, 1st Ed, vol. 1, M. Shahid, G. Chen, R. Tang (Eds.), pp. 1–27, Studium Press LLC, USA, 2018.
21. Chomean, S., Nantabut, M., Kongtia, W., Saenguthai, K., Kaset, C., Evaluation of natural dyes for human spermatozoa morphology assessment. *Acta Histochemica*, 121, 2, 227–233, 2019.
22. Zhou, Y. and Tang, R.C., Modification of curcumin with a reactive UV absorber and its dyeing and functional properties for silk. *Dyes Pigm.*, 134, 203–211, 2016.
23. Adeel, S., Kiran, S., Ahmad, T., Habib, N., Tariq, K., Hussaan, M., Bio-Mordants in Conjunction With Sustainable Radiation Tools for Modification of Dyeing of Natural Fibers. *Front. Text. Mater. Polym., Nanomater. Enzym, Advan. Modi. Tech.*, 355–367, 2020.
24. Aljuboury, D.A.D.A., Palaniandy, P., Abdul Aziz, H.B., Feroz, S., Treatment of petroleum wastewater by conventional and new technologies-A review. *Glob. Nest. J.*, 19, 439–452, 2017.
25. Haji, A., Natural dyeing of wool with henna and yarrow enhanced by plasma treatment and optimized with response surface methodology. *J. Text. Inst.*, 4, 1–9, 2019.
26. Haji, A., Mousavi Shoushtari, A., Mirafshar, M., Natural dyeing and antibacterial activity of atmospheric-plasma-treated nylon 6 fabric. *Color. Technol.*, 130, 1, 37–42, 2014.
27. Nguyen, H.P.T., Morancais, M., Fleurence, J., Dumay, J., *Mastocarpusstellatus* as a source of R-phycoerythrin: optimization of enzyme assisted extraction using response surface methodology. *J. Appl. Phycol.*, 29, 3, 1563–1570, 2017.

28. Ticha, M.B., Meksi, N., Attia, H.E., Haddar, W., Guesmi, A., Jannet, H.B., Mhenni, M.F., Ultrasonic extraction of *Parthenocissusquinquefolia* colorants: Extract identification by HPLC-MS analysis and cleaner application on the phytodyeing of natural fibres. *Dyes Pigm.*, 141, 103–111, 2017.
29. Sheikh, J., Jagtap, P.S., Teli, M.D., Ultrasound assisted extraction of natural dyes and natural mordants vis a vis dyeing. *Fibers Polym.*, 17, 5, 738–743, 2016.
30. Divakar, P.K., Crespo, A., Wedin, M., Leavitt, S.D., Hawksworth, D.L., Myllys, L., Schmitt, I., Evolution of complex symbiotic relationships in a morphologically derived family of lichen-forming fungi. *New Phytologist*, 208, 4, 1217–1226, 2015.
31. Mohamed, N.A., Ahmad, W.W., Ngalib, K., Ahmad, M.R., Ab Kadir, M.I., Ismail, A., Microwave-enzyme-assisted extraction and dyeing of lichen species: Parmotremapraesorediosum, in: *Proceedings of the International Colloquium in Textile Engineering, Fashion, Apparel and Design 2014 (ICTEFAD 2014)*, pp. 89–93, Springer, Singapore, 2014.
32. Shaheen, S., Iqbal, Z., Hussain, M., first report of dye yielding potential and compoundsof lichens; a cultural heritage of himaliyan communities, pakistan. *Pak. J. Bot.*, 51, 1, 341–360, 2019.
33. Rather, L.J., Jameel, S., Ganie, S.A., Bhat, K.A., Lichen Derived Natural Colorants: History, Extraction, and Applications, in: *Handbook of Renewable Materials for Coloration and Finishing*, 1st Ed., vol. 1, M. Yusuf (Ed.), pp. 103–114, Scrivener Publishing LLC, Beverly, USA, 2018.
34. Aishwarya, A.D., Extraction of natural dyes from fungus–An alternate for textile dyeing. *J. Nat. Sci. Res.*, 4, 7, 1–7, 2014.
35. Calà, E., Benzi, M., Gosetti, F., Zanin, A., Gulmini, M., Idone, A., Aceto, M., Towards the identification of the lichen species in historical orchil dyes by HPLC-MS/MS. *Microchem. J.*, 150, 104140, 2019.
36. Serafini, I., Lombardi, L., Reverberi, M., Ciccola, A., Calà, E., Sciubba, F., Bianco, A., New advanced extraction and analytical methods applied to discrimination of different lichen species used for orcein dyed yarns: Preliminary results. *Microchem. J.*, 138, 447–456, 2018.
37. Beecken, H., Gottschalk, E.M., Gizycki, U., Krämer, H., Maassen, D., Matthies, H.G., Musso, H., Rathjen, C., Zdhorszky, U.I., Orcein and litmus. *Biotechnic Histochem.*, 78, 6, 289–302, 2003.
38. Haddar, W., Ben Ticha, M., Meksi, N., Guesmi, A., Application of anthocyanins as natural dye extracted from *Brassica oleracea* L. var. capitata f. rubra: Dyeing studies of wool and silk fibres. *Nat. Prod. Res.*, 32, 2, 141–148, 2017.
39. Shukla, P., Nayaka, S., Tiwari, P., Natural dyes from Himalayan lichens. *Ind. J. Trad. Knowl.*, 13, 1, 195–201, 2014.
40. Casselman, K.D., *Lichen Dyes and Dyeing: The New Source Book*, Dover Publications, Mineola, New York, 2001.

41. Raisanen, R., *Dyes from lichens and mushrooms. Handbook of natural colorants. Handbook of natural colorants*, T. Bechtold and R. Mussak (Eds.), pp. 183–200, Wiley, UK, 2009.
42. Voicu, D.M.F., Mitoi, M.E., Gavriloaie, C., Helepciuc, F.E., Toma, N., Chemical investigations of lichen biomass in Usneabarbata, Cetrariaislandica, and Xanthoriaparietina species. *Extreme Life Biospeol. Astrobiol.*, 11, 1, 1–8, 2019.
43. Cardon, D., *Natural Dyes: Sources, Tradition, Technology and Science*, Archetype Publications, London, 2007.
44. Casselman, K.D., Magic, mystery and mayhem: Lichen dyes old and new, in: *Seminar 12.8.2002, Publication of the Natural Dyes Product Research and Development Project*, EVTEK Institute of Art and Design, Vantaa, 2002.
45. Brough, S.G., Navajo lichen dyes. *The Lichenologist*, 20, 3, 279–290, 1988.
46. Tiedemann, E.J. and Yang, Y., Fiber-safe extraction of red mordant dyes from hair fibers. *J. The Am. Inst. Conserv.*, 34, 3, 195–206, 1995.
47. Devkota, S., Chaudhary, R.P., Werth, S., Scheidegger, C., Indigenous knowledge and use of lichens by the lichenophilic communities of the Nepal Himalaya. *J. Ethnobiol. Ethnomed.*, 13, 15, 3–10, 2017.
48. Llano, G.A., Economic uses of lichens. *Econ. Bot.*, 2, 1, 15–45, 1948.

12

Biomedical Perspectives of Lichen-Derived Products

Muhammad Shahid[1], Arslan Rasool[1], Fozia Anjum[2*] and Muhammad Tjammal Rehman[1]

[1]*Department of Biochemistry, University of Agriculture, Faisalabad, Pakistan*
[2]*Department of Chemistry, Government College University, Faisalabad, Pakistan*

Abstract
Lichens are existing in the symbiotic relationship between a filamentous fungus and a photosynthetic collaborator, which could be either cyanobacterium or a green alga. Approximately, more than 18,000 species of lichen are estimated worldwide. Lichens are large number of valuable lower plants consider as mini ecosystem with distinctive features and various applications worldwide in the fields of textile, cosmetics, food, medicine industries, etc. There are more than 1,000 lichen metabolites that are reported and established for a diverse biological and pharmacological activities including antioxidant, antiviral, antifungal, antitumor, cytotoxic, and antigenotoxic activities. Interestingly, the last decade witnessed renewed and growing interest in lichen substances as a source of novel, pharmacologically active biomolecules, and environmental bio-indicators. Lichen-derived products exhibited a diverse nature of biological and therapeutical potential. This study explains the significance of lichens as a potential source of these bioactive secondary metabolites.

Keywords: Secondary metabolites, biological activities, antioxidant, antimicrobial, lichen extracts

12.1 Introduction

Lichens are the complex organisms having unique, self-supporting, and stable symbiotic relationship between fungi and a photosynthetic partner

**Corresponding author*: foziaanjum2008@yahoo.com

Mohd Yusuf (ed.) Lichen-Derived Products: Extraction and Applications, (263–276) © 2020 Scrivener Publishing LLC

(referred as cyanobacteria or micro algae) which play a vital role in many ecosystems. These are the pioneer colonized organisms to be found on Earth [1]. Traditionally, lichens are used as medicines in the treatment of bronchitis, dyspepsia, inflammation of bones, infections, bleeding piles, kidney diseases, and stomach disorders. Most of the traditional methods of medicines like Unani or Ayurveda include lichens [2]. These are also used as flavoring agents in spices, production of dyes, and cosmetics [3].

Lichens are widely distributed on Earth and its concentration is varied depending upon climate condition. It is estimated that 6% of Earth is covered by lichen and these act as bio-indicators for the environment. They abundantly grow on the barks, leaves, rocks, and many other substrates. They exhibit different growth forms like [4]:

- Crustose-crust like and rapidly over the surface,
- Foliose-leaf like and loosely attached to the surface,
- Fruticose-free standing branching tubes.

Moreover, lichens metabolism produce metabolites classified as primary metabolites and secondary metabolites. Primary metabolites include sugars, amino acids, vitamins, lipids, and carotenoids that are soluble in water and can be isolated by boiling. Lichen produces a large variety of secondary metabolites that are usually small but have complex structures. More than 1,050 secondary metabolites of lichens are known to date [5]. These are almost insoluble in water and can be isolated by using organic solvents. Most of them are produced by fungi [6]. Secondary metabolites ranges from 0.1 to 10% of thallus dry weight, but sometimes, it reaches more than 30% [7]. Lichens secondary metabolites exhibit a large variety of biological activities including antioxidant, enzyme inhibitory, antifungal, antiviral, insecticidal, antitumor, hepato-protective, antiprotozoal, neuroactive, antidiabetic, anti-inflammatory, and cytotoxic activities [8]. From Egyptian times, due to these effects, they are used in the treatment of different disorders or conditions like tuberculosis, injuries, burns, respiratory problems, and gastritis [9].

12.2 Coevolution of Lichens

Lichens are the stable and unique form of association between photobionts and mycobionts as shown in Figure 12.1. They are known to be the first colonizers of land habitats on Earth [10]. There are more than 25,000 species of lichens that inhabits about 8 to 10% of land surface from north pole

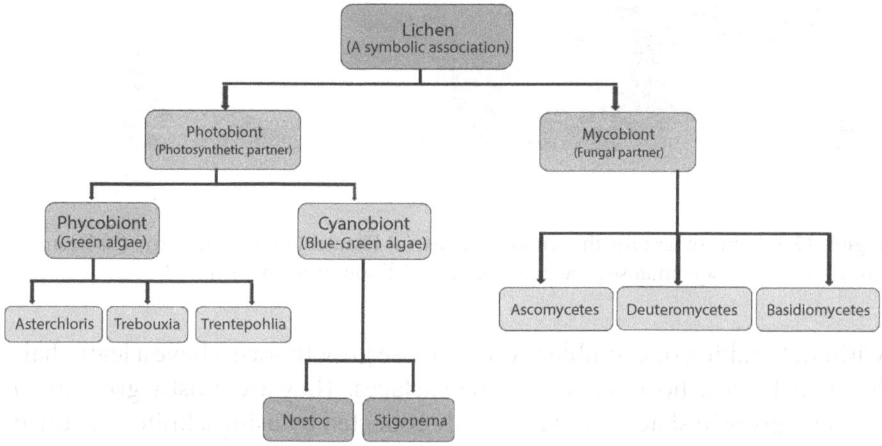

Figure 12.1 Schematic illustration of lichen association.

to south pole and from sea levels to the mountains [7]. Lichens have slow growth rate but maximum life spans up to hundreds of years and protect their metabolites from physiological effects [9].

12.3 Diversity Among Growth Forms of Lichens

The abundant type of lichens is terrestrial but some of them inhabits marine and freshwater ecosystem [5]. The most commonly growth forms of lichens are foliose (leaf like), crustose (crust like), and fruticose (branching tubes like) as shown in Figure 12.2. Crustose growth forms are attached to their habitat surface. It is difficult to remove them from their surfaces

Figure 12.2 Diversity among basic growth forms of lichens. (a) Crustose form: They are attached the surfaces like rocks. (b) Foliose form: They have a leafy thallus. (c) Fruticose form: They have bushy structures and can grow in extreme conditions.

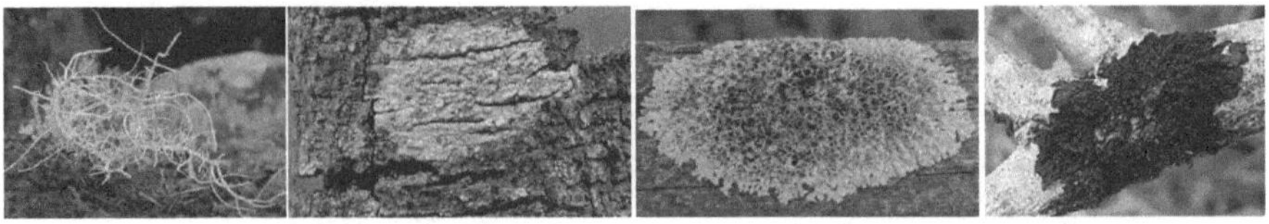

(a) (b) (c) (d)

Figure 12.3 Some other growth forms of lichens. (a) Filamentous growth form. (b) Leprose growth form. (c) Squamulose growth form. (d) Gelatinous growth form.

without breaking or crumbles away. Foliose growth forms have a leafy thallus, which grow horizontally on the surfaces. They are mostly grey-green in color, grow in shades, and can easily collected by using a knife. The third basic growth form is fruticose that are shrubby having many branching tubes. They have usually light-green or light-yellow color and can grow in extreme environment like Arctic and high mountains. These lichens can be collected by hands [11].

Moreover, the lichens also exhibit some other growth forms that include squamulose (clustered like), leprose (powder like), gelatinous (jelly like), and filamentous (thin hair like) as shown in Figure 12.3. Squamulose growth form usually consists of tiny overlapping scales. When their thallus raised off which are loosely attached to the substrate, it appear leafy like foliose form. Filamentous growth forms are unique from all other growth forms. In this type, their fungal partner form sheets or wrapped around cyanobacteria or their photosynthetic member. Leprose growth forms are mostly powdered and their surface is composed of granules. These lichens lack cortex due to which they are unable to form an organized thallus. In gelatinous growth form, carbohydrates are produce by blue-green algae which are then absorbed [12].

12.4 Lichen Secondary Metabolites and Their Structures (Lichen Substances)

Lichens produce many types of secondary metabolites which show a great variety of biological activities. They are classified as aromatic (benzene ring is present) and aliphatic (benzene ring is absent) lichen substances which have less molecular weight. These lichen substances make up 20% of the total dry weight of lichen specie [13]. These metabolites are small substances having complex structures. There are many classes of lichens secondary metabolites including aliphatic acids, chromones, depsidones, dibenzofurans, depsones, xanthones, depsides, monocyclic aromatic compounds, sugar alcohols, amino acid derivatives, steroids, terpenoids, carotenoids, and Pulvinic acid derivatives. The fungal partner (Mycobiont) synthesized lichen substances and accumulated them in its layers. Some secondary metabolites like usnic acid, atranorin, physodalic acid, physodic acid, Lecanoric acid, melanins, and physodic acid accumulate in cortex and medullary layer and their chemical structures are shown in Figure 12.4. These secondary metabolites are present in the form of small crystals outside of the cells [14].

About 85% of these lichen substances are very unique and they are absent in vascular plants. There are three main pathways that are involved in the synthesis of these secondary metabolites as shown in Figure 12.5. They include [15, 16]:

1. Mevalonate pathway
2. Shikimate pathway
3. Acetyl polymalonyl pathway

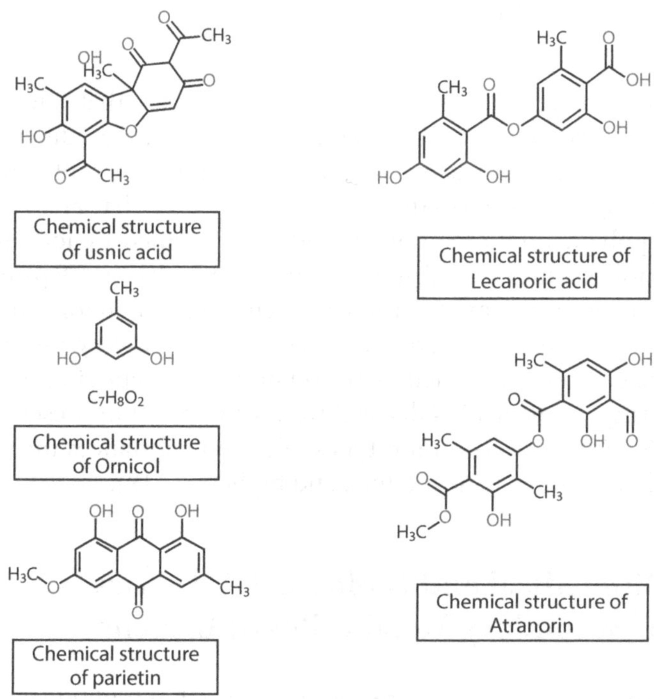

Figure 12.4 Chemical structures of some lichen metabolites.

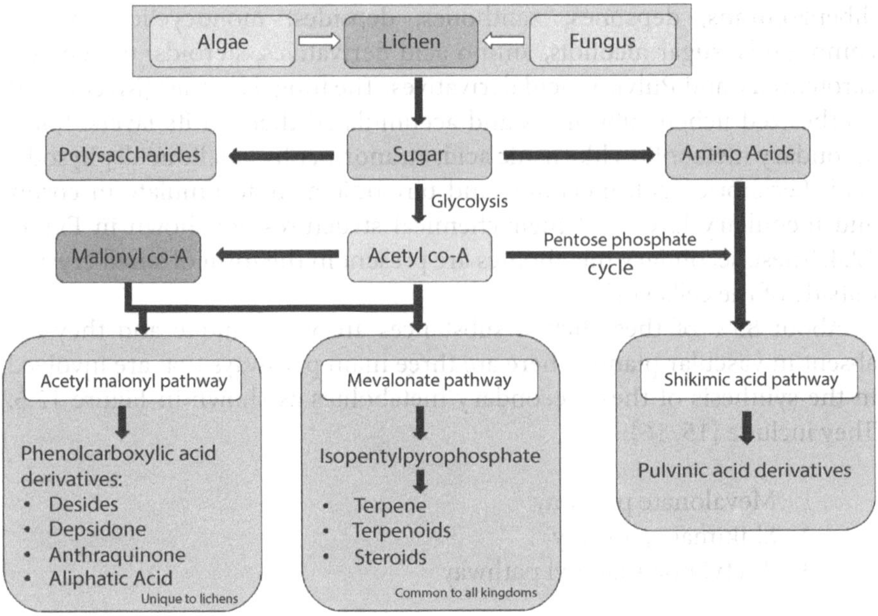

Figure 12.5 Schematic representation of outlined pathways leading to biosynthesis of lichens secondary metabolites.

There are about 1,050 known lichen substances that show definite biological activities [17]. Depsidones, anthraquinone, dibenzofurans, Chromones, Xanthones, usnic acids, and mononuclear phenolic compounds are significant bioactive lichen metabolites that are synthesized via acetyl polymalonyl pathway. Depsides are formed by joining two or more orcinol residues linked together by ester linkages. Depsidones are derivatives of orcinol consist of three benzene rings fused together by ester and cyclic ether functional group. Dibenzofuran is a very rare heterocyclic organic compound formed by fusion of two benzene rings with a central furan ring. Usnic acid is derived from dibenzofuran present in many lichens [18]. Some less common types of lichen substances includes depsones, diphenyl esters, benzyl esters, and biphenyls [19].

12.5 Biomedical and Biological Activities of Secondary Metabolites of Lichens

Lichen-derived products are more commonly used in traditional medicines to cure many animals and humans disorders. Many fruticose species of genus

Usnea like *Usnea filipendula* is used for wound healing, pain relief, and to control fever in many countries of Asia and Africa [7, 8]. In Argentina, many health diseases are treated by *Usnea densirostra* Taylor specie [20]. Similarly, some foliose lichens of genus *Umbilicaria* or *Parmelia* are useful in treatment of infections, epilepsy, rosacea (skin disease), seizers, and diarrhea in Chile (country in South America). In Finland (country in Europe), another fruticose lichen of genus *Ramalina* are used to cure many skin diseases, pain in jaws or teeth, and fungal infections [21]. Moreover, many lichen species are used in the treatment of diabetes, lungs, heart, and kidney disorders. Many studies demonstrated that many lichen-derived substances shows a great variety of biological effects like anticancer, cytotoxicity, enzyme inhibition, antiviral, antifungal, antidiabetic, and antioxidant effects, etc., as shown in Table 12.1. These biological effects support the significance of lichens to be used as traditional medicines in different countries [22].

Table 12.1 Lichen-derived compounds possess pharmaceutical activities.

Lichen metabolites	Pharmaceutical activities	Ref.
Atranorin	Anti-inflammatory, antidiabetic, antimicrobial	[8]
Zeorin	Antioxidant, antidiabetic, antifungal	[41]
Stictic acid	Anticancer, antioxidant, antibacterial	[42]
Salazinic acid	Antiviral, antitumor, antifungal	[26]
Evernic acid	Antifungal, antimicrobial, antioxidant	[8, 40]
Lobaric acid	Antimicrobial, antineurodegenerative, anticancer	[43]
Protocetraric acid	Antimicrobial, anticancer, antioxidant	[44]
Usnic acid	Antipyretic, hepatotoxic, antidiabetic, anticancer, antiviral	[44]
Diffractaic acid	Analgetic, antioxidant, anti-inflammatory, antipyretic	[45, 46]
Protolichesterinic acid	Anticancer, antimicrobial, antitumor	[44, 45]
Parietin	Antiviral, antifungal, anticancer, antibacterial	[47]
Methyl evernate	Antioxidant, antibacterial, anticancer, antifungal	[48]

12.5.1 Antioxidant Activity

Lichens show great antioxidant activities. The secondary metabolites of lichens that are phenolic derivatives are more commonly exhibit antioxidant effects. Under unfavorable environmental conditions, synthesis of polyphenolic substances increases to compensate the oxidative stress by their antioxidant properties. Phenol plays a vital role in the growth of lichens [4]. Some of phenolic compounds include dibenzofurans, depsides, pulvinic acid-derived products, and depsidones [23].

*Sphaerophorin*and pannarin are phenolic derivatives that inhibit the formation of superoxide anion and induced cell death in melanoma cells [24]. In animal models, it is reported that indomethacin is a nonsteroidal drug prepared by methanol extract of *Lobaria pulmonaria* that decreases the oxidative stress by increasing concentration of superoxide dismutase [25]. A correlation between antioxidant effect and total phenolic contents (TPCs) for *Toninia candida* specie by using different extracts (methanol, ether, and chloroform) was established [26]. In Malaysia, correlation of antioxidant effect and TFC of four lichen species from different genus (*Bulbothrix*, *Ramalina*, *Cladia*, and *Parmotrema*) were also evaluated by Stanly *et al.* [27]. Lecanoric acid is an important bioactive compound which was derived from extract of *Umbilicaria antarctica* and shows superoxide anion scavenging capacity [28].

12.5.2 Cytotoxic Activity

Most of the lichen-derived products are very useful in pharmaceutical industries that exhibit cytotoxic potential. The cytotoxic effect on four cancer cell line in humans by eight lichen species including *Evernia prunastri* (commonly known asoakmoss), *Cladonia convolute* (cup like), *Ramalina cuspidate* (strap like), *Cladonia rangiformis*, *Flavoparmelia caperata* (commonly known as green shield lichen), *Platismatia glauca*, *Usnea rubicunda* (commonly known as red beard lichen), and *Parmotrema perlatum* (commonly known as black stone lichens) was also reported [29]. In murine hepatocytes, Usnic acid develops oxidative stress by inhibit the oxidative phosphorylation in mitochondria [30].

12.5.3 Antiviral Activity

Usnic acid is an important bioactive secondary metabolite of many lichens that shows its antiviral activity against certain types of viruses like Junin virus which is an arenavirus that is responsible for causing Argentine hemorrhagic fever (AHF). It also shows its antiviral properties against Tacaribe virus which was isolated from bats and mosquitoes that also cause AHF [31]. Parietin is another bioactive secondary metabolite of lichen that also possesses its antiviral characteristics against these viruses. Parietin can be isolated from *Teloschistes chrysophthalmus* (commonly known as golden eye lichen) specie. It is also reported that *Ramalina farinacea* from genus *Ramalina* shows broad range of antiviral activity against both RNA and DNA viruses [32].

12.5.4 Antitumor Activity

In malignant melanoma (cancer cells), pannarin (depsidone) and sphaerophorin induce the apoptosis in M14-MEL cells by inhibiting its growth. These secondary metabolites can be isolated from many lichen species like *S. globosus*, *P. pulchrum*, and *P pholidotoides* [24]. These lichen substances proved very effective in the treatment of many skin cancers. Moreover, many species from genus *Ramalina* and their secondary metabolites are used to treat colon cancer cell lines by target HCT116 cells. In HCT116 cells, *Ramalin* induces apoptotic cell death by showing its anticancer activities [33].

12.5.5 Antidiabetic Activity of Lichens

Diabetes mellitus also known as diabetes is a metabolic disorder caused by elevation of blood glucose level for long period of time due to insufficient insulin supply or insulin resistance [34]. Many studies suggested that different lichens extract show effective antidiabetic activity by inducing alpha amylase inhibition. In Karnataka (state in south west india), methanol

Table 12.2 Some reported anti diabetic activities of lichens and their derived products.

Lichen species/compounds	Activities	Reference
R. nervulosa	α-glucosidase	[48]
R. pacifica		
Caloplaca biatorina		[47]
Diterpene furanoids	PTP	[9]
Usnic acid		[48]
Physcia aipolia	Alpha amylase	[38]
R. conduplicans		[49]

extracts of six lichens species were prepared to determined its alpha amylase inhibition effect. These lichen species included *P. tinctorum*, *Ramalina conduplicans*, *E. cirrhatum*, *Parmotrema pseudotinctorum*, *R. hossei*, and *U. sinensis* that are collected from Sahyadri [35]. Ethyl acetate and methanol extracts of *P. aipolia* and *F. caperata* also showed antidiabetic effects by amylase inhibition [36]. Some of the lichens and their secondary metabolites that possess antidiabetic activity are given in Table 12.2.

12.5.6 Antibacterial and Antifungal Activity

Lichens play a vital role in treatment of many bacterial and fungal infections. Some lichen secondary metabolites like protocetraric acid, atranorin, gyrophoric acid, usnic acid, lecanoric acid, and physodic acid possess effective antibacterial and antifungal effects against 6 bacteria and 10 fungi [37]. Some secondary metabolites of lichens that are collected from Spain showed effective antimicrobial activity against gram +ve bacteria [38]. Methanol extracts of five lichen species that was collected from Antarctica also showed antimicrobial effects against gram +ve bacteria [39]. Similarly, some lichen metabolites like isodivaricatic acid and usnic acid and methanol extracts of *P. poeppigii* (lichen specie) also exhibit strong antifungal effects against *Trichophyton rubrum* (dermatophytic fungus) and *M. gypseum* (soil-associated dermatophyte) [40].

12.6 Conclusion

Lichens are amazing organisms with interesting composition. Their survival abilities under unfavorable conditions depend upon their chemical composition. These are the untapped treasures and potential source for bioactive secondary metabolites that can be used in pharmaceuticals and other industries. Structures for more than thousand lichens are known but further studies are needed to discover un-explored biological metabolites. Simultaneously, these are slow growing organisms that required improved culture techniques that can be used for increased production of these secondary metabolites.

References

1. Shah, N.C., Lichens of commercial importance in India. *Scitech J.*, 1, 2, 32–36, 2014.
2. Behera, B.C., Morey, M.V., Gaikwad, S.B., Anti-lipoxygenase, radical scavenging and antimicrobial activities of lichen species of genus Heterodermia (Physciaceae). Botanica Pacifica. *J. Plant Sci. Conserv.*, 5, 1, 79–85, 2016.

3. Nguyen, K.H., Chollet-Krugler, M., Gouault, N., Tomasi, S., UV-protectant metabolites from lichens and their symbiotic partners. *Nat. Prod. Rep.*, 30, 12, 1490–1508, 2013.
4. Kumar, B., Assessment of lichen species in a temperate region of Garhwal Himalaya, India. *Am. J. Sci.*, 5, 107–12, 2009.
5. Molnár, K. and Farkas, E., Current results on biological activities of lichen secondary metabolites: A review. *Z. Naturforsch. C.*, 65, 3-4, 157–173, 2010.
6. Bačkorová, M., Jendželovský, R., Kello, M., Bačkor, M., Mikeš, J., Fedoročko, P., Lichen secondary metabolites are responsible for induction of apoptosis in HT-29 and A2780 human cancer cell lines. *Toxicol. In Vitro*, 26, 3, 462–468, 2012.
7. Muggia, L., Lichens as treasure chests of natural products. *Sim News*, 59, 3, 85–97, 2009.
8. Reddy, R.G., Veeraval, L., Maitra, S., Chollet-Krugler, M., Tomasi, S., Lohezic-Le Devehat, F., Chakravarty, S., Lichen-derived compounds show potential for central nervous system therapeutics. *Phytomedicine.*, 23, 12, 1527–1534, 2016.
9. Mitrović, T., Stamenković, S., Cvetković, V., Nikolić, M., Tošić, S., Stojičić, D., Lichens as source of versatile bioactive compounds. *Biol. Nyssana.*, 2, 1, 1–6, 2011.
10. Saini, K.C., Nayaka, S., Bast, F., Diversity of Lichen Photobionts: Their Coevolution and Bioprospecting Potential, in: *Microbial Diversity in Ecosystem Sustainability and Biotechnological Applications*, pp. 307–323, Springer, Singapore, 2019.
11. Jayanthi, S., Priya, P., Monica Devi, D., Benila Smily, J.M., Lichens: Origin, types, secondary metabolites and applications. *J. Acad. Indus. Res.*, 1, 1, 45, 2012.
12. Bhattacharyya, S., Deep, P.R., Singh, S., Nayak, B., Lichen secondary metabolites and its biological activity. *Am. J. Pharm-Tech Res.*, 6, 6, 29–44, 2016.
13. Türk, A.Ö., Yılmaz, M., Kıvanç, M., Türk, H., The antimicrobial activity of extracts of the lichen Cetraria aculeata and its protolichesterinic acid constituent. *Z Naturforsch C.*, 58, 11-12, 850–854, 2003.
14. Yamamoto, Y., Miura, Y., Higuchi, M., Kinoshita, Y., Yoshimura, I., Using lichen tissue cultures in modern biology. *Bryologist.*, 1, 384–393, 1993.
15. Shukla, V., Joshi, G.P., Rawat, M.S.M., Lichens as a potential natural source of bioactive compounds: A review. *Phytochem. Rev.*, 9, 2, 303–314, 2010.
16. Stocker-Wörgötter, E., Cordeiro, L.M.C., Iacomini, M., Accumulation of potential pharmaceutically relevant lichen metabolites in lichens and cultured lichen symbionts, in: *Studies in Natural Products Chemistry*, vol. 39, pp. 337–380, Elsevier, Amsterdam, 2013.
17. Shrestha, G. and Clair, L.L.S., Lichens: A promising source of antibiotic and anticancer drugs. *Phytochem. Rev.*, 12, 1, 229–244, 2013.
18. Musharraf, S.G., Siddiqi, F., Ali, A., Thadhani, V.M., Sensitive analysis of bioactive secondary metabolites in lichen species using liquid chromatography–mass spectrometry. *J. Pharmaceut. Biomed. Anal.*, 146, 279–284, 2017.

19. Le Pogam, P. and Boustie, J., Xanthones of lichen source: A 2016 update. *Molecules.*, 21, 3, 294, 2016.
20. Correche, E.R., Carrasco, M., Escudero, M.E., Velazquez, L., DE GUZMAN, A.S., Study of the cytotoxic and antimicrobial activities of usnic acid and derivatives. *Fitoterapia (Milano)*, 69, 6, 493–501, 1998.
21. Mitrović, T., Stamenković, S., Cvetković, V., Tošić, S., Stanković, M., Radojević, I., Marković, S., Antioxidant, antimicrobial and antiproliferative activities of five lichen species. *Int. J. Mol. Sci.*, 12, 8, 5428–5448, 2011.
22. Yousuf, S. and Choudhary, M.I., Lichens: Chemistry and biological activities, in: *Studies in natural products chemistry*, vol. 43, pp. 223–259, Elsevier, Amsterdam, 2014.
23. Paudel, B., Bhattarai, H.D., Lee, J.S., Hong, S.G., Shin, H.W., Yim, J.H., Antioxidant activity of polar lichens from King George Island (Antarctica). *Polar Biol.*, 31, 5, 605–608, 2008.
24. Russo, A., Piovano, M., Lombardo, L., Garbarino, J., Cardile, V., Lichen metabolites prevent UV light and nitric oxide-mediated plasmid DNA damage and induce apoptosis in human melanoma cells. *Life sci.*, 83, 13-14, 468–474, 2008.
25. Karakus, B., Odabasoglu, F., Cakir, A., Halici, Z., Bayir, Y., Halici, M., Suleyman, H., The effects of methanol extract of Lobaria pulmonaria, a lichen species, on indometacin-induced gastric mucosal damage, oxidative stress and neutrophil infiltration. *Phytother. Res.: Int. J. Devoted to Pharmacol. Toxicol. Eval. Nat. Prod. Deriv.*, 23, 5, 635–639, 2009.
26. Manojlović, N., Ranković, B., Kosanić, M., Vasiljević, P., Stanojković, T., Chemical composition of three Parmelia lichens and antioxidant, antimicrobial and cytotoxic activities of some their major metabolites. *Phytomedicine.*, 19, 13, 1166–1172, 2012.
27. Stanley, C., Ali, D.M.F., Keng, C.L., Boey, P., Bhatt, A., Comparative evaluation of antioxidant activity and total phenolic content of selected lichen species from Malaysia. *J. Pharm. Res.*, 8, 4, 2824–2827, 2011.
28. Luo, H., Yamamoto, Y., Kim, J.A., Jung, J.S., Koh, Y.J., Hur, J.S., Lecanoric acid, a secondary lichen substance with antioxidant properties from Umbilicaria antarctica in maritime Antarctica (King George Island). *Polar Biol.*, 32, 7, 1033–1040, 2009.
29. Bézivin, C., Tomasi, S., Lohézic-Le Dévéhat, F., Boustie, J., Cytotoxic activity of some lichen extracts on murine and human cancer cell lines. *Phytomedicine*, 10, 6–7, 499–503, 2003.
30. Han, D., Matsumaru, K., Rettori, D., Kaplowitz, N., Usnic acid-induced necrosis of cultured mouse hepatocytes: Inhibition of mitochondrial function and oxidative stress. *Biochem. Pharmacol.*, 67, 3, 439–451, 2004.
31. Fazio, A.T., Adler, M.T., Bertoni, M.D., Sepúlveda, C.S., Damonte, E.B., Maier, M.S., Lichen secondary metabolites from the cultured lichen mycobionts of Teloschistes chrysophthalmus and Ramalina celastri and their antiviral activities. *Z. Naturforsch. C.*, 62, 7–8, 543–549, 2007.

32. Esimone, C.O., Grunwald, T., Nworu, C.S., Kuate, S., Proksch, P., Überla, K., Broad spectrum antiviral fractions from the lichen Ramalina farinacea (L.). *Ach. Chemotherapy.*, 55, 2, 119–126, 2009.
33. Suh, S.S., Kim, T.K., Kim, J.E., Hong, J.M., Nguyen, T.T.T., Han, S.J., Kim, I.C., Anticancer activity of ramalin, a secondary metabolite from the antarctic lichen Ramalina terebrata, against colorectal cancer cells. *Molecules*, 22, 8, 1361, 2017.
34. Baynes, H.W., Classification, pathophysiology, diagnosis and management of diabetes mellitus. *J. Diabetes Metab.*, 6, 5, 1–9, 2015.
35. Vinayaka, K.S., Karthik, S., Nandini, K.C., Kekuda, P.T.R., Amylase inhibitory activity of some macrolichens of Western Ghats, Karnataka, India. *IJNDD*, 5, 4, 225–228, 2013.
36. Shivanna, R., Parizadeh, H., Garampalli, R.H., Screening of lichen extracts for *in vitro* antidiabetic activity using alpha amylase inhibitory assay. *Int. J. Biol. Pharm. Res.*, 6, 5, 364–7, 2015.
37. Ranković, B. and Mišić, M., The antimicrobial activity of the lichen substances of the lichens Cladonia furcata, Ochrolechia androgyna, Parmelia caperata and Parmelia conspresa. *Biotechnol. Biotechnol. Equip.*, 22, 4, 1013–1016, 2008.
38. Rezanka, T. and Sigler, K., Hirtusneanoside, an unsymmetrical dimeric tetrahydroxanthone from the lichen Usnea hirta. *J. Nat. Prod.*, 70, 9, 1487–1491, 2007.
39. Paudel, B., Bhattarai, H.D., Lee, J.S., Hong, S.G., Shin, H.W., Yim, J.H., Antibacterial potential of Antarctic lichens against human pathogenic Gram-positive bacteria. *Phytother. Res.*, 22, 9, 1269–1271, 2008.
40. Schmeda-Hirschmann, G., Tapia, A., Lima, B., Pertino, M., Sortino, M., Zacchino, S., Feresin, G.E., A new antifungal and antiprotozoal depside from the Andean lichen Protousnea poeppigii. *Phytother. Res.: Int. J. Devoted to Pharmacol. Toxicol. Eval. Nat. Prod. Deriv*, 22, 3, 349–355, 2008.
41. Kosanić, M., Manojlović, N., Janković, S., Stanojković, T., Ranković, B., Evernia prunastri and Pseudoevernia furfuraceae lichens and their major metabolites as antioxidant, antimicrobial and anticancer agents. *Food Chem. Toxicol.*, 53, 112–118, 2013.
42. Lohézic-Le Dévéhat, F., Tomasi, S., Elix, J.A., Bernard, A., Rouaud, I., Uriac, P., Boustie, J., Stictic acid derivatives from the lichen Usnea articulata and their antioxidant activities. *J. Nat. Prod.*, 70, 7, 1218–1220, 2007.
43. Thadhani, V.M., Naaz, Q., Choudhary, M.I., Mesaik, M.A., Karunaratne, V., Enzyme inhibitory and immunomodulatory activities of the depsidone lobaric acid extracted from the lichen Heterodermia sp. *J. Natl. Sci. Found. Sri.*, 42, 2, 193–196, 2014.
44. Honda, N.K., Pavan, F.R., Coelho, R.G., de Andrade Leite, S.R., Micheletti, A.C., Lopes, T.I.B., Leite, C.Q.F., Antimycobacterial activity of lichen substances. *Phytomedicine.*, 17, 5, 328–332, 2010.

45. Brisdelli, F., Perilli, M., Sellitri, D., Piovano, M., Garbarino, J.A., Nicoletti, M., Celenza, G., Cytotoxic activity and antioxidant capacity of purified lichen metabolites: An *in vitro* study. *Phytother. Res.*, 27, 3, 431–437, 2013.
46. Studzińska-Sroka, E. and Dubino, A., Lichens as a source of chemical compounds with anti-inflammatory activity. *Herba Pol.*, 64, 1, 56–64, 2018.
47. Comini, L.R., Vieyra, F.E.M., Mignone, R.A., Páez, P.L., Mugas, M.L., Konigheim, B.S., Cabrera, J.L., Montoya, S.C.N., Borsarelli, C.D., Parietin: An efficient photo-screening pigment *in vivo* with good photosensitizing and photodynamic antibacterial effects *in vitro*. *Photochem. Photobiol. Sci.*, 16, 2, 201–210, 2017.
48. Ristic, S., Rankovic, B., Stamenkovic, S., Biopharmaceutical potential of two Ramalina lichens and their metabolites. *Curr. Pharmaceut. Biotechnol.*, 17, 7, 651–658, 2016.

Index

Printed in the USA/Agawam, MA
November 19, 2025

896310.002